Financing National Defense

Policy and Process

A volume in
Research in Public Management
Lawrence R. Jones, *Series Editor*

Financing National Defense

Policy and Process

L. R. Jones
Naval Postgraduate School

Philip J. Candreva
Naval Postgraduate School

Marc R. DeVore
European University Institute and University of St. Gallen

INFORMATION AGE PUBLISHING, INC.
Charlotte, NC • www.infoagepub.com

Library of Congress Cataloging-in-Publication Data

Jones, L. R., Candreva, Philip J., DeVore, Marc R.
  Financing national defense : policy and process / L.R. Jones,
Philip J. Candreva, Marc R. DeVore.
    p. cm. – (Research in public management)
  Includes bibliographical references and index.
  ISBN 978-1-61735-677-3 (pbk.) – ISBN 978-1-61735-678-0 (hbk.) – ISBN
978-1-61735-679-7 (ebook)
  1. United States–Armed Forces–Appropriations and expenditures. 2.
United States. Dept. of Defense–Appropriations and expenditures. 3. United
States. Dept. of Defense–Finance. I. Jones, L. R. II. Candreva, Philip J.
III. DeVore, Marc R.
  UA23.F497 2012
  355.6'220973–dc23

                                                      2011045541

Printed in the United States of America

CONTENTS

ACKNOWLEDGEMENTS

The authors would like to acknowledge the value of colleagues and students who contributed to completion of this book, both directly and indirectly. In particular, we want to thank Professor of Public Budgeting (emeritus) Jerry L. McCaffery whose previous work and writing in many areas related to budgeting for national defense helped us significantly in writing this book. In addition, we wish to acknowledge the contribution of CAPT John Mutty (Ret.) for his work on earlier versions of our chapter on budget execution. Among all those we owe thanks are colleagues the Hon. Professor Douglas A. Brook, Professor Richard Doyle, Distinguished Professor Kenneth Euske, Dean William Gates, Professor emeritus Irene Rubin, Professor Philip Joyce, Professor Roy Meyers, Professor Paul Posner, CAPT Lisa Potvin, Robert Wythe Davis, LCDR Matt Jacobs, Professor Cindy King, Professor Natalie Webb, Professor Dan Nussbaum, Professor Kuno Schedler, Professor Riccardo Mussari, Professor Steven Kelman, Dr. Clay Wescott, Professor Harvey Sapolsky, Professor Eugene Gholz, Dr. Moritz Weiss, the Hon. Robert Hale, Dr. Norman Augustine, the late John Crowley, Peter Platzgummer, Sandra Eisenecker, RADM (Ret) Stanley Bozin, RADM Joseph P. Mulloy, CDR Christian A. Nelson, Mark Easton, Charlie Cook, Wes McNair, CDR (Ret) Ron Arnold, Barbara Bonessa, John Knubel, Marianna Martineau, CDR John Carty, Greg Sinclitico, Bob Donnelly, CAPT Brian Drapp, CAPT (Ret) Ed Hering, CAPT (Ret) Ken Voorhees, CAPT David Grundies (Ret), CAPT (Ret) Bob Aronson, the Hon. Sean O'Keefe, the Hon. John Raines, Paul Dissing, Ron Hass, Clinton Miles, Robert Panek, Alvin Tucker, Irv Blickstein and Charles Nemfakos.

Financing National Defense: Policy and Process, pages vii–viii
Copyright © 2012 by Information Age Publishing
All rights of reproduction in any form reserved.

Professor Jones wishes to add his thanks to RADM George F. A. Wagner (Ret) after whom the Wagner Chair he holds is named. In addition, he wishes to express his deep appreciation to two former mentors, the late Professor Aaron Wildavsky and Professor emeritus Todd La Porte, University of California, Berkeley for their support, guidance, advice, criticism and collegial interaction without which the motivation and ability to write this and other books and publications would never have emerged. Also to be thanked similarly and in particular is Professor Fred Thompson and also Professor Ilan Vertinsky, Professor emeritus W. T. Stanbury, Professor emeritus Guy Benveniste, Professor and Vice President emeritus Robert Biller, Professor emeritus Raymond Wolfinger, Jr., Al Loeb, Errol Mauchlan, the late Richard Hill, the late Robert L. Gach, the late Charles Levine, the late Martin Trow and the late Lyman Glenny.

Work by the following students whose research and consultation contributed to the research that assisted us in writing this volume includes that of Michael Owen, Dan Rieck, Paul Godek, Mark Kozar, Pam Theorgood, Chad Roum, Tiffany Hill, Erik Naley, Eric Wiese, Jim Davis, Dan Truckenbrod, Laura Kerr Nicholson, Beth Long, David Patton, Jason Woodruff, Sharon Holcombe, Nate Johnson and William E. Philips. Presently, all of these former students are military officers or DOD civilians who have dedicated their lives in service to the defense of our country and we thank them for that as well.

We would also like to thank the dozens of Department of Defense, Department of the Navy, and congressional staff experts in defense and federal budgeting and financial management who have visited NPS, taken our telephone calls and responded to our email messages, written and in some cases cited published works that inspired us, and to all those who have served as guest lecturers in our classes over the years.

Finally, we would like to thank our wives and families for their patience and support before, during and after the writing of this book. You provided safe harbors during many a storm.

L. R. Jones
Philip J. Candreva
Marc R. DeVore
Monterey, California, USA

LIST OF ACRONYMS

ACAT Acquisition Category
ADM Acquisition Decision Memorandum
AOR Area of Responsibility
APB Acquisition Program Baseline
APN Aircraft Procurement Navy
ASD (SO/LIC) Assistant Secretary of Defense for Special Operations and Low Intensity Conflict
ASN (RD&A) Assistant Secretary of the Navy for Research, Development and Acquisition (CAE for the Navy)
ATN Alliance Test Network
BA Budget Authority
BAH Basic Allowance for Housing
BAM Baseline Assessment Memorandum
BCP Budget Change Proposal
BES Budget Estimate Submission
BGM Budget Guidance Memorandum
BOR Budget OPTAR Report
BOS Base Operating Support
BR Concurrent Resolution on the budget
BRAC Base Reorganization and Closure Committee
BSO Budget Submitting Office
C4I Command, control, communications, computers and intelligence
CAD Computer Aided Design
CAE Component Acquisition Executive
CAIG Cost Assessment Improvement Group
CAIV Cost as an Independent Variable

Financing National Defense: Policy and Process, pages ix–xiv
Copyright © 2012 by Information Age Publishing

CAP   Critical Acquisition Position Description
CNDI   Commercial or Non-Developmental Items
CBO   Congressional Budget Office
CCR   Concurrent Resolution on the Budget; also CBR, BR.
CEB   CNO Executive Review Board
CENTCOM   Commander in Chief, Central Command
CFE   Commercial Furnished Equipment
CFFC   Commander Fleet Forces Command
CINC   Commander in Chief
CIVPERS   Civilian Personnel
CLF   Commander, U.S. Atlantic Fleet
Clinger-Cohen Act of 1996   Information Technology Reform Act of 1996
CNDI   Commercial or Non-Developmental Items
CNO   Chief of Naval Operations
COMNAVAIRPAC   Commander Naval Air Forces Pacific
COMNAVSUBPAC   Commander Naval Submarine Forces Pacific
COMNAVSURFPAC   Commander Naval Surface Forces Pacific
COMOPTEVFOR   Commander, Operational Test and Evaluation Force
COTS   Commercial off the Shelf
CP   Capability Plan
CPAM   CNO Program Assessment Memorandum
CPF   Commander, U.S. Pacific Fleet
CPA   Chairman's (of the Joint Chiefs) Program Assessment
CPR   Chairman's (of the Joint Chiefs) Program Recommendation
CRA   Continuing Resolution Appropriation
CVN 68   NIMITZ Class Nuclear Powered Aircraft Carrier
DAB   Defense Acquisition Board
DDG 51   Arleigh Burke Class Aegis Destroyer
DFAS   Defense Finance and Accounting Service
DIT   Design Integration Test
DOD   Department of Defense
DoDD   Department of Defense Directive
DODIG   Department of Defense Inspector General
DON   Department of the Navy
DPG   Defense Planning Guidance
DUSD   Deputy Under Secretary of Defense
DW   Defense-wide
EA   Executive Agent
EMD   Engineering, Manufacturing and Development Phase
ESPC   Energy Savings Performance Contracts
EUSA   Eighth United States Army
FAD   Funding Authorization Document
FARA   Federal Acquisition Reform Act of 1996

FASA   Federal Acquisition Streamlining Act of 1994
FASAB   Federal Accounting Standards Advisory Board
FFMIA   Federal Financial Management Improvement Act
FHCR   Flying Hour Cost Report
FHMP Family Housing Master Plan
FHP   Flying Hour Program
FHPS   Flying Hour Projection System
FMB   Navy Budget Office
FO   Flying Hours Other
FP   Force Protection
FY   Fiscal Year
FYDP   Future Years Defense Plan
GAO   General Accounting Office
GFE   Government Furnished Equipment
GPRA   Government Performance and Results Act:
HAC   House Appropriations Committee
HASC   House Armed Services Committee
HQ   Headquarters
H.R.   House Resolution
IA&I   Industrial Affairs and Installations
IDTC   Inter-deployment Training Cycle
IG   Inspector General
IMD   International Institute for Management Development
IPDE   Integrated Product Data Environment
IPPD   Integrated Product and Process Development
IPT   Integrated Process Team
ISPP   Integrated Sponsor Program Proposal
IT   Information Technology
IWAR   Integrated Warfare Architecture
JCS   Joint Chiefs of Staff
JP   Joint Publication
JROC   Joint Requirements (of JCS) Oversight Committee
JSF   Joint Strike Fighter
LAN   Local Area Network
LBTE   Design Integration in a Land Based Test Environment
LPD 17   Marine Amphibious ship used for embarking, transporting and supporting troops
LRIP   Low Rate Initial Production
MDA   Milestone Decision Authority
MDAP   Major Defense Acquisition Program
MEB   Marine Expeditionary Brigade
MFP   Major Force Program
MHPI   Military Housing Privatization Initiative

MILCON Military Construction
MIPR Military Interdepartmental Purchase Request
MOCAS Mechanization of Contract Administration Services system
MOU Memorandum of Understanding
MS Milestone
MSA Master Settlement Agreement
MUHIF Military Unaccompanied Housing Improvement Fund
MWR Morale, Welfare, and Recreation
NAVAIR Naval Aviation Systems Command
NAVSEA Naval Sea Systems Command
NDI Non-Developmental Item
NHBS Navy Headquarters Budgeting System
NMCI Navy and Marine Corps Intranet
NMSD National Military Strategy Document
No. Number
NOR Net Offsetting Receipt   Collections from the public
NSS -National Security Strategy
OAC Operating Agency Code
O&M Operations and Maintenance
O&MN Operations and Maintenance, Navy
O&MMC Operations and Maintenance, Marine Corps
O&S Operation and Support Costs
OBAD Operating Budget Activity Document
Obligation legal set aside funds for a future payment, as in letting a
 contract
ODC (P/B) Office of the Deputy Comptroller (Program/Budget)
OFPP Office for Federal Procurement Policy
OGA Other Government Activity
OFPP Office for Federal Acquisition Policy
OMB Office of Management and Budget
OMN Operation and Maintenance, Navy
OPNAV Office of the Chief of Naval Operations
OPTEMPO Operational Tempo
ORD Operational Requirements Document
OSD Office of the Secretary of Defense
Outlay An Expenditure or Liquidation of Obligations
PACFLT U.S. Pacific Fleet
PACNORWEST Pacific Northwest
PACOM U.S. Pacific Command
PBAS Program Budget and Accounting System
PBCG Program Budget Coordination Group
PBD Program Budget Decision
PCP Program Change Proposal

PDRR   Program Definition and Risk Reduction
PE   Program Element
P.L.   Public Law
PM   Program Manager
POM   Program Objective Memorandum
PPBS   Planning, Programming and Budgeting System
PPBES   Planning, Programming, Budgeting and Execution System
PR   Program Review
PRESBUD   President's Budget
QDR   Quadrennial Defense Review
R3B   Resource Requirements Review Board
RBA   Revolution in Business Affairs
Reappropriations   Extending previously appropriated funds
RDT&E   Research, Development, Training and Engineering
Rescissions   Canceling new Budget Authority or Unobligated Balances
RMA   Revolution in Military Affairs
ROE   Return-on-equity
SAC   Senate Appropriations Committee
SASC   Senate Armed Services Committee
SECDEF   Secretary of the Defense
SECNAV   Secretary of the Navy
SGL   Standard General Ledger
SLAN   Secure Local Area Network
SO   Special Operations
SOC   Special Operations Command
SOF   Special Operations Forces
SPI   Single Process Initiative
SPP   Sponsor Program Proposal
SPS   Standard Acquisition System
SRM   Sustainment, Restoration, and Maintenance
SSBN   Strategic Ballistic Missile Submarine
SSGN   Tomahawk Launch Capable Converted Strategic Ballistic Missile
      Submarine
Sub-unified   Subordinate Unified
SV   Service
TOA   Total Obligational Authority—value of the direct defense program
      in a given year from this year and previous years.
TOC   Total Ownership Costs
T-POM   Tentative Program Objective Memorandum
TSOC   Theater Special Operations Command
TSP   Thrift Savings Plan
TYCOM   Type Commander
UFR   Unfunded Requirement

UMD Unmatched Disbursement: payment that can not be matched to an existing obligation

U.S. United States

USARSO United States Army South

USC United States Code

USCENTCOM United States Central Command

USCINCSOC Commander in Chief, United States Special Operations Command

USD (AT&L) Under Secretary of Defense for Acquisition, Technology and Logistics

USD(C) Under Secretary of Defense (Comptroller)

USEUCOM United States European Command

USFK United States Forces Korea

USJFCOM United States Joint Forces Command

USNORTHCOM U.S. Northern Command

USPACFLT United States Pacific Fleet

USPACOM United States Pacific Command

USSOCOM United States Special Operations Command

USSOUTHCOM United States Southern Command

LIST OF FIGURES

Financing National Defense: Policy and Process, pages xv–xvi
Copyright © 2012 by Information Age Publishing
All rights of reproduction in any form reserved.

LIST OF TABLES

Financing National Defense: Policy and Process, pages xvii–xviii
Copyright © 2012 by Information Age Publishing
All rights of reproduction in any form reserved.

CHAPTER 1

BUDGETING IN THE FEDERAL GOVERNMENT

INTRODUCTION: BUDGETS AS MULTIPURPOSE INSTRUMENTS

A myth from the colonial period was that Americans could defend themselves by keeping a rifle in the closet, grab it, and march off to battle in times of crisis. Unfortunately, defense is more complicated that that; indeed it was more complicated even during the Revolutionary war. General George Washington's struggle to form a standing army supported by a logistics and supply organization and to get funding for both from the Revolutionary Congress are well known. Defense requires planning and resourcing in advance. Reacting at the instant of crisis is too late. Moreover, production of defense goods has long lead times and involves decisions that have consequences for decades. This includes selecting and training personnel as well as designing, buying and fielding a vast array of ground weapons, ships, aircraft and other weaponry. Moreover, one decision to buy a major defense asset sets in motion a chain of decisions likely to endure for decades. For example, buying an aircraft carrier for five billion dollars presupposes the purchase of support ships, force defense ships, aircraft, and training of pilots and crews that will cost an additional $50 billion or more. And, it is not unusual for an aircraft carrier and other ships in a carrier battle group to have a service life of forty years or more. Such decisions are resourced through the budget process, a planned yet somewhat disor-

Financing National Defense: Policy and Process, pages 1–32
Copyright © 2012 by Information Age Publishing
All rights of reproduction in any form reserved.

derly system for deciding how to allocate scarce resources in a manner that culminates with congressional and presidential approval. In this chapter we examine the concept and practice of budgeting, provide an overview of the policy making process, and analyze the most significant features of the federal government resource management system. We then describe what makes defense budgeting different.

DEFINING BUDGETING AND THE BUDGET

In sending his proposal to create an executive budget system to Congress in 1912, President Taft said, "The Constitutional purpose of a budget is to make government responsive to public opinion and responsible for its acts." (Burkhead, 1959: 19) In the proposal, it was noted that a budget served a number of purposes, from a document for congressional action, to an instrument of control and management by the President, to a basis for the administration of departments and agencies. The multiple purposes of the budget have been noted, but no one has been more eloquent in describing budgetary complexity than Aaron Wildavsky. In his classic 1964 book *The Politics of the Budgetary Process*, Wildavsky explained that a budget is:

1. "Concerned with the **translation** of financial resources into human purposes."
2. A mechanism for making choices among alternative expenditures ... **a plan**; and if detailed information is provided in the plan, it becomes a **work plan** for those who administer it.
3. An instrument to attempt to achieve **efficiency** if emphasis is placed on obtaining desired objective at least cost.
4. A **contract** over what funds shall be supplied and for what purposes:
   - Between Congress and the President
   - Between Congress and the Departments and Agencies
   - Between Departments and Agencies and their subunits
   These 'contracts' have both legal and social aspects. Those who give money expect results; those who are due to receive money expect to have the funds delivered on time to execute their programs effectively. Both superiors and subordinates have rights and expectations under such contracts, and mutual obligations are present.
5. A set of both **expectations** and **aspirations are contained in proposed budgets** from submitting agencies. Agencies expect to get money, but they may aspire to much more than they are given. The budget process regularly allows they to ask for what they aspire. What they are given in dollars reveals the preferences of others about the agency's budget. This is important information for the next budget cycle.

6. A **precedent:** something that has been funded before is highly likely to be funded again (this is defined as budgetary incrementalism).
7. A tool to **coordinate and control:** to coordinate diverse activities so they complement each other, to control and discipline subordinate units, e.g., by limiting spending to what was budgeted or by providing money to or taking it from pet political projects.
8. A **call to clientele** to mobilize support for the agency, when programs appear to be underfunded or losing ground to other programs.
9. "A **representation** in monetary terms of governmental activity. . . ." (Wildavsky: 1964, 1–4)

In the American context under the Constitutional separation of powers between the executive, congressional and judicial branches of government, the budget process begins in the executive branch of government, where the budget plan is developed, and proceeds into the legislative branch where it is reviewed, reformulated, (sometimes even rejected in total), amended and enacted. The process usually concludes in the executive branch where the President, as Chief Executive of the executive branch (and Commander-in-Chief of the armed forces), signs the budget bill or bills into law. At the federal level, the chief executive may veto budget bills he does not like, but eventually he must sign some sort of compromise bill. In contrast, a majority of state governors has an "item veto" authority that allows them to change bills in different ways depending upon the precise nature of the veto power, e.g., in California the Governor can use the line-item veto to cut money from an appropriation for a program approved by the legislature, but cannot add money through the veto.

Budgetary power is shared power, employed under a system of checks and balances made possible by the separation of powers, although the Constitution is clear that the "power of the purse" rests with the Congress. Article I, Section 9, clause 7 of the United States Constitution requires that "No money shall be drawn from the treasury, but in consequence of appropriations made by law; and a regular statement and account of receipts and expenditures of all public money shall be published from time to time." This clause flows from the basic "power of the purse" granted in Article I Section 8, authorizing Congress to "pay the debts and provide for the common defense and general welfare of the United States." Alexander Hamilton, the first Secretary of the Treasury, said: "The House of Representatives cannot only refuse, but they alone can propose, the supplies requisite for the support of the government. They . . . hold the purse. . . . This power over the purse may, in fact be regarded as the mot complete and effectual weapon with which any constitution can arm the immediate representatives of the people . . ." (McCaffery and Jones, 2001: 53)

Under the Articles of Confederation, Congress did try to organize itself to run the government, but this was largely a failure. U.S. history from 1790 to 1921 reflects the gradual growth of executive competence and power over the budget, capped by the Budget and Accounting Act of 1921, which directed the President to submit an annual budget and equipped him with a staff office, the Bureau of the Budget, for assistance. Thus, the President submits budgets each year as required, but Congress still holds the power of the purse and feels free to put its own imprint on that budget. Congress also feels a Constitutional obligation to exercise precise and detailed oversight of how programs are administered as part of its budgetary responsibility. It does this through authorization of programs and program review in the appropriation process.

In aggregate terms, Congress may not seem to make great changes in the President's budget. While it often appears that Congress is a "marginal modifier," of the President's budget, Congress may take a position that is very different from the President with respect to agencies and programs, notwithstanding that the final dollar numbers are not very far apart. This is particularly true when control of the Congress and the Presidency is divided between the two parties. When one party holds the executive branch and the other party holds one or both houses in the legislative branch, the budget process can be both heated and extended. In the 1980s, several Presidential budgets were termed 'dead on arrival' when submitted to Congress, because Congress did not even consider them as a base for spending negotiation. This might have led to a vote of no confidence and a general election in a parliamentary system. Instead, it led to late appropriation bills, summit meetings between leaders of the executive and legislative branches, and, to some extent, policy gridlock, sometimes followed by budget process reform efforts.

Once the President has signed the appropriation bills, it is the function of the executive branch agencies to execute the budget as enacted, and not as submitted. This is not as simple as it seems. Changing conditions may lead the executive branch to try to defer spending or rescind (cut) programs; sometimes emergency supplemental appropriations are sought to fund emergencies, e.g., natural disasters or military action. A continuing traffic exits in reprogramming (moving money within appropriations) and transfers (moving money between appropriations), some of which agencies can do on their own and some that Congress must approve. In short, budgeting seems to be a never-ending activity. While the current budget is executed, the budget for the next year is under review in the legislature, and the budget for the year following that is under preparation by agencies in the executive branch. Some reformers have suggested that the federal government pursue a biennial budget process where budgeting is done every two years in order to put more planning and analysis into the budget process. Biennial budgeting is a form of budgeting found in 19 American states in the period 1990–2011.

At its heart, the budget process is a planning process. It is about what should happen in the future. For non-defense agencies, this planning process may involve agency estimation of the amount of services to be provided in the next year, and to whom services will be supplied. For income security and welfare programs, planning may involve estimation of what it will take to provide a decent standard of living for the poor. For defense agencies, it may involve estimation of the consequences of U.S. foreign policy commitments and defense resource planning in terms of threat response capacity and the personnel and support resources necessary for threat management and deterrence. While numbers and quantification give the budget document the aura of precision, it is still a plan; this is most clearly evident in budget execution, where agencies struggle to spend the budget they have received in an environment inevitably changed from the one for which the budget was developed. Consequently, all budget systems provide some capacity to modify the enacted budget during budget execution, e.g., fund transfers and reprogramming, emergency bills and supplemental additions of new funding.

Finally, it is important to recognize that budgeting is not done within government in a vacuum. Both in formulation and execution, various stakeholders outside of government attempt to influence budget decisions and outcomes. In the defense environment, these stakeholders range from corporations that do defense business, state and local governments where the corporations are located, hire people, make purchases, and pay taxes, to employees and employee unions, lobbyists and legislators who represent these and more general interests and those who would either like to share in the defense spending pie or diminish it in order to have it spent on other policy areas. Each policy area seems to have major players; in defense, the major players exist in the Department of Defense and the Military Departments, in Congress in the defense authorization committees and in the appropriations committees and sub-committees. Program advocates focus on these critical players. They articulate demands and show support for those demands. A 'good idea' that gathers little or no support has no chance of making it through the policy-making apparatus, whereas an 'average' idea that has widespread support is very likely to become policy. Voters, citizens, lobbyists, political action committees (PACs) all help articulate demands and gather support. The various outlets of the news media are very important in seizing on issues and helping the public understand what is at stake, even if they sometimes prefer the relatively unimportant but titillating, to the important but obscure and complex.

Most observers of the American system assume that nothing is written in stone, thus information about current policy outcomes may be immediately fed back into the policy-making mechanism to help correct flaws in current policy. This is not as easy to do, as it is to say. Moreover, some things do seem written in stone; subsidy programs and entitlements are very difficult to change. Tax laws, especially when they increase taxes, usually only get the necessary support if they are written to occur at a point far enough in the future that a majority of the potential taxpayers feel that they can arrange their affairs so the tax will not affect them. Thus, flawed policies do endure and are hard to change, particularly when a small, but intensely vocal group favors the current arrangement. Some focused arrangements have developed historically where the interests of a particular company or clientele, an executive branch agency and a Congressional Committee or Committees combine to make and sustain policy favorable to those specific parties, perhaps at the expense of the public. This is true to some extent in all policy areas. These relationships are commonly referred to as 'iron triangles' to denote their power.

AN OVERVIEW OF THE FEDERAL BUDGET PROCESS CYCLE

Government budgeting is a process that matches resources and needs in an organized and repetitive way so that collective choices are properly funded. The product of this process is the budget document- an itemized and programmatic estimate of expected income and operating expenses for a given unit of government over a set time period. Budgeting is the process of arriving at such a plan and executing it. Once a fiscal year has begun, the budget becomes a plan for tracking and managing the collection of taxes, fees, and other revenues, and for distributing and disbursing these revenues to attain the goals specified in the budget. A variety of financial management functions are performed throughout the fiscal year, in conjunction with taxing and spending in attempt to coordinate the financial activities of government, and to insure accountability, safety, legality and propriety in the raising and expending of public monies. At the end of the year, the budget process produces reports that allow for comparison of the achievements of government relative to the commitments made when the budget was enacted. In democratic systems, these commitments represent the will of the people as expressed during the politics of the budget process.

Budgeting and financial management are not performed in a vacuum. They are part of a public policy cycle in which (a) public service demands and preferences are articulated, (b) public policy is developed to respond to these demands and preferences by elected officials, (c) resources are generated and allocated to various public and private purposes, (d) programs and implementation strategies and tactics are developed and executed, (e) spending is incurred in the delivery of services and benefits, (f) the outcomes of policies and programs are reported and analyzed. Citizens consume this information and respond to the manner in which services are delivered and the amounts of services supplied, and again articulate their service demands and preferences to their representatives in government. In democratic political systems, it is assumed that the role of government in large part is to meet the demands and preferences of citizens with resources afforded relative to the condition of the national economy, and to do so in a manner that promotes social equity, economic efficiency, and social and economic stability.[1] Below is a simple graphic to help explain the public policy process.

This graphic helps us understand that budgeting is not done within government in a vacuum. Both in formulation and execution various stakeholders outside of government attempt to influence budget decisions and outcomes. This environment is depicted in the "stakeholder space" graphic in which the Boeing Corporation (non-governmental organizations or public interest lobby groups, etc. could be substituted in this diagram to reflect their input) is shown as an example of how private firms play a role in the budget process. What this diagram shows is that before policy can be made, demands and support must be articulated. A 'good idea' that gathers little or no support has no chance of making it through the policy-making apparatus whereas an 'average' idea that has intense support from relatively small groups may well become policy. Voters, citizens, lobbyists, political

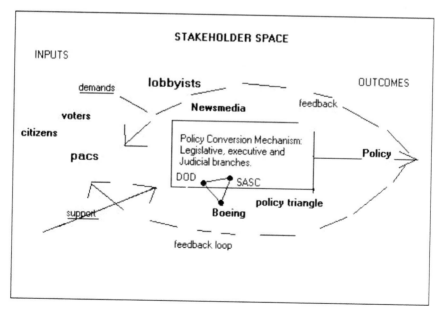

Figure 1.1 A Stakeholder Space Model of Public Policy. *Source:* Authors, 2011

action committees (PACs) all help articulate demands and gather support. The various outlets of the news media are very important in seizing on issues and helping the public understand what is at stake, even if they sometimes prefer the relatively unimportant but titillating to the important but obscure and complex.

Most observers of the American system assume that nothing is written in stone, thus feedback about current policy outcomes may be immediately fed back into the policy-making mechanism to help correct flaws in current policy. This is not as easy to do as it is to say. Moreover, some things do seem written in stone; subsidy programs and entitlements are very difficult to change. Tax laws, especially when they increase taxes usually only get the necessary support if they are written to occur at a point far enough in the future that a majority of the potential taxpayers feel that they can arrange their affairs so the tax will not affect them. Thus flawed policies do endure and are hard to change, feedback loop or not, particularly when a small, but intensely vocal group favors the current arrangement. The triangle in the diagram is meant to indicate that there are some focused arrangements that have developed historically where the interests of a particular company or clientele, an executive branch agency and a Congressional Committee or Committees combine to make and sustain policy favorable to those specific parties, perhaps at the expense of the general public. This is true to some extent in all policy areas. These triangular relationships are sometimes referred to as 'iron triangles' to denote their power. (The one in our diagram shows the Boeing Corporation, the Department of Defense, and the Senate Armed Services Committee, but it could as well show almost any large company or group of commercial interests or pensioners, their relevant executive agency and their relevant committee.)

The budget process is often portrayed as a process where disaggregated groups of experts make policy, experts on the appropriations sub-committees and their staffs, experts on the authorization committees and their staffs, experts in OMB, and experts in the agencies. Their decisions are later tested and ratified—or rejected—by others in the budget process. This is another way of describing the 'iron triangle' mechanism. The anchoring of the triangle in public space to represent interest groups, lobbies, citizens, and corporations only indicates that they too can be experts in a policy area and bring to lawmakers information about impacts or consequences that may not be available to or provided by participants within the administration or by the policy experts in Congress. This is the positive case for 'iron triangles.' On the negative side is the worry that the iron triangle mechanism will result in unfair or inefficient or ineffective decisions. For example, in early 2007, the Pentagon's former acquisitions chief, Kenneth Krieg was asked "How do you break the so-called "iron triangle" of the Pentagon, the Congress and the defense industry? Does it force you to make unwise decisions?" Krieg responded: "I worry about it greatly. I worry about getting the best product I can for the war fighter at the best price in a changing strategic environment, which means you need to be able to move in different directions. And that is really hard when people are anchored in a different world view. I have a requirement for a Lamborghini; I drive a 104,000-mile Town and Country minivan. The reality is we make trades all the time between what we want and what we buy." (Bennett and Muradian, 2007: 30) Some might think the iron triangle mechanism might make the job of the Assistant Secretary of Defense for Acquisition, Technology and Logistics (ASD AT&L) easier, but the reality is that the military departments have their own allies at different points in iron triangle space and their influence in any particular case might be strong enough to overturn a 'DOD' perspective or recommendation. The Department of Defense rarely speaks with a single voice. For example, earmarks buying more aircraft of ships than requested or improving bases where no improvement was requested in the annual budget submittal are examples of iron triangle mechanisms subverting the main DOD budget submission.

To a limited extent, the struggle over the deficit in the 1990s, with imposition of spending caps on discretionary expenditures and pay-go provisions (pay as you go, i.e., to increase entitlement spending either new revenues had to be identified or existing programs cut) may have changed the dimensions of this struggle so that players within each triangle are forced to limit their aspirations to taking money away from other players within the triangle, rather than preying upon the treasury at large, or another policy area. This is, however, to a limited extent and the growing use of earmarks in the 2000s illustrates a different aspect of the iron triangle game; that is how it can be driven by self-concerned interests in Congress, even to the extent that budget control is lost. For example, in the early 2000s spending caps and pay-go provisions were suspended by a Republican dominated Congress, which led some experienced budget scholars to conclude that the congressional process was "broken." (Posner, 2002; Joyce, 2002; Meyers, 2002) Also, a policy area that is blessed by the President or majority interests in Congress can be assured of an increasing budget, as has been the case with national defense under two Presidents since 9/11/2001. This is all part of the politics of the budgetary process and serves to remind that real benefits are distributed through the budget process. Citizens who are not aware of this will end up paying for benefits enjoyed by other people as they wonder where their taxes went or how long it will take their children to pay down the national debt.

In a formal sense, public policy consists of a set of goals or objectives, strategies and tactics to achieve these goals, a commensurate commitment of resources, an implementation plan based on strategy and limited by resource availability, and a means for measuring, reporting and evaluating the extent to which goals have been met and other outcomes achieved. If these basic components are not present, then policy may be viewed as not properly formulated. Unfortunately, policy analysis often determines that important components are omitted in government policy development and application. Analysis of policy outcomes may reveal that the costs of service development and delivery by government are lower than the benefits achieved. However, in some cases, benefits are exceeded by costs, i.e., net negative benefits have resulted from government action. Where this occurs, policy and service delivery may be questioned relative to their effectiveness. In democratic systems of governance, it is intended that such questions be addressed regularly in the public budget process.

Within stakeholder space as described above, the budget process may be understood to operate in two main phases, formulation and execution. The model of the budget process shown in Figure 1.2 is useful for delineating the major phases and activities of the budgetary process.

The American governmental system is an open system, thus almost all of the stages of the budget process above must be understood to be open to the impacts of stakeholders and would-be stakeholders who seek to maximize their own goals and aspirations. Some are much more powerful than others, of course, and not all stages are equally open to all participants, but it would be incorrect to view any model of the budget process as a closed model immune from what is happening in the broader society. The audit stage is perhaps the most technical and most closed, but this stage is also open to public impact as a result of the way the other executive branch participants react to an audit or in terms of the way Congress uses audit find-

---

I. Budget Formulation
  A. Preparation of Estimates (Executive branch)
  B. Negotiation
    (1) Executive: departments, agencies, budget offices
    (2) Legislative: budget, authorizing and appropriating committee analysis and hearings, amendment and voting
  C. Enactment
    (1) Legislative: debate, amendment, conference committees, regular and continuing appropriations voting
    (2) Executive: lobbying, signing or vetoing appropriations
II. Budget Execution (Executive branch)
  A. Apportionment of appropriations to departments and agencies, Allotment within agencies and Spending
  B. Monitoring and Control of Spending
  C. End Year Accounting and Reconciliation to Appropriations
  D. Financial Audit, Management Audit, Program Evaluation and Policy Analysis

---

Figure 1.2 Phases of the Budget Process Cycle. *Source*: Authors, 2011.

ings as a basis for future public policy decisions. Thus we offer our model of the stages of the budget process as a way to interpret what the process looks like when extracted from its environment; it is offered as a starting point.

Also, while the exhibit is linear and based on a single year evolution, readers must understand that at least three budget years are in play at any one time. As agency and department leaders prepare estimates for one year, e.g., FY 2013 in the summer of 2011 (IA above), they also execute the current budget (IIA to C) and testify in Congress in support of the FY 2012 budget (IB). The budget under execution (FY 2011) has various reporting dates and deadlines, quarterly plans, and reports to both OMB and Congress as funds are allotted and obligated, and transfers and reprogrammings are made. The budget under scrutiny in Congress is the focus of hearings in a number committee and subcommittee venues, including the budget, authorization and appropriations committees in each chamber. To some extent, what members of these committees say about budget proposals under review must be considered by the agency as it begins preparation of the budget for the following year. Hearings usually begin in February and last through May, and should be understood as part of the negotiation process noted above. As appropriation bills progress through Congress, departments and agencies begin to prepare their initial obligation plans for OMB in mid-August, with the final plan due by October 1 or within 10 days of passage of their appropriation bill. Agencies within the department have to prepare financial and staffing plans for the upcoming fiscal year for the department budget office The budget preparation process for the following fiscal year (BY+1-2013) runs concurrently with the execution of the current budget (CY-2011) and testimony on the next budget (BY-2012).

More elaborate time horizons exist within defense. For example, the defense PPBE system is focused around a multiple year array of budget data provided in the Five Year Defense Plan or FYDP and other documents, e.g., the Quadrennial Defense Review or QDR. The FYDP displays show the current year, the budget year, and five future years. The FYDP allows defense planners to examine resource-planning profiles over an extended time horizon. Any budget changes that have future year consequences must be carefully tracked and changed appropriately over the course of the FYDP. DOD budget insiders have remarked that the first two years of the FYDP are of 'budget quality,' but the 'outyears,' the last two or three years, are not of budget quality. Even so, they serve as target planning estimates.

THE PRESIDENT'S BUDGET

Budget preparation begins with a Guidance letter from OMB in late winter, is followed up with a Spring Preview session where the Department and

OMB may work certain issues at dispute, and concludes with the routines of budget preparation as dictated in OMB circular A-11, the budget preparation circular. In 1999, A-11 was issued on July 12 and contained over 580 pages of explanation, definitions, and instructions. In 1999 and 2000, efforts were made to rewrite A-11 in plain language to make it more accessible to the departments and agencies that must use it. This has been largely successful, but the budget process remains a stunningly complex process, where confusion abounds and the necessity for negotiation is obvious. Various amendments and clarifications are added to A-11 during the year as is necessary.

The executive budget process timetable has five discernable stages:

1. **April–June**: Agencies begin development of their budget requests based on prior year programs, problems and issues, and new initiatives. The President, assisted by OMB, reviews the current budget and makes decisions to guide policy for budget decisions. These may be conveyed to agencies at the Spring Preview or through budget instructions.

2. **July–August:** OMB issues policy directions to agencies and provides guidance for the agencies formal budget decisions. Agencies usually prepare and submit budgets for agency budget review. This may be done at several levels in large agencies and involve various hearings and negotiation stages. Ultimately, issues will be decided by the head of the agency and a final budget will be prepared.

3. **Early fall:** Agencies will submit their budget requests to OMB. In defense the Military Departments go through a formal budget preparation process in the summer with hearings and cuts and appeals, arrive at a final budget and then submit it to DOD in the early fall where it is diligently reviewed. Both at the Military Department level and the DOD level, most issues are negotiated and solved by analysts, but some major issues can only be settled at the Secretary of Defense level. The Defense budget is then submitted to OMB.

4. **November–December:** OMB and the President review and make decisions about agency requests and "passback" their decisions to the agencies. Agencies then revise their budget submission for inclusion in the President's budget. Agencies may appeal analyst decisions to the OMB Director or even to the President when they cannot negotiate a satisfactory solution with line budget analysts. Only major issues go to the President. When the major agencies, e.g., DOD, appeal issues to the President, they are expected to bring suggested solutions with them to the table.

5. **January–early February**: OMB and the President continue to make decisions on the budget. The President is required by law to submit

the budget to Congress by the first Monday in February. At some point during this period, the decision process must give way to the needs of the government printing office; hence, the budget database is locked and the budget is printed and subsequently presented to Congress.

Each new Presidential regime requires a transition year with basically ad hoc procedures. For example, in 2000, departments prepared their current services budget (with very little or no policy change) until an early amendment to A-11 was issued by Mitchell Daniels, President George W. Bush's OMB Director. On February 14, 2001, Daniels ordered the departments to carry on with the Government Performance and Results Act (GPRA) initiative and to surface the Bush Administration priorities. This transmittal letter said, in part:

> Most agencies submitted an initial version of their FY 2002 performance plan to OMB last fall. The performance goals in this initial plan were set using a current services funding level, and did not anticipate policy and initiative decisions by the new Administration. You should immediately begin making all necessary changes to your FY 2002 performance goals to reflect both the agency's top-line allowance and any applicable policy and initiative decisions. The top-line allowance will need to be translated into goal target levels for individual programs and operations. Your FY 2002 performance plans and budget materials should reflect the focus on bringing about a better alignment of performance information and budget resources...These plans should be sent to OMB at least two weeks prior to being sent to Congress to ensure that the President's decisions, policies, and initiatives are appropriately reflected. (Daniels, 2001)

We should note that this letter sent in February 2001 was intended to have an impact on the appropriation bills that would be passed in Congress in the summer of 2001 to fund departments beginning in October 2001 (FY 2002). This shows the desire of a new administration to get its priorities stamped on the budget as soon as possible.

Normally, for the United States federal government, each year the programs and spending departments initially propose to begin the budget cycle are prepared and reviewed in close detail first by agency and then department budget staff. These budgets are based upon instructions from the executive so departments will have some notion of how much they may ask for in total, thus the spring preview by OMB and the Mid-session reviews. The instructions include policy guidance and directions about the form and format of the budget. The federal budget process did not always operate in this top-down manner, but it has done so since the early 1980s. After this review, budgets are sent to the President's Office of Management and

Budget (OMB) where hearings are held, decisions made and passed back to the departments and appeals are heard. December is spent preparing the multiple products that comprise the President's budget, updating and locking-up electronic databases, and preparing congressional justifications. Even in normal years, this process lasts until the end of January.

By the first Monday in February, the President is required by law to submit to Congress his proposed budget for the next fiscal year. As part of his submittal, he delivers a myriad of exhibits, tables, graphics and thousands of pages of text that show where revenues come from and on which programs he proposes they be spent. For example, for fiscal year 2002, the federal budget was composed of four volumes, the Budget of the U.S. Government: FY 2002; Analytical Perspectives; Historical Tables; and A Citizen's Guide to the Budget. These are presented to Congress and widely disseminated in the media because they are supposed to make the gargantuan sums of money collected and distributed by the federal budget comprehensible to the average citizen. Unfortunately, given the complexity of the data and the general absence of knowledge and interest of the citizenry in budgeting and what the government does with money, it is doubtful that this objective is achieved to any extent.

THE CONGRESSIONAL BUDGET PROCESS

It is important to understand that Congress never appropriates the President's budget proposal exactly as proposed. This is because the Constitution provides the power to enact taxes and budgets to the Congress, and not the President. An old and true budget aphorism is, "The President proposes and the Congress disposes." Congress spends eight or more months of effort each year scrutinizing the President's proposal in great detail in numerous committees and sub-committees, debating alternatives and amendments of their own origination, asking questions of witnesses they call, and listening to testimony from the President's administration and a variety of other advocates and interests group lobbyists in hearings on the budget. Congress then writes separate authorization and appropriation bills that may include substantial changes to what the President proposed, votes to approve these huge bills (typically, there are thirteen separate appropriation bills alone) and sends them to the President for his signature or veto.

In addition to appropriation bills, Congress may also have to pass other legislation to complete the budget plan, e.g., bills that change existing tax laws, enact new ones, or modify benefit structures for entitlement programs, e.g., social security.

Legislative products that affect the budget include:

The Concurrent Resolution on the Budget: The budget resolution (BR) sets aggregate spending and taxing totals and estimates the resulting deficit or surplus. It also sets spending totals by functional area, e.g., defense, transportation and so on. The budget resolution is a plan; once it is adopted, Congress tries to stick to it through 'scorekeeping' mechanisms enforced by points of order. The appropriation committees take the amounts allotted to them by the budget resolution and divide them up among the sub-committees that produce appropriation bills. As these bills progress through Congress, members of the budget committees assisted by the Congressional Budget Office keep score to ensure that the functional total does not exceed the amount assigned to it in the budget resolution. The rules of Congress call for the budget resolution to be reported out of the Senate budget committee by 1 April and passed both chambers by 15 April, though it rarely is. Budget resolutions usually set targets for the budget year and a number of future years, usually three or four, but this has been extended to as many as seven years to coincide with some key electoral strategy. Without doing too much violence to the concept, it may be said that the budget resolution is Congress's plan for spending and taxing, and just as much a budget as is the president's budget, although at this stage the President's budget is much richer in details. At the end of the process, the Budget Resolution and the appropriations bills comprise a comparable level of detail, and indeed, govern in detail how agencies will operate, at least for the discretionary parts of the budget

Reconciliation Bills: A reconciliation instruction may be added to the budget resolution to affect tax or mandatory spending changes. When this is done, it results in a reconciliation bill drafted by various committees at the direction of the budget committee and submitted to Congress by the Budget Committees. Almost all of the major budget and tax changes of the last two decades have come as a result of reconciliation bills, starting with the Reagan administration sponsored and congressionally approved tax cut of 1981. Usually reconciliation bills do not affect defense, but remain focused on tax and entitlement matters.

Appropriation Bills: Discretionary spending for the federal government is provided by annual appropriation bills, including the defense appropriation bill. The budget resolution comes first. Once it has been passed, serious work may begin on the appropriation bills. The leaders of this process are the appropriation sub-committees for each bill, e.g., defense, transportation, agriculture and so on. They hold hearings, question witnesses from the agency or department as well as independent experts and lobbyist or other interested parties (e.g., defense contractors, GAO), review the departmental request, listen to committee staff, and react to the chair's mark (suggested list of changes). When the chairman provides a mark, the committee generally supports that mark; after all, the chairman holds his position because his party controls a majority of votes on the committee. The bill must then stand for full committee scrutiny and then pass on to floor deliberation, debate, and amendment. When each chamber has passed its version of the bill, a conference committee is appointed to resolve differences between the two versions. As a guideline, appropriation bills are supposed to be out of the House by the end of June and enacted before the new fiscal year begins. In general, the House often meets the end of June test, but appropriation bills are seldom passed by 1 October.

Continuing Resolution Appropriation (CRA): When no new appropriation has been passed and the fiscal year is about to begin, Congress passes a Continuing Resolution Appropriation (CRA) to cover the gap. The CRA provides agencies with budget authority to operate in the interim. The amount of money provided may be the current rate or an amount set in a bill passed by one chamber or one committee in one chamber. It is usually set at the current operating or rate. For example, for FY2011, the initial CRA was clear to stipulate the FY2010 rate, or the lower of the bills passed the House or Senate, or the lowest of the FY2010 rate or the bills passed the House or Senate. The intention is clearly to fund at the lowest possible level. This means that no new personnel can be hired, no new programs started, no new equipment purchased and so on. The purpose of a CRA is meant to be quite restrictive, with no or minimal new activities. This is seen in how the Department of Defense was treated by a September 30, 2005 CRA. This specified for DOD "...no new production of items not funded in FY2005 (the preceding year fiscal year), no increase in production rates sustained with FY2005 funds or the initiation, resumption, or continuation of any project or activity... for which appropriations... were not available during fiscal year 2005." The CRA also advised that no multi-year procurement programs could be entered into. The CRA did give the Secretary of Defense the authority to initiate projects or activities required for 'force protection purposes, using funds from the Iraq Freedom Fund, following notification of the Congressional Defense committees (normally the House and Senate Armed Services Committees and the Appropriations sub-committees on Defense of each chamber). In allowing purchases for force protection needs, this CRA did allow for "new" items/projects not in the current budget base, if the Secretary of Defense deemed it necessary, so the blanket statement can not be made that new programs are never permitted.. In general CRA's are not meant to be controversial. When it votes a CRA, Congress picks the appropriate period for it, a morning, a day, a week, a month, or whatever it decides is necessary. The time chosen indicates roughly how long Congressional leadership thinks it will take to come to a compromise and pass the remaining appropriation bill or bills. As individual bills pass, each ensuing CRA may cover fewer and fewer appropriations, until finally all appropriations have been provided. In some years, compromise is very difficult and an omnibus continuing resolution appropriation may be passed to include all remaining appropriation bills for the remainder of the fiscal year. This has been the rule, rather than the exception during the last decade.

Authorization Bills: Authorization bills create or modify programs, providing program authority. They establish the department and its mission in the first place and make any changes to it subsequently. Defense has an annual authorization cycle, but other policy areas may have different cycles for the authorization process, from three to five years to permanent authorization. In defense, annual authorizing bills may set limits on what appropriators may appropriate for the program created in the authorizing bill, but appropriators do not have to follow authorization dictates. An authorization bill does not make money available; only the appropriation bill does this. The Defense authorizing committees see themselves as helping inform the appropriators on major defense policy issues, thus they try to keep the authorization bill ahead of or even with the appropriation bill in the congressional budgeting cycle. This does not always work; sometimes the authorizers get involved with treaties, test ban limits, when to commit American troops and other controversial issues, with the result that the authorization bill is passed after the appropriation bill. When this happens, it is good to remember that it is the appropriation bill that provides the money. Recently the authorization bill has established policy for military pay, benefits, health care, retirement and a number of quality of life areas for uniformed personnel in defense and it is always of interest to defense contractors and weapon system suppliers hoping to get additional systems authorized for procurement.

For example, in the summer of 2006 the House and Senate approved a three-year buy of 60 F-22 Raptors in their versions of authorization bills. Usually aircraft are bought annually; what the two authorization bills did was provide for buys of 20 a year for three years. Multiple year buys are doable, but there are tests the weapons system must meet, related to cost savings, a stable weapon system, and a stable mission. According to the auditing agency for Congress, the Government Accountability Office (GAO), the F-22 did not meet four of the six requirements to qualify for multiple year purchase. In fact, GAO said that the multiyear buy appeared to drive costs up, not down. What made this debate particularly interesting was that the Senate approved its version of the bill 70 to 28, over the resistance of the Chairman of the Senate Armed Services Committee, Senator Warner (R-VA) and that of the ranking minority member Senator Levin (D-MI) and Senator McCain (R-AZ) the chair of the Air Land subcommittee of the Senate Armed Services Committee. Senator McCain then held a hearing in late July during the conference committee time period to question the multiyear buy, a hearing at which Senator Warner appeared and said that he fully supported McCain's actions. The multiyear buy was inserted into the authorization bill during Senate floor debate on the authorization bill, lead by Senator Saxby Chambliss (R.) from Georgia where the F-22's are built. The Armed Services Committee opposed the purchase, but it should be noted that not only was Senator Chambliss a member of the Senate Armed Services Committee, he was also a member of McCain's Air Land sub-committee. Some observers felt that this was a triumph of constituent interest over party and committee discipline. They also noted that the F-22 buy had been capped by Secretary Rumsfeld at 183 and that the multiyear buy would extend the program into 2011, into a new administration and a new Secretary of Defense, thus creating an opportunity for the Air Force to renegotiate its goal of 381 F-22's. (Defense News, 2006: 1). The point is that this congressional guidance process is not as simple as it might appear. In this case the contractor, Lockheed Martin, and Senator Chambliss from Georgia and the Air Force appeared to have formed a coalition intended to counter both the authorization committees, political party discipline and a decision already made by the Secretary of Defense.

Supplemental Appropriation Bills: Supplemental appropriations occur when emergency needs dictate, for natural disasters and for defense needs, such as the $48 billion supplemental passed after September 11, 2001. In defense, they generally supply funding to replenish accounts drawn down in response to mission tasking generated by the President that was not foreseen in the annual budget such as evacuation of American citizens or embassy personnel or providing aid and comfort to victims of earthquakes, floods, and other natural disasters in foreign countries. Supplementals are meant to be largely non-controversial; they allow for quick response to an unpredictable emergent need, the money used out of current funds and then reimbursed later, but still within the current fiscal year. However, the 'war supplementals' passed after 2003 are somewhat different. In 2006, despite supplementals amounting to more than $100 billion for war on terrorism, in July the Army was in the uncomfortable position of freezing travel and hiring and laying off temporary employees, at some bases, while other bases were running at full capacity, these inequities seemingly due to the way the supplementals were accounted for. The dynamics of supplemental appropriations are analyzed in chapter 6.

While no two legislative sessions are identical, benchmarks do exist. The following are suggested key dates for monitoring the legislative budget and appropriation processes:

1. First Monday in February: President sends budget to Congress
2. 1 April: Senate Budget Committee reports out the budget resolution
3. 15 April: Conference Committee report on the budget resolution passes both chambers.

4. 30 June: all appropriation bills passed by the House.
5. 1 October: all appropriation bills passed.
6. Anytime: the defense authorization bill precedes the defense appropriation bill.
7. By mid-August: defense supplemental is passed (if any). If passed later than this, the supplemental may be caught up in the end of fiscal year politics.

The final steps to coordinate bills in Congress involve appointing a conference committee of leaders from each chamber to meet and reconcile the provisions that are different in each bill. When a bill is passed in the House or Senate, the different constituencies will result in different provisions in a bill, thus it is the job of the conference committee to iron out the differences and get a unified version that will be supported by both chambers. The conference committee only exits for that time period it takes to meet and hammer out a compromise that will stand in both chambers. If it is a defense appropriation bill, the conference committee will include the appropriations sub-committee Chairman and ranking member (senior member from the other party on the committee) other members of the defense appropriations sub-committee from each chamber, and these will be supplemented by key party leaders from the authorization committees, and perhaps the budget committees or the party leadership. Both parties are represented. Interestingly, the conference committee is not necessarily bound to what is in either bill before them; if a solution takes an idea or proposal not in either bill, the conference committee can include it in the report. Conference committees are the focus of intense lobbying efforts. By law, when the House and Senate versions of the DOD appropriation bill differ, DOD may submit an appeal to the conference committee favoring its position. This may or may not be successful. For example, in 2000, the House had cut $48 million from the $305 million request for DD-21 destroyer class ships and the Conference Committee allowed $292 million. Other interests also lobby the Conference committee. In another example from 2000, the Presidents budget asked for 4 C-130 cargo planes, the House and Senate gave 5 and the Conference Committee allowed 6 (Congressional Quarterly Almanac 2000: 2–51). The Conference Committee Report does have to gain a majority vote of each chamber in an all or nothing vote process. Conference committees are very powerful, but they are disciplined by the full membership when it votes on the conference committee report.

The budget and appropriations process is described in the graphic below. In general, the budget committees should finish their work before the appropriation committees. Both the appropriation committees and authorization committees may send views and estimates letters to the budget committees to help them decide how much to set aside for the programs under their jurisdiction. The budget resolution conference report includes 302a allocations for the appropriations committees. When these are passed on to the appropriation subcommittee for separate bills (defense, agriculture) they become 302b allocation targets. The Senate Appropriations Committee issued the information below in a press release on July 19, 2003. It contains the 302b allocations for the appropriations bills measured against what was enacted the previous year and what the President requested for the current year.

When reconciliation is called for in the budget resolution, the conference report also contains reconciliation instructions for the authorizing committees advising each committee of how much it is expected to save in its programs to meet the reconciliation changes. While the Budget Resolution is not signed by the President and does not become law, the reconciliation bill does become law and therefore must be signed by the President.

It is a fact of life that because Congress rarely passes budgets before the beginning of the fiscal year, departments and agencies often begin each fiscal year under a Continuing Resolution Appropriation (CRA).

BUDGET EXECUTION

After the appropriation bill has been passed by Congress and signed by the President, the process for providing spending authority to departments and agencies begins. OMB and the Department of the Treasury apportion money to the Departments which in turn allot money to their sub-units. Each agency head then uses allotments to delegate to subordinates the authority to incur a specific amount of obligations. These allotments may be further subdivided into allocations for lower administrative levels. Following these allotments and allocations, obligations can be incurred (e.g., a contract issued) and outlays are paid when the work or service is completed or supplies and equipment delivered.

Apportionment, allotment, and allocation processes are guided by department planning for when funds will be spent, by quarter and month, and by administrative level. This process also requires departments to resubmit their budgets to OMB for approval, indicating how actual appropriations, rather than the proposals included in the President's Budget, will be spent. Department requests must be approved jointly by OMB and the Treasury before money is approved for expenditure and made available for obligation in department and agency accounts maintained by the Treasury. In effect, the apportionment/allotment process represents a separate mini-budget cycle within the executive, although its major focus is on when dollars will be spent within the fiscal year and to a lesser extent, what the mix of consumables will be within the categories approved in the appropriations bill.

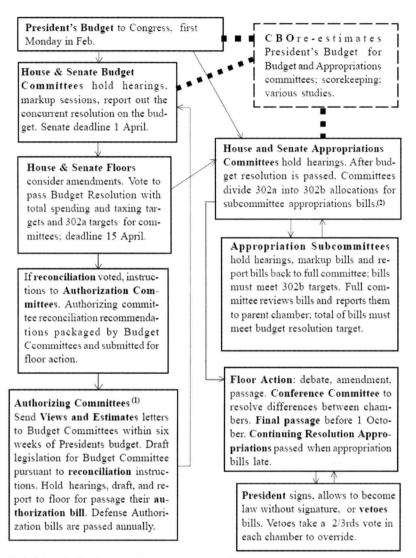

President's Budget to Congress, first Monday in Feb.

House & Senate Budget Committees hold hearings, markup sessions, report out the concurrent resolution on the budget. Senate deadline 1 April.

CBO re-estimates President's Budget for Budget and Appropriations committees; scorekeeping; various studies.

House & Senate Floors consider amendments. Vote to pass Budget Resolution with total spending and taxing targets and 302a targets for committees; deadline 15 April.

House and Senate Appropriations Committees hold hearings. After budget resolution is passed, Committees divide 302a into 302b allocations for subcommittee appropriations bills.[2]

If **reconciliation** voted, instructions to **Authorization Committees.** Authorizing committee reconciliation recommendations packaged by Budget Ccommittees and submitted for floor action.

Appropriation Subcommittees hold hearings, markup bills and report bills back to full committee; bills must meet 302b targets. Full committee reviews bills and reports them to parent chamber; total of bills must meet budget resolution target.

Authorizing Committees [1]
Send **Views and Estimates** letters to Budget Committees within six weeks of Presidents budget. Draft legislation for Budget Committee pursuant to **reconciliation** instructions. Hold hearings, draft, and report to floor for passage their **authorization bill.** Defense Authorization bills are passed annually.

Floor Action: debate, amendment, passage. **Conference Committee** to resolve differences between chambers. **Final passage** before 1 October. **Continuing Resolution Appropriations** passed when appropriation bills late.

President signs, allows to become law without signature, or **vetoes** bills. Vetoes take a 2/3rds vote in each chamber to override.

Exhibit 2.2: Budget and Appropriation Process

(1) **House Authorizing Committees:** Agriculture; Banking & Financial Services; Housing & Urban Affairs; Commerce, Economic & Educational Opportunities; Government Reform & Oversight; House Oversight; International Relations; Judiciary; National Security; Resources; Science; Select Intelligence; Small Business; Transportation & Infrastructure; Veterans' Affairs; Ways & Means. **Senate Authorizing Committees:** Agriculture, Nutrition & Forestry; Armed Services; Banking, Housing & Urban Affairs; Commerce, Science & Transportation; Energy & Natural Resources; Environment & Public Works; Finance; Foreign Relations; Government Affairs; Indian Affairs; Judiciary; Labor & Human Resources; Rules & Administration; Select Intelligence; Small Business; Special Aging; Veterans' Affairs.

(2) **House & Senate Appropriations Bills:** Agriculture & Rural Development; Commerce, Justice, State; Judiciary; District of Columbia; Energy & Water Development; Foreign Operations; Interior; Labor, HHS, Education; Legislative Branch; Military Construction; Defense; Transportation; Treasury and General Govt.; VA, HUD, and Independent Agencies.

Figure 1.3 Congressional Budget and Appropriation Process. Source: U.S. Senate, Budget Committee, 2011.

Once the appropriations have been allotted, departments allocate their budgets to their sub-unit agency budget staff, which then prepare and issue spending authority and guidance to the various program components where spending obligations are incurred, services delivered and resources consumed. It is important to recognize that some agencies are very large and have huge budgets. In the Department of Defense (DOD), the Office of the Secretary of Defense and the DOD Comptroller receive and allocate the budget for national defense appropriated by Congress. Among the agencies to which funding authority is provided are the Departments of the Army, Navy and Air Force.

In allocating the budget, the central budget staffs of departments and large agencies do not completely free the programmatic side of their enterprises to spend as they wish, despite the desire for such flexibility on the part of those who spend. Rather, spending is accompanied by constant monitoring and control by central budget office staff as well as by budget, accounting, and audit staff internal to the program units. Spending is monitored in terms of actual rates versus those projected, and by other variables, including legality and purpose of expenditure (of utmost importance, spending has to conform to the appropriation and other attendant control language), schedule and timing, location, measures of production and volume, and other variables. In essence, monitoring looks for variances between planned and actual spending that then have to be accommodated through management and control.

Where funding is not available in places most needed, reprogramming or transfers (defined subsequently) are requested. Where funding is exhausted due to exceptional circumstances (e.g., natural disasters), supplemental funding requests are sent to and approved by Congress. Moving money to the highest priority and executing the full amounts appropriated are the key tasks of budget execution. Execution is often taken for granted, but it is a very important phase of the budget cycle. Execution is where services are either provided or not, are provided efficiently or not, and where needs are met or neglected.

While the act of budgeting is a planning process, budget execution is a management process. This process is treated in detail in following chapter. In budget execution, agencies obligate or commit funds in pursuit of accomplishing their program goals. Following plans made in the budget preparation cycle, employees or contractors are engaged, materials and supplies purchased, contracts let, and capital equipment purchased. The basic assumption is that the budget will be executed as it was built once it has been approved. In many jurisdictions, this concept has legal backing; the summary numbers that appear in the budget documents stand for all the little numbers behind them. Executing the budget simply means going back to the detailed spreadsheets. Many also assume that this execution phase is a relatively simple task compared to preparing and passing the budget, since all it involves is returning to the budget building documents and executing the plans described in them, as modified by the final version of the appropriations bill. The reality of administrative life is somewhat different.

A substantial portion of budget execution is driven by the necessity of rescuing careful plans from unforeseen events and emergencies and unknowable contingencies. At the close of the fiscal year, in the aggregate and on average, budget execution may appear to have been a matter of uninteresting routine dominated by financial control procedures, but it is unlikely to have appeared so uneventful to the department budget officer and his staff or the manager charged with carrying out the program. The usual occurrence is for most of the budget year to unroll as planned, but for execution of a small percentage of the budget to consume a major investment of managerial and leadership effort. All jurisdictions have set rules and procedures to help guide agencies through the budget execution process. The tensions in this process involve issues of control, flexibility, and the proper use of public funds.

Audit and Evaluation

The final phase of the budget execution process involves audit and evaluation. In this phase the disbursement of public money is scrutinized to assure officials and the public, first, that funds were used in accordance with legislative intent and that no monies were spent illegally or for personal gain and, secondly, that public agencies are carrying out programs and activities in the most efficient and effective manner pursuant to legal and institutional constraints.

The first type of audit is called a financial audit. It concentrates on reviews of financial documents to ensure that products and services are delivered as agreed on, payment is accurate and prompt, that no money is siphoned off for personal use, and that all transactions follow the legal codes and restrictions of the jurisdiction. Such audits are generally carried out or supervised by agents external to the entity undergoing audit. These agents include the General Accounting Office and agency Inspectors General at the federal level as well as internal auditing agencies (for example, the Navy Audit Service), elected or appointed state auditors at the state level and by private sector accounting firms at all levels of government. Some accounting occurs between levels of government–the federal government audits use of federal monies in state and local programs while state governments audit certain local fiscal practices.

Financial audits evaluate honesty and correctness in handling money. As the role of government expanded after World War II, more effort was spent to measure the efficiency and effectiveness of government. This led to experiments with performance budgeting at different levels of government and ultimately to performance auditing. Here the auditor attempts to ensure that the agency is conducting programs in a manner consistent with the laws and regulations that authorize the program and to ensure that the agency has taken judicious action in resource deployment to attain programmatic ends. Basically, the auditor is attempting to judge efficiency in resource use and effectiveness in program delivery. In theory at least, the findings from such audits help policy makers enhance program outcomes while minimizing the resources required to operate programs. Often the same agencies responsible for financial audits also do program audits, but since the audit focus differs, personnel with different skills are required. The reports often become part of the budget process, especially during the legislative stage, and may lead to changes in the laws and rules that guide the program and to the managerial practices of the agency that administers it.

Audit findings may also be reviewed in special hearings by oversight committees, outside the budget process, who want to ensure that agencies behave responsibly. Generally, these audits are published and available to the public. Both financial audits and performance audits help ensure that a jurisdiction is getting the most for its tax dollars. In so doing, they help maintain the vital element of trust in government and in those who disburse its monies, and create and administer its programs. In general, the same entities who audit for propriety may also be called upon to audit for performance, including GAO at the federal level and legislative audit agencies at the state level. OMB also has a management review staff and at state and local levels, the budget agencies often require that budget analysts combine managerial analysis work with budget analysis.

After all of this activity concludes at the end of the fiscal year and accounts are closed-out to prevent further obligation, budget accounts are audited internally and externally by agency auditors, Inspectors General, the General Accounting Office and other audit agents. In some cases, private firms conduct parts of or all of these audits. After or in conjunction with auditing, programs are evaluated to see to what extent they have met the objectives and commitments promised to Congress and the executive, and policies are analyzed for their value. All of the information thus developed becomes grist for the mill for preparation of future budgets, as the cycle operates continuously.

Revenue and Taxation Legislation: A Different Cycle

A final note is required regarding the revenue side of budgets. In general, it is only at the top level of the budget staff of Congress, the Budget

Committee and Congressional Budget Office (CBO) and the executive Office of Management and Budget (OMB) for the President, where revenue and expenditure estimates for the upcoming year are totaled based upon what Congress passes in appropriations. Simultaneously, as the President and Congress propose and enact the spending plan for the upcoming year, expenditure budgeting management and control of the current year budget is the pervasive activity that occurs throughout the executive branch departments and agencies as organizational sub-units compete for their share of funds. Ordinarily, departments and agencies do not concern themselves with revenues other than fees and charges they are allowed to retain in current or future year budgets, and others they collect incidental to their own operation, e.g., National Park admission fees or federal license fees.

For the most part, departments and agencies compete for their share of general revenues accumulated from the tax mechanism. For example, the Department of Defense and the Department of Health and Human Services do not concern themselves much with general government tax policy and revenue generation. Instead, they rely on Congress and the President to tell them if their needs can be accommodated within revenues approved for the current and next budget year.

BUDGET FORMATS

The budget of the federal government, indeed of all governments, has many forms. It can be rendered simply as a single line with a few descriptors (e.g., agency name) and funding proposed or available for spending, or as complexly as is imaginable, containing all kinds of programmatic and even performance–related information and associated funding. Essentially, the only constant in budgets is that funding (dollars in the U.S.) is shown in nominal values, i.e., not discounted over time.

All of the different budget formats have their particular uses, as is the case for the myriad of ways in which expenditure plans and programs are displayed. For example, the display by function for Defense includes expenditures in other departments that are defense-related, e.g., nuclear energy expenditures in the Department of Energy, civil defense programs in the Federal Emergence Management Administration (FEMA), and defense-related activities in other agencies such as the Department of Homeland Security including the Coast Guard and the FBI.

That the format of the budget depends on its intended use comes as no surprise; what is surprising is that the budget is used in so many different venues and contexts. To understand why this is so requires considerable familiarity with the decision process cycle employed to determine and manage resource allocation. Why are so many formats—probably too many—used in federal budgeting? Technically, the fundamental task in budgeting is twofold; it lies in predicting the future and evaluating the past. Visions about the future are complicated by differing visions of what the future should be, as well as arguments about what it will be given current understandings of causal relationships. Since government is a coercive arrangement and can extract money and time from its citizens and deliver goods and services unequally, say for example more to those who are more successful in lobbying and legislating, thus budgeting tends to be a conservative process where players examine, in excruciating detail, the evidence on decisions that have been made and information about decisions that are about to be made. Consequently, the great problem in budgeting is information overload. Reformers have attempted to address this problem in a variety of ways, with the executive budget movement, the creation of central budget bureaus and legislative staff agencies, public hearings, and published documents describing the budget. Much of this effort involves a rationalistic sense that a better budget process results from better information.

Reformers have also attempted to solve the information overload problem by creating better budget systems. Historically, the basic budgeting system has been an object of expenditure system, which features objects of expenditures, arrayed in lines of items in spreadsheet format with their names and proposed expenditures, hence the name line item budget. The items are usually personnel and supporting expenses, like travel, telephone, and offices supplies. In its early form, this type of budget format was closely identified with the accounting system, thus making it relatively simple to execute and audit. In small jurisdictions, with few functions and a limited and homogeneous role for government, this type of budget is still simple to construct, review, execute, and audit.

The format of a line item budget presentation includes the line items, what each category cost the previous year, its authorized level in the current year, and the amount requested for funding in the budget year. If this is a biennial presentation, two budget years will be shown. Usually, these documents will also show the percentage of change from the current year to the budget year, thus enabling reviewers to scan the presentation and select objects for review, for example, that which increased the most. This type of budget emphasizes control and fiscal accountability, rather than management or planning (Schick, 1966). Usually the budget document will have introductory narrative explaining the purposes of the unit; this usually changes minimally from year to year and is almost universally referred to as 'boilerplate.' Budget officers feel that once they have it right, all they need to do is fine-tune it with this year's emphasis. Most, but not all, line item systems also will have some explanation of changes, usually very brief, and very carefully written. Writers know that readers will not read long justifications, but they also know that the justification will focus how the budget request is perceived. In many budget shops, the less experienced budget analysts write the boilerplate and the experts write justification paragraphs.

At the agency level, the line item display usually carries with it a list of all personnel employed in the unit and their salary cost. This total is transferred to the personal services line on the budget display. Supporting expense categories will be displayed by item, e.g., postage, travel, office supplies, and so on. A list of minor capital outlay items (desks, chairs,) will be constructed; this may or may not accompany the budget, but the agency budget analyst will use the exhibit to monitor the agency purchases during the budget year. It is easy to see how this type of budget lends itself to control purposes, both for accounting and for managerial control.

TYPES OF BUDGETS

Budgets may be divided into three basic types: operating, capital, and cash. The operating budget contains funds to be spent on short-term consumables including the personnel payroll and the goods and services that keep government agencies in business on a day-to-day basis. The capital budget plans for and purchases long-term assets such as land for parks and buildings. The cash budget is used to manage the cash flow of a government or government agency.

The Operating Budget

The operating budget includes salaries and benefits for employees, funding for utilities, money to contract for trash management, maintenance and supplies, space rental and lease payments, pens, pencils, copy machine paper and so on. Typically, employee salaries and benefits are the highest cost item in the operating budget, representing anywhere from 60% to 90% of spending. Personnel are usually listed by position by type and grade and perhaps seniority, with a certain percentages added on to cover the cost of fringe benefits including vacation, sick leave, health, life and disability insurance and retirement plan costs.

The operating budget funds the annual operational needs of the jurisdiction, ranging from road maintenance, to park and recreation supervision, tax collection, education, welfare, public safety and defense. The time period for spending the funds in the operating budget is almost always one year at state and local levels. At the federal level, the time period of obligational availability varies depending on the type of appropriation.

The operating budget usually includes minor capital outlay amounts for desks and office machines and the like, below a specified item cost threshold, e.g., $10,000. Consumable items range from computer parts, to telephone installations, to pencils. In line-item budgets, the line includes the name of the item, the budgeted amount for the current year and a requested amount for the next year. Many jurisdictions also include what was spent on that item the previous year. Revenue sources may be broken into equally excruciating detail, not only general taxes, e.g., sales, property and income, but also revenue from fees, charges and miscellaneous sources, including parking fines, dog and bicycle licenses, and garage sale permits. When the line-items are re-organized into programs, the budget emphasizes program accomplishment and is called a program budget. A performance budget changes the mix so that consumables are attributed to activities accomplished and a cost may then be attributed to each activity output, such as the cost to construct a mile of road or the cost to collect a ton of refuse. The same budget may be displayed in different ways; the defense budget is commonly displayed by appropriation for Congressional purposes and force structure and mission within the Department of Defense.

The Capital Budget

The capital budget contains funding for purchase of long-lived physical assets such as buildings, bridges, land for parks, and high cost equipment such as aircraft, ships, etc. Many, but not all, state government and other jurisdictions in the U.S. have separate capital and operating budgets. The federal government does not have a separate capital budget, but it has distinct capital accounts, mostly used to pay for acquisition of military hardware in the Defense Department. The capital budget is usually used for long-term investment type functions, including buying parklands and constructing buildings, bridges, and roads, where consumption spans over a period of years or decades. Thus, the capital budget is appropriated and consumed on a multi-year basis. In state and local governments, capital projects are usually paid for through issuance of bonds whose principal provides the money for the capital budget projects. These bonds normally are paid back over a time period that approximates the consumption of the asset, e.g., thirty years. In contrast, the U.S. federal government appropriates money for capital consumption out of the annual operating budget, and obligates and outlays it over multiple year terms as the project is executed. This is how the interstate highway system was funded and built in the 1950's and 60's. On the defense side, weapons systems such as aircraft carriers and certain aircraft have enjoyed lifespans measured in decades, thanks to outstanding initial design and continuous modification processes. Both initial purchase and subsequent modifications were funded by annual appropriation bills.

It also should be noted that capital and operating budgets must be linked to provide services. Where the capital budget pays for construction of an office building, it is the operating budget that pays for the furniture and equipment to make the building ready for use, and for daily activities of maintenance in and around the building. This is true both in the defense and non-defense sectors. In defense, a new aircraft will have to have pilots, a pilot training program, weapons, hangar space, and spare parts funded out of the annual operating budget. On the non-defense side, the capital budget will prepare a project that will be taken over and operated out of the operating budget, including maintenance personnel, paint and other maintenance and other items, such as snow removal for highways. Good budget analysis always means checks are made to ensure that capital projects scheduled to come on line are fully funded on the operating side.

The Cash Budget

Cash budgets are used to manage the cash flow demands of operation on a daily, weekly, monthly and annual basis. The objective of cash budgeting is to insure the liquidity and solvency of government and its agencies. Liquidity may be defined as having the cash and readily convertible assets that are used to pay the government's bills, i.e., its short-term liabilities. In contrast, solvency refers to the ability of government to sustain operations over the long-term, i.e., greater than one year. Typically, the cash budget is managed in close coordination with the investment functions of the treasury department or other cash controlling entity, e.g., the comptroller or department of finance. The task for proper cash management is to have enough cash available to cover current liabilities, but not too much, so that opportunity to invest and earn interest from surplus cash is wasted. In the private sector, what is termed the "quick ratio" is used as a measure of cash flow management. A ratio of 1.1 or 1.2 cash on hand to 1.00 liabilities is appropriate. Ratios lower than this imperils the ability to meet payment obligations; a higher ratio wastes investment opportunity.

PERSPECTIVES ON BUDGET THEORY

Scholars in political science and public administration tend to examine budgeting in terms of who gets what, why, and under what conditions. Those in public administration typically are concerned with how the budget is proposed, decided upon, how services are produced, and how budget information is presented and analyzed. They take great interest in the internal workings of government and governance, a propensity shared by political science from which public administration was born. Matters of policy and program effectiveness are of primary concern from this perspective. Public administration attempts to improve the welfare of citizens through attempts to make governance more responsive and government more effective. Often, their prescription for improving effectiveness includes employing more staff and spending more money to achieve unequivocally reasonable and desirable objectives.

Criticism of the public administration perspective on budgeting is that it is too focused on process, overly concerned with function versus output and outcomes, and obsessed with the details of congressional, presidential, and agency decision-making and service production, to little consequence. However, it may be pointed out that many of the reforms that have made governance more responsive and governments more effective in the past one hundred years or so are wholly or in part attributable to reformist efforts led by scholars in public administration.

On the other hand, economists tend to examine budgeting from the perspectives of equity (who pays and who benefits, i.e., the distributive consequences of tax and allocation decisions), allocative efficiency, and stability. Our interpretation of this perspective, generally speaking, is that it assumes that the role of government is to assess and determine the validity of arguments made by various claimants for shares of the distribution of public money, i.e., to define equity in practice. Equity in the context of budgeting means what is "fair" given that how we define "fairness" is always a debatable question in a democracy, and that some degree of competition for resources is inevitable in any socio-economic system, because demand always exceeds supply. How we define what is fair varies over time and is always a primary consideration in the numerous forums of government decision-making.

To some, distributional equity refers to policies that transfer income from one set of citizens to another through tax policy, spending, and other government actions. Others vehemently disagree with this perspective and generally do not acquiesce to the redistributive role of government. Clearly, the definition of what is fair in a democracy is virtually always up for grabs— a work in progress—ever changing relative to the political will of the people and their elected representatives. We are quite aware that fair is sometimes, but certainly not always, defined as equal. But, given variances in need, income, wealth and other variables, distribution of spending based on an "equal share for all" rule would, in many instances, not be judged as fair.

To economists, efficiency means how the decisions of the government affect the productivity of the private sector and the economy as a whole. Efficiency in this context is not concerned with whether the internal operations of government operate in a managerially efficient manner. Rather, efficiency is determined in essence by economic decisions about what should be produced, how and by whom. The means of production and what is produced are determined through the interaction of the public and private sectors of the economy, and in choices over which goods and services should be provided by government and which are better provided by the private sector. To some, the government is regarded as a net drag on the private economy, causing the sacrifice of efficiency in pursuit of equity objectives. From this view, the best government is that which governs least. To others, the essence of the role of government is to supply equity, a good that cannot and will never be supplied purely as a function of the pursuit of efficiency in the private sector. The mixed capitalist-socialist form of economy that prevails in the U.S. represents a compromise between these polar perspectives. As such, the trade-offs that are forced constantly between equity and efficiency are the very stuff that makes the government budget process intensely competitive.

From an economic perspective, stability refers to policies pursued through the budget to stabilize the economy, in conjunction with the fiscal and monetary policies of the government and the actions and productivity of the private sector. Stability is measured in terms of prices, employment, growth of the gross domestic product and other indices. Just as trade-offs are made between equity and efficiency, inevitably potential trade-offs may be considered between each of these and stability under some circumstances. Balancing these exchanges is in part a result of decisions made in the ongoing cycle of government budgeting and financial management. We subscribe to the perspectives of both political science—public administration and economics. Our hybrid view attempts to draw in an interdisciplinary manner on both perspectives, that is to say that neither view is wrong—but that each emphasizes different aspects and ways of understanding budgeting and spending outcomes. Now that we understand some of the basics of budgeting, we need to understand something about the history of federal government budgeting and financial management. Then, in chapter 3 we consider the intricacies of defense budgeting.

NOTE

1. Different schools of thought exist on how individuals and groups gain and exert power in society. In this book we pursue an institutional model embedded in a pluralist model of democracy. This is our way of saying that groups and individuals have access to the institutions of governmental power and use that access to govern through the distribution of budget (and tax) policy. We believe there is no set of dominant groups that always controls American society and that it is not controlled by the elite few who control

and profit from the efforts of the many. We believe laws, institutions, and individual behavior matter and that individuals are well-advised to gather into like-interested groups to affect political action. There are other schools of thought on these matters, from those who would argue that a focus on institutions is too narrow, that groups control everything and the individual is powerless, or that elites rule behind the scenes and what is seen publicly is a charade. There are also schools of thought about the policy process, that it is rational or incremental or that game theory or public choice theory provide better explanations for how decisions get made or can be explained. In these debates, we tend to be mid-level rational focusing on the science of muddling through, the virtues of incremental behavior and the necessity to work the numbers. We suggest that this is typical for budgeteers. For those interested in the larger debates about society and power, we suggest starting with the work of Thomas Dye, S. M. Lipsett, or the classical political theorists, e.g., Plato, Aristotle, Machiavelli, Hobbes.

CHAPTER 2

HISTORY AND DEVELOPMENT OF FEDERAL GOVERNMENT BUDGETING

Executive and Legislative Branch Competition

INTRODUCTION

Defense spending has been at the center of budgetary history before this country was born. Taxing for defense of the colonies against the French and along the borderlands led to resentment of the King and charges of taxation without representation. Staggering debts were accrued during the Revolutionary War and the existence of the new nation was threatened if it could not develop a stable currency and pay off its war debt. Debate occurred over what the new President's budgetary power should be and Congress first tried to run the country and provide for its defense without an executive branch. Later, with the beginnings of a budget system, arguments transpired over whether lump sum or line item appropriation was better, over the idea of control and how much was appropriate, particularly for important functions like the Army and Navy. Presidents did not have a budget office until 1921 and different Presidents took different views of the

Financing National Defense: Policy and Process, pages 33–69
Copyright © 2012 by Information Age Publishing
All rights of reproduction in any form reserved.

budget power. This notwithstanding, crises continued to occur and with them came stress for the fiscal system, especially after WWI and WWII that resulted in budget reforms and responsibilities. It is clear that the Constitution gives the President great latitude to act as commander in chief when the nation is attacked or threatened, but Congress also has been given the power of the purse as a check on the power of the President. By and large this sharing of power has resulted in satisfactory outcomes as the president was able to act swiftly in the nation's defense (e.g., President Truman's commitment of troops to Korea in 1950) and then explain and justify his actions later to Congress. There have been times however when this sharing of power has been tested, in the latter years of the Viet Nam conflict and in 2006, 2007 and subsequently when Congress sought to direct how the President should bring the war in Iraq to a successful conclusion by including detailed guidance in the supplemental appropriation bills, especially in 2007. In this chapter we review some of this history to explain how we got where we are presently.

BUDGETING IN THE EARLY YEARS OF THE NATION

A review of the history of budgeting in the United States reveals debate over two prominent questions: how the budgetary power should be divided between Congress and the President, and how the budget is to be employed as a tool to better govern and manage. (McCaffery and Jones, 2001: chapter 2.) The question of how power is shared between Congress and the executive branch is particularly important for defense, since confusion arises over the correct division of power between the president who must function as commander in chief and Congress which has been given the power of the purse by the Constitution. The struggle for power (or over how best to discharge their lawful responsibilities) between the executive and the legislative branches has been a recurrent theme in the American budget process since the founding of the nation. Shortly after the Revolutionary War, Congress appeared to have taken the initiative in competition over the proper role of the two branches in debating whether the executive was legally empowered under the Constitution to make budget estimates, and whether the Secretary of Treasury could or should submit a budget framework to Congress (Burkhead, 1959: 3). Members of Congress were opposed to having the Secretary of Treasury even submit plans to Congress for the following fiscal year (Browne, 1949: 12). In contrast, beginning with the Budget and Accounting Act of 1921, by the late 1960s the steady accretion of power in the executive branch, and within it in the Bureau of the Budget, provided the Chief Executive significant leverage in the budget process. Presidential use of impoundment power and the politicization of the Bureau of the Budget essentially consolidated gains for the Presidency over Congress and over the executive branch as well.

However, just as nature abhors a vacuum, the American federal system abhors an imbalance of power. The Congressional Budget Impoundment and Control Act of 1974 re-established Congress's role in the budget process. In 1980, Congress used the provisions of this Act to rewrite the President's budget. While President Ronald Reagan used the Reconciliation Instruction from the 1974 Act to seize a great victory in Congress in 1981, Congress again rewrote the President's budget in 1982. The years since 1974 have witnessed a reversal of form of most of the twentieth-century practices that had seen Congress increasingly relegated to the role of making marginal and incremental changes in the President's budget. Since 1976 Congress has had its own budget plan to contrast to the President's budget request. Notwithstanding this, any enlightened observer would have to cede the weight of power to the executive branch and within it the President's budget office, the Office of Management and Budget. Such an observer would also have to conclude that Congressional procedures have not led to a timely or orderly appropriations process and that no two years are alike. Allen Schick has termed this 'improvisational budgeting.' (Schick, 1990: 159–196) Indeed in recent years, Congress sometimes has not been able to agree on a budget plan. However, even given all this turbulence, there is no doubt but that Congress has legitimate power over the purse and intends to exercise it. In retrospect, the Republican Congresses of 1995, 1997 and to some extent in 2010 seized much of the initiative from the President and drove much of the budget planning for a balanced budget with seven and five year budget plans.

In 1789, arguments over the balance of power centered on the extent to which the President should prepare a budget; and although this is no longer debatable, the limits to the advantage that preparation gives the executive branch still is arguable. This theme is still as current as it was in 1789. Within this recurrent theme is an idea central to American democracy: power is balanced at the national level between the executive and legislative branches. When a serious imbalance occurs, corrective action ensues to restore and ensure the balance so that when one side is a leader the other remains a powerful modifier.

During the first century of the nation's existence, simple forms seemed sufficient for simple functions, a premise that held true through the opening decades of the twentieth century. Then, as the functions and responsibilities of government expanded, changes were made in budget technology and technique. This seems to be a linear and expanding process, with more reforms attempted in the last 50 years than in the first 175 of the American experience. Nonetheless, those in charge at the earliest stages of budgeting in this country recognized the need for different budget forms. As early as

1800, civilian agency budgets were presented in carefully detailed object-of-expenditure form, while military expenditures tended to be appropriated as lump sums not unlike specific program categories. Early debate on budget development focused on flexibility and program accomplishments rather than on strict agency accountability.

Early American budgetary patterns were both part of and separate from their predominantly English colonial heritage. They were part of that heritage in that the American colonies inherited the full line of English historical experience with a limited monarchy and expanded legislative powers. This historical legacy may be dated to 1215, when a group of dissident nobles forced the King of England to accede to and sign the Magna Carta. Of the sixty-nine articles in this document, the most important is that which stated, "No scutage on revenue shall be imposed in the kingdom unless by the Common Council of the Realm." (Caiden and Wildavsky, 1974: 25) The Common Council preceded Parliament, and the statement that revenue could be raised only with the consent of a legislative assembly remained constant. This is often hailed as a beginning of popular government, but it is useful to note that this was basically a sharing of power between the king and the most powerful nobles in the realm, the two top tiers in an elite-dominated society where status was conferred mainly by birth. Nonetheless, by the end of the thirteenth century, the principle was established that the Crown had available only those sources of revenue previously authorized by Parliament.

In England, the Magna Carta was only the beginning of a long process of movement toward popular government, a process completed in the 20th century when the House of Lords lost the power to reject money bills. By the middle of the fourteenth century, the House of Commons was established and its leaders realized that a further check upon the power of the king would result from legislative control over appropriations. At first, revenue acts were phrased broadly, and once the money authorized was available the king could spend it as he wished. However, Parliament began to insert appropriation language in the Acts of Supply and other similar legislation, stating that the money be used for a particular purposes. Moreover, rules were made for the proper disposal of money, and penalties were imposed for noncompliance (Thomson, 1938: 206). Consequently, by the middle of the fourteenth century, fiscal practices included a check on the Crown's right to tax and spend; bills from Parliament carried notice of intent designating what money was to be used for, rules for disbursement of money, and penalties when rules were not followed.

Refinement of this system would take centuries, and its progress was not linear. Some kings were more skillful, personable, or powerful than others, and Parliament's role manifested steady evolution only in the most general terms. Wildavsky suggests that if a benchmark is needed, formal budgeting can be dated from the reforms of William Pitt the Younger (Wildavsky, 1975: 272). As Chancellor of the English Exchequer from 1783 to 1801, Pitt faced a heavy debt burden as a result of the American Revolution. In response to this, Pitt consolidated a maze of customs and excise duties into one general fund from which all creditors would be paid, reduced fraud in revenue collection by introducing new auditing measures, and instituted double-entry bookkeeping procedures (where each transaction is entered twice, as a credit to one account and a debit to another). Moreover, Pitt established a sinking fund schedule for amortization of debt, requiring that all new loans made by government impose an additional 1 per cent levy as a term of repayment (Rose, 1911a; 1911b; 1912). Pitt raised some taxes and lowered others to reduce the allure of smuggling. The legacy Pitt left was a model that encompassed a royal executive with varying degrees of strength and a legislative body attempting to exert financial control over the Crown by requiring parliamentary approval of sources of revenue and expenditure. Approval was provided through appropriations legislation. In this way, administrative officials ultimately were held accountable to Parliament (Browne, 1949: 15).

The history of the American colonies has been described as a replication of the struggle between Parliament and the Crown, with the colonies, like Parliament, gradually winning a more independent position (Labaree, 1958: 35), even before the Revolutionary War. For example, the colonies turned the power of the purse against the English Royal Governors. Colonial legislatures voted the salaries of Governors and their agents, appropriating them in annual authorizations rather than for longer periods. Indeed, one colonial governor's salary was set semiannually. In theory, the Royal Governors had extensive fiscal powers, but in fact these powers were often exercised by colonial legislative assemblies. These included raising taxes, appropriating revenues, and granting salaries to the Royal Governors and their officers. Caiden and Wildavsky remark that the colonists were thoroughly in the English tradition of denying supply (budget dollars) to the colonial governors to force compliance with the will of colonial legislatures. Not only were salaries voted annually, but taxes were also often reenacted annually. "Royal Governors were allowed no permanent sources of revenue that might make them 'uppity.'" (Caiden and Wildavsky, 1974: 25) Appropriations were specified by object and amount and appropriation language was used to specify exactly what funds could and could not be used for, e.g., "...no other purpose or use whatsoever." (Caiden and Wildavsky, 1974: 26) Royal Governors even were prevented from using surplus funds or unexpected balances; these were required to be returned to the treasury. Various mechanisms were used to impose further restrictions on the power of the Royal Governors. In some colonies, independent Treasurers were elected to manage funds. Several colonies required legislative approval prior to disbursement of funds; in emergencies this might necessitate a special appropriation. Some colonial legislatures appointed special commissioners accountable not to the Royal Governor but to the legislature as a further check on the power of the Governor. Caiden and Wildavsky conclude that "... power, not money was the issue" (Caiden and Wildavsky, 1974: 32). Thus, from the earliest days, budget decisions in the American colonies focused around the issue of the correct balance of power between the colonial legislatures and Royal Governors, a discussion that continues to occur every year at the national level, but rarely at the state level.

Generally, neither expenditures nor taxes were heavy during the colonial period; England did not extract much revenue from the colonies except in periods of war (Browne, 1949: 16). What was troublesome to the colonists was that the Crown could and did impose duties and excises intended to regulate trade and navigation without the colonists' approval, hence the revolutionary complaint of "taxation without representation." Up to the Revolutionary War, the colonies followed the British pattern of a gradually developing budget power. Decisions about taxation were paramount, and the exercise of budget power was basically sought as a check upon Royal power. Development of the instruments of taxation, appropriations, and accounting were all evidenced in this pattern, but a formal budget system did not yet exist.

The colonies departed from English tradition when they gained independence after the Revolutionary War. Like Pitt, the founders of this new nation faced a heavy debt burden; unlike Pitt their primary concern seemed to be focused on creating a country that could operate without an executive branch in a decentralized format almost dependent upon the voluntary contributions of the individual colonies. Under the Articles of Confederation, a fiscal system was created in which there was no executive branch. Power was vested in various legislative arrangements. As a result of fear of central government inherited from their experience with the British, the powers of the first Congress established under the Articles of Confederation were very weak. In fact, this fear was evident in the manner in which powers were delegated to both the legislative and executive branches as the Constitution was drafted. Also the colonists were averse to a system of' national taxation. Taxation imposed a special hardship on the colonies because hard coinage was scarce and bills or letters of credit were used irregularly. Consequently, the colonies were chronically short of cash and coinage schemes abounded. During the Revolutionary War, borrowing and promising to pay either with bills of credit or by coining paper money became endemic as the colonists pursued the war and made expenditures without tax revenues. Washington's continuing struggle to adequately equip his armies is well known, with the winter at Valley Forge standing for all time as a symbol of heroic efforts to contend with a new nation's ineffectual and rudimentary governmental systems.

The costs of war were great. Thomas Jefferson estimated the cost at $140 million from 1775 to 1783. By contrast, the federal government operating budget in 1784 was $457,000. Bills of credit were issued both by the states and by Congress from 1775 to 1779. Bills of credit rapidly depreciated. In 1790 Congress was forced to admit that a dollar of paper money was worth less than two cents and passed a resolution to redeem bills of credit at one fortieth their face value (Dewey, 1968: 36). Paper currency did not become legal tender again until after the Civil War.

The Articles of Confederation provided that revenues were to be raised from a direct tax on property in proportion to the value of all land within each state, according to a method stipulated by Congress. These limitations upon congressional taxing power left it dependent upon the states. Congress was not disposed to provide effective fiscal leadership to states, in part because Congress was debating issues related to its own budgetary procedures and its leverage over states regarding fiscal power. Congress was attempting to act as both the executive and legislative branch in a system where the preponderance of power was held by the individual states. Not only was this a departure from the English tradition but, it was a model of government that would be short-lived in this country.

Constitutional government began, then, with a long history of British practices further shaped by both the inefficiencies of the Confederation and the cost of the Revolutionary War. If American institutions were influenced by an anti-executive trend, they were also affected by the chaotic nature of legislative government under the Articles of Confederation. This period was marked by extraordinary negligence, wastefulness, disorder, and corruption, as Congress in its committees prepared all revenue and appropriation estimates, legislated them, and then attempted to exercise exacting control over accounts (Bolles, 1969: 358). As Vincent J. Browne observes:

> "Until the framing of the Constitution, the future of the States was almost as much imperiled by financial indiscretions as it had been previously jeopardized by the forces of George III." (Browne, 1949: 17)

Legislative dominance began to give way when the Continental Congress created the post of Superintendent of Finance in early 1781. Robert Morris, the first Superintendent of Finance, was charged with oversight of the public debt, expenditures, revenues, and accounts to the end that he would, "... report plans for improving and regulating the finances, and for establishing order and economy in the expenditure of the public money," (Powell, 1939: 33) as well as perform oversight of budget execution, purchasing and receiving, and collecting delinquent accounts owed the United States.

The enabling legislation has been referred to as "a bit radical for the times," because of the vast authority it delegated to one man (Browne, 1949: 21–22). Morris's pressure for revenue collection seemed to have angered some members of Congress. Consequently, in 1784, a Treasury Board or committee was established. However, the benefits derived from a single executive, albeit not the President, equipped with broad powers was evident and this pattern would reappear. As a practical matter, the whole period of Confederation was a time of experimentation within the context of anti-monarchical rule. The events of this period seem somewhat confusing, but there existed no model financial system to follow. Let it be remembered that William Pitt was the contemporary of Morris and the Founding Fathers. Pitt did not take office until after two years after Morris had been appointed, and the system Pitt created operated in a system where law and tradition still gave the balance of power to the Crown. Pitt was the King's minister. The Americans were busy negotiating the mechanics of representative government, influenced by the Confederation model of strong legislative assemblies.

THE CONSTITUTIONAL BASIS FOR POWER OVER BUDGETS AND SPENDING

The Constitution provides the basis of congressional power over the budget and spending as indicated below:

Article I. The Legislative Branch, Section 8: Powers of Congress

The Congress shall have Power To lay and collect Taxes, Duties, Imposts and Excises, to pay the Debts and provide for the common Defence and general Welfare of the United States;

To raise and support Armies, but no Appropriation of Money to that Use shall be for a longer Term than two Years;

To provide and maintain a Navy;

To make Rules for the Government and Regulation of the land and naval Forces;

Article I. The Legislative Branch, Section 9: Limits on Congress

No Money shall be drawn from the Treasury, but in Consequence of Appropriations made by Law; and a regular Statement and Account of the Receipts and Expenditures of all public Money shall be published from time to time.

Article II. The Executive Branch, Section 2: Civilian Power over Military

The President shall be Commander in Chief of the Army and Navy of the United States, and of the Militia of the several States, when called into the actual Service of the United States..."

Source: U.S. Constitution, 1798.

However, by the time of the Constitutional Convention, it was clear that this experiment in representative democracy operated through the legislature and by committees within it was not a practical solution to administering government, whatever its virtues in representing the people, thus the founding fathers created the Presidency (despite their fear of kings) and laid out the design for the taxing and spending power. The Constitution provided the right to tax to Congress and set forth four qualifications on spending power:

1. No money shall be drawn from the Treasury but in consequence of appropriation.
2. A regular statement and account of all receipts and expenditures must be rendered from time to time.
3. No appropriations to support the army shall run for longer than two years.
4. All expenditures shall be made for the general welfare. (Article 1, Sec. 8, U.S. Constitution)

The first two points contained in Article 1, Section 8 are the cornerstones of the budget process. On the revenue side, all money bills were directed to originate in the House of Representatives because of its proportional and direct representation of the people. The role of the Senate was debated, with the compromise that the Senate could concur with the House, or it could propose amendments to revenue bills. Fiscal power would be developed within an environment where the Congress was expected to be supreme at the federal level, and the states were expected to be jealous guardians of their powers.

THE RISE IN EXECUTIVE BRANCH BUDGETARY POWER

Under the impact of the American Revolution, planning had been nonexistent, management had been a legislative responsibility, and control for propriety was honored more in its absence than in its presence. However, the period 1789 to 1800 marked the beginning of the movement toward executive management and perhaps could be called the first stage of U.S. budget reform. The talents of Alexander Hamilton strongly influenced development in this period (Seiko, 1940: 45; Caldwell, 1944; Fesler, 1982: 71–89).

Hamilton was a man of great achievement. He learned applied finance at the age of 11 while a clerk in a counting house on the island of St. Croix in the West Indies. He learned quickly and was promoted to bookkeeper and then to manager. Before Hamilton was 21, he had impressed friends with his abilities to the extent that they sponsored him in a course of studies, first at a preparatory school and then at the predecessor to Columbia University in New York. Here he quickly gained a reputation as an adroit protagonist for the cause of the American colonies (Miller, 1959; Caldwell, 1944: 71-89). In 1776, he had won George Washington's eye with his conspicuous bravery as an artillery captain at the Battle of Trenton. Washington used him as a staff officer until 1781 when Hamilton, chafing under the limitations of staff routine, seized upon a trivial quarrel to break with Washington and leave his position. Washington seemed to have understood his impetuous subordinate well. He gave Hamilton command of a battalion that attacked a British stronghold at the siege of Yorktown in October of 1781, a siege that ultimately became the decisive battle of the Revolutionary War.

During the 1780s, Hamilton practiced law in New York City and was active in congressional politics, arguing for a strong central government. Hamilton believed that English government, as then constituted under George III, should be the American model. Hamilton proposed a President elected for life, who would exercise an absolute veto over the legislature. The central government would appoint the state governors, who would have an absolute power over state legislation. The judiciary would be composed of a supreme court whose justices would have life tenure. The legislature would consist of a Senate, elected for life, and a lower house, elected for three years. In this system the states would have virtually no power.

Hamilton's ideas seem to have had little influence upon the Constitutional Convention. However, when opponents attacked the document brought forth by the convention. Hamilton, with James Madison and John Jay, authored *The Federalist Papers,* a collection of eighty-five essays that were widely read and helped mold contemporary opinion; they became one of the classic works in American political literature. This was the man Washington appointed as Secretary of the Treasury in September of 1789.

Hamilton fused his own goals for a strong central government with the new nation's fiscal needs. His first efforts were directed toward establishing the credit of the new government. His first two reports on public credit urged funding the national debt at full value, the assumption by the federal government of all debts incurred by the states during the Revolutionary War, and a system of taxation to pay for the debts assumed (Hamilton, 1790; 1791). Strong opposition arose to these proposals, but Hamilton's position prevailed after he made a bargain with Thomas Jefferson, who delivered southern votes in return for Hamilton's support for locating the future nation's capital on the banks of the Potomac near Virginia.

Hamilton's third report to Congress proposed a national bank, modeled after the Bank of England. Through this proposal, Hamilton saw a chance to knit the concerns of the wealthy and mercantilist classes to the financial dealings of the central government. This was a very controversial proposal, not only because it was a national bank, but because banks of any sort were almost unknown in colonial America until the 1781 when the Confederation Congress set up a bank of North America, unlike England where there were "dozens and upon dozens of private and county banks scattered all over..." (Wood, 2006: 133). Despite heated opposition, Congress passed and Washington signed the bill creating what became the Federal Reserve Bank into law, establishing this national bank based, in part, on Hamilton's argument that the Constitution was a source of both enumerated and implied powers, an interpretation he used to expand the powers of the Constitution in later years. Hamilton's fourth report to Congress was perhaps the most philosophic and visionary. Influenced by Adam Smith's classic work "The Wealth of Nations" (1776), Hamilton broke new ground by arguing that it was in the interest of the federal government to aid the growth of infant industries through various protective laws and that, to aid the general welfare, the federal government was obliged to encourage manufacturing through tax and tariff policy. Hamilton's contemporaries seem to have rejected the latter view; Congress, at least, would have nothing to do with it. Nonetheless, in little more than two years Hamilton submitted four major reports to Congress, gaining acceptance of three that funded the national debt at full value, established the nation's credit at home and abroad by creating a banking system and a stable currency, and developed a stable tax system based on excise taxes to fund steady recovery from the debt and to provide for future appropriations. Indeed, Hamilton opposed the popular cry for war with England in the mid-1790s, at a time when France and England were at war, and England was seizing American ships in the Caribbean. He believed that commerce with England and the import duties it provided were crucial.

Hamilton's essays, published in New York newspapers in 1795, helped avoid war with England, thus helping to save his revenue system. Also, Hamilton was an admirer of the English system with its strong central government, fiscal systems and professionalized standing Army; in the latter respect, he advocated a strong standing Army for the U.S. to allow it to subdue any 'refractory state' and "to deal independently and equally with the warring powers of Europe." (Wood, 2006: 130) While neither of these conditions would come to pass in his lifetime, Hamilton was recalled to service by George Washington as second in command of the Army under Washington in 1798 when it seemed that France might invade the U.S.

The elements of the new nation's monetary and fiscal policy were bitterly contested issues, and groups coalesced around various positions. Hamilton became the leader of one faction, the Federalists, and because Washington supported most of Hamilton's program, in effect he became a Federalist. The two most prominent individuals in opposition were James Madison in the House of Representatives and Thomas Jefferson in the Cabinet. Madison and Jefferson were the Republican leaders. Hamilton and Jefferson feuded for several years beginning in 1791, as each tried to drive the other from the Cabinet. Finally, tired, stung by criticism of his operation of the Treasury Department and needing to repair his personal fortunes, Hamilton announced his intentions to resign his post as Secretary of the Treasury at the end of 1794. Hamilton did not, however, retreat to obscurity. He still held presidential ambitions that were narrowly frustrated; he was appointed to high military command, and he remained within the inner circle of the nation's political elite before departing center stage, killed in a pistol duel with Aaron Burr in 1804. Hamilton had made both great accomplishments and bitter enemies. Moreover, all of his suggested reforms became settled and accepted policy, thought he was perhaps fifty years too early on using tariffs to protect infant industries.

Hamilton's role in establishing a system for debt management, securing the currency, and providing a stable revenue base make him perhaps the founding father of the American budgeting system. Without faith in the soundness of the nation's currency and credit system, and a productive revenue base, it is difficult to make any budget system work. The federal taxing power alone was a dramatic change from the system envisioned under the Articles of Confederation, which approximated a contributory position by the separate states, hectored by the central government. Only a sure and certain revenue base, providing predictable revenue collections, allows the creation and maintenance of the modern nation-state. It was Hamilton's genius to direct the United States to that pathway.

To Congress, Hamilton represented a transitional figure. Before his appointment the House of Representatives had a tax committee, the Committee on Ways and Means, established in the summer of 1789. But this committee fell into disuse when a Secretary of the Treasury was appointed. In fact, from 1789 to 1795 when Hamilton resigned as Secretary of Treasury, Congress discharged its Committee on Ways and Means and stated that it would rely on Hamilton for its financial knowledge (Wood, 2006: 129). At this juncture in history, Congress viewed the Treasury Department as a legislative agency and the Secretary of Treasury as its officer (Browne, 1949: 34). The first appropriations bill for an operating budget came about because the House ordered the Secretary of Treasury to ". . . report to this House an estimate of the sums requisite to be appropriated during the present year; and for satisfying such warrants as have been drawn by the late Board of the Treasury and which may not heretofore have been paid." (Annals of Congress, no date, 1: 929)

When the articles of the Constitution were being debated, Hamilton wrote:

The House of Representatives cannot only refuse, but they alone can propose, the supplies requisite for the support of the government. They, in a word, hold the purse....This power over the purse may, in fact, be regarded as the most complete and effectual weapon with which any constitution can arm the immediate representatives of the people for obtaining a redress for every grievance and for carrying into effect every just and salutary measure (Hamilton, as reprinted in Miller, 1959).

Whatever the flaws of the act creating a Department of the Treasury, it seems clear that its intent was to make the Congress alone responsible for the budget process. That there was little room for executive leadership is demonstrated in the fact that the Act mentions the President only in connection with the appointment and removal of officers. Furthermore, while the act was being debated, opinion was divided over the wording of the duties of the Secretary of the Treasury with respect to whether he was to digest and *report* revenue and spending plans or whether he was to digest and *prepare* plans. Those Congressmen hostile to strong executive power believed that giving the Secretary the power to digest and report plans would take the fiscal policy initiative away from the House. The Secretary would report only what he had already done; this would deprive the House of its ability to exercise a prior restraint on the actions of the Secretary. The word *report* was deleted from the legislation and *prepare* was inserted and carried by the majority (Browne, 1949: 31).

Some observers have mistaken Hamilton's approach to appropriation as having established an executive budget system. The traditional model of an executive budget system would encompass a presidential review of departmental documents, revision of estimates, and a unified submission by the President or his agent of those estimates to Congress for approval. As Secretary of the Treasury, Hamilton did not wish the system to function in this manner, and personally he acted as an agent of Congress. The development of an executive budget system occurred in more gradual process, with a steady line of evolution leading to the authorization of a formal presidential budget—but not until 1921.

THE FIRST APPROPRIATIONS

The first appropriation act of Congress was brief and general:

That there be appropriated for the service of the present year, *to be* paid out of the monies which arise, either from the requisitions heretofore made upon the several states, or from the duties on impost and tonnage, the following sums. *viz.* A sum not exceeding two hundred and sixteen thousand dollars for defraying the expenses of the *civil* list, under the late and present government: a sum not exceeding one hundred and thirty-seven thousand dollars for defraying the expenses of the department of war; a sum not exceeding one hundred and ninety-six thousand dollars for discharging the warrants issued by the late board of Treasury and remaining unsatisfied; and a sum not exceeding ninety-six thousand dollars for paying pensions to invalids. (I Statutes at Large, U.S. Congress, Ch. XXIII, Sept. 29, 1789: 95)

Although salaries are the largest single item in this list (civil list), mandated expenditures—bills and pensions—comprised 45 per cent of the budget, defense 21 per cent and entitlements 14.8 per cent. De facto uncontrollability was high. True to modern practice, this appropriation bill was not the only money bill passed by Congress. Between the summer of 1789 and May of 1792, numerous bills were passed to provide for a variety of expenses, including defense, Indian treaties, debt reduction, and establishment of the federal mint[1] (Annals, no date: 3, 1: 259)

The first three bills were written as lump sum general appropriations, for the civil list, the department of war, invalid pensions, the expenses of Congress, and contingent charges upon government. Appropriating by lump sum caused resentment among some congressmen. One wrote of the appropriations bill of 1790 in his diary:

> The appropriations were all in gross, and to the amount upward of half a million. I could not get a copy of it. I wished to have seen the particulars specified, but such a hurry I never saw before ... Here is a general appropriation of above half a million dollars—the particulars not mentioned—the estimates on which it is founded may be mislaid or changed; in fact it is giving the Secretary the money for him to account for as he pleases. (Wilmerding, 1943: 21)

Notwithstanding their general nature, appropriation bills were linked to estimates of expense as specified in other bills. Expenditures for salaries were generally governed by laws enumerating the salary and number of the officers stipulated; for example, five associate Supreme Court justices at a salary no more than $3,500 per year. Estimates for the military were assumed to control the appropriations voted for the military. Therefore, even though the appropriations were voted in gross, the calculations adding up to the total were assumed to control the total. By 1792, Congress was appropriating money in gross but stipulating what the money was to be used for, with "that is to say" clauses: for example, $329,653.56 for the civil list, with a "that is to say" clause followed by specific sums attached to an enumeration of the corresponding general items (Wilmerding, 1943: 23; Fisher, 1975: 61).[2]

Congress was now planning in detail and the executive branch accepted that detail, although knowing full well that the dictates of administering might make it imperative to depart from the detailed plans expressed in the appropriation acts. Budgeting by lump sum was not a characteristic of the routine of American government except in case of emergency appropriations (Browne, 1949, Wilmerding, 1943). Congress increasingly specified the itemization of appropriation bills, in part as a strategy to control Secretary of the Treasury Hamilton, who was viewed by some as a member of the executive branch. In 1790, the House had sixty-five members, and most of its business could be carried out as a committee of the whole, but by 1795 it was clear that the Treasury Department could not serve the needs of Congress as well as it could serve the needs of the executive. Therefore, Congress reinstituted the Committee on Ways and Means, initially as a select or special committee, and by 1802 as a standing committee. Also, during this period, Woolcott, Hamilton's successor at Treasury, was embroiled in an increasingly bitter argument with Congress over the transfer of appropriations. Although Congress could appropriate in very specific terms, it could not stop the administration from transferring from one account to another when the situation seemed to warrant such transfers. The War and Navy Departments seemed particularly able to transfer funds, thereby dissolving the discipline of detailed itemization.

THE JEFFERSONIAN PERIOD

In 1801, when the Federalists were defeated and the Republicans took office, Thomas Jefferson spoke to the need for increased itemization of expenditure in appropriations. Nonetheless, the transfer of appropriations was an accepted practice in the administration, albeit an illegal one (Wilmerding, 1943: 48). Jefferson himself made the Louisiana Purchase after a liberal interpretation of executive authority to issue stock when government revenues were insufficient to cover necessary expenditures. Thus Congress's insistence on itemization led to deficiencies in accounts. Later, budget practices also became part of developing party politics. The Federalist Party believed in a strong executive, thus they preferred lump sum appropriations for activities that would give administrators as much flexibility as possible in managing programs. Meanwhile, the Republican party favored specific line item appropriations that limited department heads to doing specifically what Congress intended. However, Congress found it could not control everything and, beginning with appropriations for the Army and the Navy, line item controls were relaxed and other controls not invoked, e.g., penalties for unauthorized transfers.

The tension between delegating power to the executive and retaining appropriate congressional control of the power of the purse has remained an issue to current times. Generally, when relationships between the executive and legislative branch deteriorate, because of divided party control or because an executive agency becomes either too aggressive in its budget practices or not aggressive enough, Congress changes the rules to ensure that its intent is preserved, e.g., by changing reprogramming within appropriation thresholds (ceilings), requiring advance notification of Congress for reprogramming changes from one category within an appropriation account to another, or by attaching a legislative interest note to an item to ensure that money is spent in a special way, or that a program is executed within the fiscal year.

By 1800, the initial pattern had been set. Appropriation bills were passed and linked to specific estimates for specific purposes, military and civilian expenditures were treated somewhat differently, and it was recognized that transfer of funds between categories necessary to meet contingencies unforeseen at the time of appropriations, if technically illegal. The House held major control of the purse. By 1802, it had developed a standing committee to deal with revenues and appropriations, while on the executive side; the Secretary of Treasury had become more and more the President's agent in shaping appropriations bills, collecting revenues, and debt management. Both the legislature and the executive were elected by and responsible to the people. Rudimentary and disconnected as it seems from modern perspectives, no other country had such a budget system. The themes that surfaced during these years still appear: correct balance between executive and legislative, lump sum versus line item, correct use of funds during emergencies, even to the need to vote large sums on the basis of limited information. Still, this was democracy in action, and we reiterate that no other nation had such a budget system. Great Britain also had a system, but ultimate power resided in a king, not the people.

By the early 1800s, the transfer of power to Thomas Jefferson and the Republicans marked the end of the period of creation of the Republic, the end of the process of separating from England and the setting up of a new government. Much remained to be done, but many of the basic mechanisms of governance, and of monetary and fiscal policy, were now in place. In fiscal affairs the federal government had established its powers to tax and to budget, as well as to issue notes of credit when revenues did not match expenditures. Budgeting was in the main a legislative power. The Department of the Treasury was originally conceived as Congress's assistant. Budgeting power in the Congress was held in the House, which was small enough so that it could operate as a committee of the whole. As the Treasury increasingly served the President, Treasury's power in Congress declined, and Congress chose its own internal review and enactment body, the Committee on Ways and Means, and this committee would gain great power. During the course of the nineteenth century, appropriation bills would be sent to other committees as the Ways and Means Committee work load became heavier, or as political factors dictated.

FROM THE CIVIL WAR TO THE 20TH CENTURY

After the Civil War, a committee on appropriations was created. When it used retrenchment powers to reach into substantive legislation under the jurisdiction of other committees, Congress reacted against this expansion of the appropriation committees' powers and diminished the power of the appropriations committee. It seems that whenever a committee role became too important, Congress changed its procedures to move power away from it. Thus, legislative procedure changed, but budgeting maintained eminence as a vital legislative process.

There was an executive branch component to this process, but different presidents chose different profiles. Some were quite involved, others not. Fisher suggests that during the nineteenth century a number of Presidents revised departmental estimates before they were sent to Congress, including John Quincy Adams, Martin Van Buren, John Tyler, James K. Polk, James Buchanan, Ulysses S. Grant, and Grover Cleveland, and were assisted in this task by a number of Secretaries of the Treasury. Some ascribe an even larger role to the executive in this period.[3] (Smithies, 1955: 53, White, 1951: 68–69, Fisher, 1975: 269–270)

The first decade of the twentieth century was pivotal in terms of the balance of budgetary power. Government revenues based on customs and excise taxes were insufficient for the task of achieving the nation's "manifest destiny". Although the budget had been in a surplus position from the conclusion of the Civil War, after 1893, the economy and the budget ran into trouble and some policy makers worried that an antiquated revenue system prevented government from meeting new needs. The Spanish American War and the expense incurred in building the Panama Canal created budget deficits. Moreover, customs revenues began to decline. The federal budget was in a deficit position for eleven of the seventeen years from 1894 to 1911, including five of the seven years from 1904 through 1910. In addition to these debts arising from emergencies, some also felt that the revenue system was not up to the task of funding America's new and expanding world role. Passage of the sixteenth Amendment by Congress in 1909, ratified by the states in 1913, authorizing a federal income tax was a major milestone event in U.S. fiscal policy, and was intended in part to remedy this dilemma.

The debate over strengthening presidential spending power was essentially completed by 1912, with issuance of the report of the Taft Commission on Economy and Efficiency. Taft submitted this report to Congress, along with a plan for a national budget system, but his party did not control the House during that session of Congress and the two branches of government could not agree on a new budget process. The Commission's position was succinctly stated:

...the budget is the only] effective means whereby the Executive may be made responsible for getting before the country definite, well-considered, comprehensive programs with respect to which the legislature must also assume responsibility either for action or inaction. (Taft, 1912: 62–63, 138)

Budget reform was further delayed at the national level by the First World War, but reform continued apace at the local and state levels. Indeed, some observers have suggested that budget reform during this period in the American context began at the local level. Reform efforts resulted from indignation over corruption, graft, and mismanagement prevalent in local governments, exposed by journalists and good government movements, and supported by the Progressive party. Budget reform complemented other innovations including establishment of city manager and commission government forms, and the initiative, referendum, recall, and short-ballot electoral procedures. Budget reform in this period may be considered a local affair that eventually carried over to the federal government (Burkhead, 1959: 15; see also Schick, 1966: 243-258).[4] The fiscal stress caused by the American commitments to World War I, and President Woodrow Wilson's own interest in budget and administrative reform also precipitated the adoption of the executive budget process.

In his 1917 annual message to Congress, President Wilson stressed his party's platform on budget reform. Although reform seems to have been possible in any of these years, Wilson chose to wait until the end of World War 1. While he waited, the nation incurred a large deficit. In the three years from 1917 through 1919, federal debt grew from $1.2 billion to $25.5 billion. This gave urgency to the case for budget reform. After the peace treaty had been signed, Wilson argued that budget reform would give him a better grasp of the continuing level of defense spending, the effect of the disposal of surplus military property, and the impact of demobilization upon the economy (Fisher, 1975: 33).

THE BUDGET AND ACCOUNTING ACT OF 1921: THE BUREAU OF THE BUDGET

In 1918 and 1919, a series of bills intended to reform the distribution of budget power were passed, and in 1921 Congress passed the Budget and Accounting Act (Burkhead, 1959: 26–28).[5] This bill created the Bureau of the Budget (BOB), to be located in the Department of the Treasury with a Director appointed by and responsible to the President. The Bureau was given the authority to, "...assemble, correlate, revise, reduce, or increase" departmental budget estimates (42 Stat. 20, 1921). The intention of the writers of this law was to avoid unnecessary friction between the President and his cabinet officers over budget matters by locating the budget review power within the BOB in Treasury. This was intended to avoid setting the Bureau against the more powerful cabinet officers. Also, placing the Bureau in the Department of the Treasury facilitated the coordination of expenditures and revenues (Fisher, 1975: 34). Later it was moved to the executive office of the president. Under any interpretation, establishment of the BOB in 1921 and the crucial tasking of the President to prepare and submit a budget to Congress shifted power to the executive.

However, in passing reform legislation that increased the power of the executive, Congress also took something back by creating the General Accounting Office (the GAO, now the Government Accountability Office) to audit and account for expenditures, led by a Comptroller General of the U.S. responsible to Congress and appointed for a 15 year term. Any perusal of reports and testimony generated by GAO indicate how important this office has become in providing information for Congress to use in reviewing budgets and making financial management decisions. At the time, creation of the GAO was overshadowed by the attention directed at the Bureau of the Budget.

The halcyon days of the Bureau of the Budget existed from 1939 through the end of the 1940s. During this time, the Bureau built and held a reputation for unsurpassed excellence as a neutral, analytic power operating as a staff instrument for the executive. The reputation for excellence gained during these years of depression and war would mantle the Bureau into the late 1960s, but then its function changed to match the politics of the time. The Bureau was renamed the Office of Management and Budget (OMB) in 1969. Further, OMB would become tainted somewhat by the politics of Watergate but more by the aggressiveness of Richard Nixon in using and abusing his Presidential impoundment authority by refusing to spend money appropriated to executive agencies by Congress. OMB would be accused of exerting too much power, of resistance to change in a world interested in policy analysis instead of budget examination, and failure as an intergovernmental program manager for the multiplicity of programs resulting from President Lyndon Johnson's quest for the Great Society (Fisher, 1975: 58; see also Davis and Ripley, 1967, p: 749-769). However, during this entire period, for better or worse, the BOB functioned increasingly as the instrument of executive budget and policy making power.

One way to conceptualize budgetary control of the type wielded by the BOB and all central executive budget control agencies in government is to envision it as a tool that operationalizes fiscal values. These fiscal values are basically economizing values. As identified by Appleby, they include fiscal sense and fiscal coordination: "Fiscal sense and fiscal coordination are certainly values. The budgeting organization is designed to give representation in institutional interaction and decision-making to this set of values."

(Appleby, 1957: 134) Appleby argued that the budget function is inherently and preponderantly negative because it is against program expenditure and expansion. He explained that this is proper because program agencies and pressure groups are so extensive that there is no danger the values they represent will be overlooked or smothered by budgeter control agents.

Appleby conceded that a budget control agency cannot always be negative, for there are ways to save money by spending money, and the controllers have to be on the lookout for these occasions. In the main, however, budget control agencies will be the aggressors, pushed to cut, trim and squeeze spending. Spending agencies, on the other hand, will temper their requests by their judgment of what is wise and practical, and what policy makers and the budget bureau will accept. The executive budget bureau is at the center of this struggle, and yet it is removed from direct contact with many if not most of the political pressures of the politics of budgeting due to its isolation within the executive and the fact that it works for only one political party (the President's) at a time. Consequently, the budget bureau should act as a counter-weight to ensure that economizing fiscal values are entered into the decision making calculus.

Wildavsky (1964) characterized the budget process as a competition between the "spenders" and the "cutters," with BOB and agency budget control offices as the primary cutters in the executive branch of government, and the appropriations committees as the cutters in Congress. However, when the appropriations committees play this role, it is often to cut one program so as to add funding to another.

With respect to spenders, program agencies are expected to advocate for their programs and constituencies both within and outside of government. Members of Congress and the President, as elected officials, generally are expected to play the role of spending advocate most of the time. Otherwise, how would they get reelected? In a democracy, what do people send their elected officials to Washington, D.C. to do? The answer in large part is to solve or resolve problems, and to do so *requires* Congress to spend—from the perspective of the clients of governments and many stakeholders in the economy. Consequently, according to Wildavsky (1964), the spenders vastly outnumber the cutters and this creates a pro-spending bias in government. An understanding of the roles, duties and expectations related to the players in the budget process is critical to comprehension of budgetary competition for power. It is also important in attempting to understand the proper functioning of budget control agencies.

Neutral competence was the keystone of the philosophy of the BOB. This was typified by the folklore the BOB perpetuated about itself. As Berman noted, "BOB officials often told the story that if an army from Mars marched on the Capitol, everyone in Washington would flee to the hills, except the Budget bureau staff, who would stay behind and prepare for an orderly transition in government." (Berman, 1979: 29)

FROM THE BOB TO THE OFFICE OF MANAGEMENT AND BUDGET

The Bureau of the Budget was reorganized in 1969–1970 and became the Office of Management and Budget (OMB) in part to add political acumen by layering political appointees over the career staff (Reorganization Plan No. 2, 1970). After 1970, OMB's representation of fiscal values would be filtered through nets of political values before they reached the President, a change that may have improved the advice the Bureau could give the President but probably changed the character of neutral competence. Gone was the pure budget technician, lost in part to the era of policy analysis where the ability to detail the consequences of alternative budget decisions was the task at hand.

What happened to merely cutting the budget through close examination and intimate knowledge of the program? A reorientation of the role of BOB to become OMB probably was a necessary change. Schick observes that the Bureau as a simple representative of fiscal values could serve every President with, "...fidelity, but it could effectively serve only a caretaker President. It could not be quick or responsive enough for an activist President who wants to keep tight hold over program initiatives." (Schick, 1970: 532) As the functions and responsibilities of the Presidency changed, so did the role of the budget bureau.

Highly respected budgetary scholar Jesse Burkhead judged the institution of the executive budgetary system in the United States to be a revolutionary change. Burkhead argued: "The installation of a budget system is implicit recognition that a government has positive responsibilities to perform and that it intends to perform them." (Burkhead, 1959: 28–29) To do this would require reorganizing administrative authority in the executive branch, said Burkhead, and an increase in publicly organized economic power relative to privately organized economic power. Thus, the institution of executive budgetary systems in the United States clashed with customary doctrine about public versus private economic responsibility, but more importantly it was fundamentally at odds with the basic organizing precepts of the founding fathers. The budget system after 1921 and particularly in the post-World War II years through 1970 was an integrating system that allowed positive movement toward goals by relatively small groups of participants within the political system. It had to work this way, or it could not be an efficient system.

However, this kind of organizational efficiency appears to run counter to the Constitutional doctrines of separation of powers and checks and balances. Consequently, Burkhead suggested that not only would the practices of government have to be altered before budget systems could be installed and operated, but their development and installation alone were "revolutionary" in the context of American society. Burkhead concluded that although budget systems need not be synonymous with an increase in governmental activities (budget systems can be used for retrenchment), their installation is synonymous with a clarification of responsibility in government.

THE CONGRESSIONAL BUDGET AND IMPOUNDMENT CONTROL ACT OF 1974

In reaction to a number of what it saw as abuses of executive power in the early 1970s, Congress reasserted its power with the passage of the Congressional Budget and Impoundment Control Act of 1974.[6] (Schick, 1980) This Act sought to correct certain abuses of presidential impoundment powers, but more importantly, it also sought to reorganize the Congressional budget power to give Congress a better chance at full partnership in budgeting for the modern welfare state. If the Full Employment Act of 1946 gave the president responsibility for managing the economy, the Congressional Budget and Impoundment Act of 1974 extended the same opportunity to Congress. In addition, Congress equipped itself with more analytic power by creating the Congressional Budget Office, comprised of a neutral staff imbued with a sense of high calling and professionalism similar to that found in the Bureau of the Budget of the 1940s but in a somewhat more complex fiscal world.

The 1974 congressional Budget Impoundment and Control Act centralized the planning function of the budget in the House and Senate Budget committees. These two committees have the responsibility to develop a target resolution in the spring of each year containing detailed appropriation, spending and other targets (e.g., lending) to guide the work of the appropriations committees and subcommittees. The target resolution shows the overall situation, total spending, including the level of spending by function, taxing projections, and the level of surplus or debt forecasted. Then, in September, the budget committees were to shepherd a second resolution through Congress that matched the early planning target to the final appropriation bills. Through the reconciliation process, these committees may also ask Congress to tell its appropriation and taxing committees what and where reductions are appropriate in order to reconcile the final bills against the target resolution. In June of 1981, Congress attached reconciliation instructions to the first resolution and in effect dictated what would be done later that year in appropriations committee work.

The reconciliation instruction of June 1981 marked a turning point in the American budgetary process. For the first time in the history of the United States, the Congress set budget targets for taxing, spending, and debt, and thereby had a sense of what the national budget ought to be before it started enacting appropriation bills. At no other time since 1789 was this done. After 190 years of titular vesting of the power of the purse in Congress, Congress organized itself to pursue a budget prospectively, rather than adding up the total appropriations and expenditures and calling it a budget.

The 1981 budget was essentially an executive budget due to the fact that the Republican party controlled Congress to a much greater degree after the 1980 elections and tended to support the budget proposals of recently elected President Ronald Reagan, but still it was endorsed in Congress only after a bitter struggle. However, Congress used its newly developed budget power to develop congressional budgets that were different from the proposed executive budget of 1980, the final Carter budget. Congress asserted itself again in 1982, in deliberations over the second Reagan budget (Peckman, 1983: 19). As we note in analysis of the separation of powers, Congress has exercised its budget power both in support of and against the executive. Although the power to prepare and submit budgets remains with the executive, and a formidable power it is, Congress has evolved into a powerful and systematic modifier of budgets.

From 1945 to 1970, congressional scrutiny of budgets was characterized by students of budgeting as one of incremental review and marginal adjustments by appropriation committees to whom the other members of Congress deferred. Incremental behavior was rational, according to Aaron Wildavsky, because in reviewing that in which he or she was most interested, members allowed individual self-interest to protect the public good. Fenno documented the success of final adoption of appropriations committees' recommendations as being 87 per cent. Sharkansky observed that congressional behaviors could be summarized as the concept of contained specialization-elite status, specialized expertise, deference to the acknowledged experts, and conflict management (Wildavsky, 1979; Fenno, 1966; Sharkansky, 1969).

During this era of stability of review, enormous changes were taking place within society, and the composition of the budget reflected it. Social service and especially entitlement expenditures increased dramatically. Although each bill was intensely scrutinized, no one in Congress knew what all the appropriations bills would total in terms of actual spending (outlays) until the end of the fiscal year. As macroeconomic management became more important to the nation (Jones and Wildavsky, 1995), Congress had no apparent forum of its own to make and enforce economic policy through the budget. Thus, the budget process became less and less useful to the realities of managing a modern welfare state.

Seizing on the Nixon abuses of impoundment power to reorganize the budget process and to make itself a full partner in the process once more was an outcome not totally anticipated by Congress. Some thought that Congress had merely changed the fiscal year in an attempt to give itself more time to process appropriation bills to offset its difficulty in passing them on time. Others saw the 1974 Budget Act as a weighting of the budget power in favor of Congress at the expense of the executive. Although the Act did improve congressional potential, it need not be said that it usurps executive prerogatives. If it did it would have been ruled unconstitutional by the U.S. Supreme Court. In fact, there is more than enough budget power for both branches to share in attempt to satisfy constituents. Congress did become a powerful critic, however, because as a result of passage of the 1974 Budget Act it has the power to redevelop an executive budget when the President's original submission does not match congressional interpretations of the needs of a particular year, and it has the assistance of the Congressional Budget Office to help members exert their will and power in negotiations with the President and OMB.

SEPARATION OF POWERS AND CHECKS AND BALANCES: THE U.S. AND OTHER NATIONS

The United States is lawfully set up with three branches of government, legislative, executive and judicial, with each having certain powers and each equipped with powers to check the other. A short excursion on the internet to such sites as wikipedia, the State Department website or the CIA Fact Book shows how different other countries are from the U.S. in formal government processes (more on this in chapter 12). In particular, the countries that provide the most students to the Naval Postgraduate School are parliamentary democracies where the President is titular head of state, but his main functions are ceremonial and whoever controls Parliament exercises the power of the legislative branch. These countries use mostly the parliamentary system with a unicameral legislature and a long electoral cycle. Except for Lithuania, all of these systems result in one-party dominance in the Parliament even if there are two or more parties competing at elections. For example, in a past election in Greece the winning party held 45% of the popular vote and the second party had 40%; in number of seats won, the winning party held 55% of the seats and the second party 39%. Thus when the Vice President casts the tie-breaking vote in the US Senate, he is doing something which is not going to happen in these countries. First there is no

Senate. Secondly an officer from the executive branch is casting the deciding vote. Thirdly, the second house in this issue is in effect forcing the other chamber to compromise with whatever the first house wants, whereas in all these other systems the lower house is the only house and in most cases it is ruled by one party. Fourth, the party which holds the Presidency gets to vote twice in the instance cited above; the vice president votes in the Senate and then the President may sign or veto the bill as he wishes. Admittedly this does not happen often, but it does happen and it is legally permissible and has to be understood to be a part of the American government mechanism. The countries under the parliamentary system basically are designed to promote stability (long electoral terms, one party dominance) and efficiency (Prime minister runs country, holds majority in parliament, picks cabinet of ministers to administer departments, dominant party wins all votes, no negotiation with other branch of government, other chamber, other party or parties.) Both the strengths and weaknesses of the separation of powers systems have long been recognized. In fact some believe that the US system is qualitatively better than that of other countries because it results in a unique political structure with an unusually large number of interest groups, because it gives groups more places to try to influence, and creates more potential group activity. Opponents of separation of powers indicate that it also slows down the process of governing, promotes executive dictatorship and unaccountability, and tends to marginalize the legislature. This is somewhat of a modern view given the tendency of the press to build up the power of the presidency, while founding fathers were clear that the legislative power was the most important one.

Much of the US design was designed and has evolved to limit the power of the other branches of government while in many other democratic countries power is organized to govern efficiently following the doctrines of the party which has won the most recent election. Also, elections tend to be at least four years apart and some are longer, thus guaranteeing a longer time period for the dominant party to implement its programs. Turkey for example has a five year electoral cycle, unlike the US where the House gets elected every two years, along with one-third of the Senate. Thus in the US, all the House members and one-third of the Senate are almost always trying to make a track record for re-election. Here our foreign neighbors have established the value of stability over longer terms, rather than responsiveness to what may be of concern to the public at the moment. Typically, this allows for a more statesman-like stance by legislators in these countries and in some cases less legislative churn. In the US it is necessary to distinguish between what candidates say when they run for election as opposed to what they say and do when they have to govern. To international observers, it may seem like there is a lot of wasted motion in the US political apparatus, because someone is always running for election.

Another very important difference is that most of parliamentary systems are heavily invested in one chamber; in the five countries cited above, there is only one chamber, thus there is no fundamental appreciation for the dynamics between two chambers, on different electoral cycles and representing different constituencies, even when they are controlled by the same political party. In the US, both the House and the Senate have legitimate roles and political power and their perspectives differ, even when they are controlled by the same party. This is different for many other countries, where even in two-chamber systems like Canada and Great Britain, only one house governs. Thus, in addition to negotiation between executive and legislative branches over appropriation bills, there is an area of negotiation between legislative branches that has to be understood.

Thus, it is not only the separation of powers that international students have to understand; they also have to understand checks and balances. For example, the powers and responsibilities of the different branches of government intentionally were designed to overlap. Congressional authority to enact laws can be checked by a presidential veto, but that veto can be overridden by a two-thirds majority in both houses. The President serves as commander in chief, but only Congress has the authority to declare war and raise funds to support and equip the armed forces. The president has the power to appoint ambassadors, federal judges, and other high government officials (e.g., Director of OMB, Secretary of Defense), but all appointments must be affirmed by the Senate, and sometimes these appointments are denied, if rarely. Also, the Supreme Court has the power to overturn both presidential and legislative acts. This checking and balancing was designed to ensure that no one branch of government would grow too powerful, and in the early days, that the national government would not grow too powerful at the expense of the states.

In the world of budgeting specifically this means a variety of things such as that the President proposes budgets, but Congress passes the appropriation bill which the President can veto, but Congress can override. The President can impound, delay, and defer budgeted funds, but only within certain limits and is subjected congressional review. The President may veto bills, but he does not have an item veto since that would make him a legislator. The President appoints judges, diplomats, top level civil servants and top level military officers, but Congress has to confirm them. The President can commit troops into immediate military action, but only Congress has the power to declare war and fund the military. The Vice President chairs the Senate in a role which is largely ceremonial, but he may cast tie-breaking votes which can be very important. The House can impeach the President and the Senate then tries him, thus the legislative branch has control over the executive for gross dereliction of duty. The House is also given the responsibility of choosing the President if there is no majority in the Electoral College, which is in itself another check and balance in the electoral system. Finally the courts constitute a check on the actions of both Congress and the President through the power of judicial review, which may happen long after an issue has lost the public's attention. It is no wonder that international officers have some difficulty starting out with the public policy course. This checking and balancing is a large part of what makes the budget process so complicated. Later we discuss in more detail the ways Congress modifies, controls and generally 'improves' presidential actions.

CONGRESS AS ARBITER OF CIVIL–MILITARY RELATIONS

Another way to examine the roles of Congress and the executive branch is to assess their roles as arbiters of civilian and military relationships. Historically, the challenge of civil-military relations (CMR) in democratic societies has revolved around the dilemma of raising a military strong enough to deter and defeat a state's enemies while at same time controlled sufficiently so as not to threaten its own government. As Peter Feaver put it, the "... challenge is to reconcile a military strong enough to do anything the civilians ask them to with a military subordinate enough to do only what civilian authorities authorize them to do." (Feaver, 1996:149) Over the past half-century, the dialogue on CMR has been dominated by the views of Samuel Huntington and Morris Janowitz, but important new perspectives have emerged.

Huntington proposed that maximizing the professionalism of the military is the key to assuring a subordinated military. In *The Soldier and the State*, he argues that a professional officer corps is the key to maintaining the balance between military superiority and civilian control. An autonomous, professional military will subordinate itself to legitimate civilian control (Huntington, 1957). Janowitz (1960) also focused on the officer corps, but saw them as more politicized than Huntington. Janowitz likened Cold War missions to a constabulary and saw threats to CMR from the centralization of power in the Department of Defense. He viewed the budget process as an important tool of civilian control, but not one effectively used. He argued for greater oversight by civilian authorities and the establishment of not simply a professional military, but a professional military ethic (Janowitz, 1960).

The views of Huntington and Janowitz were foundational to the management of the military during the Cold War; for example, they were influential in the design of officer training programs. However with the end of the Cold War came a rethinking of these ideas. Roles and missions—one of the control mechanisms advocated by Janowitz—were changing. Emerging democracies have created a fertile subject for research (Cottey, 2002, p: 31–56). Newer conceptualizations have also recognized that mature democracies have not suffered military coups. Thus, if the extreme end of the CMR continuum is so unlikely to occur, then what is the relevant range of the continuum? (Feaver, 1996) The answer is not necessarily found in the sociological literature on the military such as we have seen in the Janowitz tradition. (Burk, 1993:167–185) "Their findings…are of great sociological import but seem less relevant to political scientists concerned with the exercise of power between institutions." (Feaver, 1996: 157)

Feaver argues for a more nuanced view of CMR that incorporates "interest-based and external control mechanisms" and "changing patterns of civilian control" (Feaver, 1996: 167). In developing a set of benchmarks for the development of future theory, he suggests that the new problematic is "…about the delegation of responsibility from the notional civilian to the notional military. It is about increasing or decreasing the scope of delegation and monitoring the military's behavior in the context of such delegation. And it is about the military response to delegation, desire for more delegation, and even occasional usurpation of more authority than civilians intended." (Feaver, 1996: 168–169)

Burk presents a compelling criticism of traditional CMR theory and summarizes recent trends in CMR theory-building. While not offering a theory of his own, he argues for a model based on protecting and sustaining democratic values in the context of the post-Cold War geopolitical situation and the realities of the blurring of the military and civilian spheres in the United States. He suggests in the conclusion that the answer perhaps lies in a federalist-like division of responsibilities across levels or units of government (Burk, 1993: 7–9). In their review of emerging democracies in Eastern Europe, Cottey, Edmunds and Forster came to similar conclusions: the control of the military is not necessarily an executive function, it is more democratic. They specifically address the role of the legislature or parliament in policy setting, oversight and resource allocation decisions. "Effective parliamentary oversight of the armed forces and defense policy, however, depends on both the formal constitutional or legally defined powers of the legislature and the capacity of the legislature to exercise those powers in an effective and meaningful way in practice." (Cottey, 2002: 44)

Since publication of Huzar's classic book *The Purse and the Sword*, there have been numerous empirical studies on the power of the purse as a tool to control the military. Most studies focused on the Cold War years (Kanter, 1983), the relationship of military spending to economic activity, (Kollias, et. al., 2004: 553–569; Goldsmith, Flynn and Goldsmith, 2003: 54–65) or the management of the defense establishment (Gansler, 1989; Thompson and Jones, 1994; McCaffrey and Jones, 2004). There is also a substantial literature on congressional control of the purse that explains why Congress exercises control through the budget and the mechanisms they use to do so.

DELEGATION OF BUDGET AUTHORITY BY CONGRESS

Since the birth of the nation Congress has chosen to delegate authority to the military departments and later to DOD in cases where it has appeared to be required, i.e., in times of war or imminent armed conflict. However, in the 1960s the debate in Congress over policy and funding for the war in Vietnam divided members into those willing to delegate increased authority to President Lyndon Johnson and those who sought to restrict the use of money to fight the war under any condition. This debate parallels the dialogue of the past few years in Congress over funding the war on terror and military operations in Afghanistan and, particularly, in Iraq. While most members of Congress support a war on terrorism, many have reservations about how and where this war should be fought. The defense budget is a tool for showing support and addressing reservations.

The degree of resource decision flexibility and delegation of authority from Congress to the Department of Defense has been examined by observers of congressional defense budgeting and management for decades (Augustine, 1982; Kanter, 1983; Fox and Field, 1988; Jones and Bixler, 1992). The advantages of increased delegation of resource management authority by Congress have long been asserted by DOD leadership. For example, DOD Comptroller Robert Anthony developed an extensive plan for reorganizing defense accounting and budgeting under Project Prime in the mid-1960s, but Congress rejected the proposal (Jones and Thompson, 1999). Defense Secretary Frank Carlucci asked Congress for increased resource and managerial powers at the end of the Reagan administration in 1988. Defense Secretary Dick Cheney proposed six acquisition programs in the 1991-1994 time frame for execution without congressional oversight as a test of the DOD ability to operate efficiently independent of external micromanagement. Despite considerable congressional lip service to the effect that these proposals would increase program management and budget execution efficiency and better "bang for the buck" in defense, Congress supported neither Carlucci's nor Cheney's request.

In April 2003, Defense Secretary Donald Rumsfeld offered a set of legislative proposals under the umbrella title, *Defense Transformation for the 21st Century Act.* Included among the proposals were requests for greater autonomy in budget execution. Congress largely ignored the budgetary proposals and devoted most of their attention to the more public proposal (creation of the National Security Personnel System). Given this longstanding absence of trust of DOD management and budgetary judgment it is significant when Congress deviates from traditional patterns of control.

Those traditional patterns of control have been well described in the literature. Research has explained the incentives for and the means by which Congress oversees the Defense Department through the budget and authorization processes. In summary, Congress has traditionally exercised control as constitutional prerogative, (Fox, 1988; Owens, 1990:131–146, Lindsay, 1990: 7–33) to shape defense policy, (Jones and Bixler, 1992: 293–302; Mayer, 1993: 293–302) in response to media publicity of defense mismanagement (Fox, 1988; Lindsay, 1990; Jones and Bixler, 1992), and due to partisan congressional-executive branch competition and occasional mistrust (Blechman, 1990; Mayer, 1993; Thompson and Jones, 1994). Since the defense budget has historically accounted for at least half of all federal discretionary spending, it is a favorite target for those legislators seeking to influence federal spending even for non-defense matters. (Fox, 1988; Halperin and Lomasney, 1999: 85–106, Wildavsky and Caiden, 2001) A final reason is the advocacy or protection of constituent interests, e.g., military installations, labor, or defense contractors (Lindsay, 1990; Hartung, 1999; Jones and Bixler, 1992: 29–84).

CONGRESSIONAL MONITORING AND CONTROL OVER THE DEFENSE BUDGET

Congress exerts constant effort to monitor and control the budget and spending of the Department of Defense. It is not only important to examine why Congress tends to control defense, it is useful to examine the manner in which that oversight occurs. The literature provides us with a framework of tools including restrictions on the use of funding, tools for gathering information, and accounting requirements. Regarding the use of funding, members of Congress may make line-item adjustments to the budget; (Lindsay, 1990; Blechman, 1990; Mayer, 1993; Thompson and Jones, 1994; Halperin and Lomasney, 1999) earmark funds for specific purposes; (Owens, 1990; Mayer, 1993) place restrictions on the reprogramming and transfer of funds between accounts; (Fox, 1988; Jones and Bixler, 1992; Wildavsky and Caiden, 2001) and restrict funds pending executive compliance with provisions of law or committee reports (Owens, 1990). They gather information formally in congressional hearings and informally between congressional and DOD staff outside of hearings or through the use of reviews, audits, and investigations by committee staffs, GAO, and CRS (Owens, 1990). Finally, Congress may place requirements on program execution or may specify reports be provided to members on a myriad of topics (Owens, 1990; Jones and Bixler, 1992).

Strong incentives are present for Congress to try to actively manage or micromanage defense policy and budgets, and congressional rules and procedures provide many means by which to control DOD through authorization, appropriation and oversight as shown in the exhibit below. The fundamental tension between the constitutional roles of the executive and

TABLE 2.2 Why and How Congress Exercises Authority Over the Defense Department

	Fox 1988	Owens 1990	Lindsay 1990	Blechman 1990	Jones & Bixler 1992	Mayer 1993	Thompson & Jones 1994	Halperin & Lomasney 1999	Hartung 1999	Wildavsky & Caiden 2001
Why Does Congress Control Defense?										
Legitimate exercise of constitutional power	X	X	X			X	X			
Policy influence					X	X				
Response to media publicity of DoD mismanagement	X		X		X					
Partisan politics, competitiveness, mistrust	X			X	X	X	X			
Direct federal spending (Defense budget is half of discretionary spending)	X	X			X			X		X
Advocate or Protect Constituent Interests	X	X	X	X	X			X	X	X
How Does Congress Control Defense?										
Line item adjustments to the budget		X	X	X	X	X	X	X		X
Earmarked funds		X				X				
Reprogramming and transfer restrictions	X	X			X					X
Restricting access to funds pending compliance		X								
Formal and Informal Information Gathering	X	X	X	X	X	X	X			X
Reviews/audits/investigations by committee staffs, GAO, CRS, etc.			X		X					
Structural requirements placed on programs		X			X	X	X			X
Reporting requirements	X	X			X					

Source: Authors, 2011

legislative branch over control and flexibility is longstanding and has no definitive means of resolution is present other than to seek a balance appropriate for the time and situation. This is by design, instilled into our political process by our founding fathers in the Constitution.

CONCLUSIONS

Budgeting began in this country as a legislative enterprise, and specifically as a check against executive power. The people exercised the power of the purse through elected representatives. Effective representation of demands was emphasized over the needs of executive efficiency. When this was seen as not efficient enough, we created an executive branch and have gradually released power to it. By the 1960s the executive power had taken the lion's share of power. Since 1974, the legislative branch has gradually organized itself to be a more effective partner in the resource allocation process. Thus, our experience with a prominent executive budget power is relatively limited. In periods of extreme crisis, Congress has tended to cede power to the executive and reclaim it after the crisis has passed. Although there is a good deal of conflict in the budget process, there is also a good deal of reconciliation and adaptation.

The United States was a country born poor. There were no crown jewels, no colonies to exploit. As a consequence of the Revolutionary War, the currency was a wreck and the colonists owed a substantial debt. Alexander Hamilton's efforts have long been recognized for paying down the debt and stabilizing the currency, in effect, putting the country on a sound, but not rich, fiscal footing. This uncertain fiscal history may have been the start of the balanced budget ethic which dominates so much of American budgetary discussion and practice. A country born poor, or just out of debt, must live within its means[7] (See table 1.1). How this has turned out is interesting. Aaron Wildavsky compared American National budgets to their European counterparts. He found that U.S. budgets have been consistently balanced and that per capita revenue and expenditure ratios consistently lower than that of our European counterparts. What could explain these differences? Wildavsky suggests it comes about as a social legacy rising out of the revolutionary period (Wildavsky, 1964).

Wildavsky observes that the winning side Revolutionary America was composed of three groups. The first, the *social hierarchs*, wanted to replace the king with a native variety better suited to colonial conditions. The second were *emerging market men* who wanted to control their own commerce. The third were *egalitarian Republicans*- a legacy of the continental Republicans. They stressed small, egalitarian and voluntary association. What allowed these three groups to co-exist, create, and operate a successful government was agreement on a balanced budget at low levels of expenditure, except in wartime. Wildavsky suggests that there was no formal declaration of this agreement, nor was it done on a single day. However, this agreement lasted for 150 years until a new understanding was forged in the 1960s. The social hierarchs would have preferred a stronger and more splendid central government, and the higher taxes and spending that went with it. But the emerging market men would have had to pay, thus they preferred a smaller government, except where taxing and spending provided direct aid. Together this coalition led to spending on internal improvements, like railroads, canals, and harbors. However, it was the egalitarian Republicans who did not believe that government spending was good for the common man, meaning the small property-holder and/or skilled artisan. They suggested that unless the scope of government spending was limited, groups of these common men would withdraw their consent to the union. Limiting, not expanding, government was their aim.

Out of these three strains came the impetus for a balanced budget at low levels of taxing and spending. Egalitarian Republicans were able to place limits on central government. Market men won the opportunity to seek economic growth with government subsidy, but the extent of this was limited by an unwillingness to raise revenues. The social hierarchs obtained a larger role for collective concerns provided they were able to convince the others to raise revenue. Wildavsky suggests that no group got all it wanted, but all got something. The Jacksonian belief that equality of opportunity would lead to equality of outcome helped cement this outcome observed Wildavsky. The result has been an ethic of a balanced budget at comparatively low levels of taxing and spending. Moreover, what is balanced in this equation, according to Wildavsky, is not only the budget, but also the social orders and their supporting viewpoints that helped found the country. This impetus towards balance continues to surround the decisions we make about taxing and spending.

It is clear that during the early and mid-twentieth century the budget power assumed burdens that made it different in kind from anything that had gone before. The expansion of responsibilities of the U.S. government for social welfare beginning in the 1930s under President Franklin D. Roosevelt and expanded by President Lyndon Johnson and Congress in the 1960s (for example through establishment of Social Security, Medicare and other large and expensive entitlement programs), along with macroeconomic management beginning with the Full Employment Act of 1946, and the use of the defense budget predicated on winning the Cold War (1948 to 1989) precipitated the modification that strengthened executive power in the budget process. The effect of these factors may be seen in the budget trend lines of the early 1970s when the rising trend for human service expenditures surpassed the declining trend for discretionary and defense expenditures in the federal budget.

Since 1940, the U.S. has spent a substantial share of its wealth on human resources. In 1940 the human resources share was approximately 44% and 4.3% of GD: By 2011 this spending had increased to approximately 67% by 2006 and almost tripled to 13.8% of GD: However, the way this spending was distributed changed dramatically from 1940 to 2011. Payments to individuals increased from 17.8% of budget share to approximately 63% and about 15% of GDP. As the vast majority of those payments go directly to individuals, they give individuals a sense of entitlement and ownership, thus making these patterns hard to change. Defense, on the other hand, is not linear; it responds to the threat as we have demonstrated elsewhere. It has held to about the same budget share as payments for individuals in 1940 and increased only slightly by 2011, while payments for individuals more than tripled. Thus, in the last 70 years there has been a definite change in American society and federal spending, and while few would argue that it is not for the better, there are serious questions about its sustainability. With respect to national defense, the relevant question for policy makers is not its share of the budget or percent of GDP but the extent and nature of the threat and whether what is appropriated and spent is enough to support the mission.[8]

Hormats traces US fiscal history and war financing through the Revolutionary Period and to the present (Hormats, 2007). He observes, "...looking back over this nation's more than two hundred years, one central, constant theme emerges: sound national finances have proved to be indispensable to the country's military strength. Without the former, it is difficult over an extended period of time to sustain the latter." (Hormats, 2007: xiii) Hormats notes that during the Revolutionary War Congress had no taxing power and was frequently unable to provide the money General Washington needed, how printing money quickly devalued the worth of the continental dollar, and how in the end loans from France and the Netherlands saved the day.

Hormats also observes that since then military necessity often resulted in new and expanded fiscal powers, like the income tax during the Civil War. Hormats notes that it is not enough to have well-trained troops, good generals and a good strategy, that the country also needs a sound financial strategy and skillful leaders at the Treasury, in the White House, and in Congress to ensure that sufficient money is available to meet military expenses. He concludes that under the duress of war, the nation's policymakers have produced "...dramatic innovations in the nation's tax and borrowing policies, innovations that lasted long beyond the conflict during which they were introduced..." (Hormats, 2007: xiv). He believes that America's wars have been fiscally as well as politically transforming events, leading to changes that would not have been envisioned or accepted in quieter times, and the result has been that in the twentieth century America was able to generate, "...colossal amounts of tax revenues, conduct massive bond drives, and produce great volumes of weapons" which resulted in victory in two world wars and the Cold War. (Hormats, 2007: xv) However, he warns that leadership is not only about raising huge amounts of money, but it is also about uniting the country behind the war effort and raising money without weakening the economy. This takes skillful political as well as fiscal leadership. He and others have worried that currently the US is living in a post-9/11 world with a pre-9/11 fiscal policy where a, "...heavily debt-laden, over obligated, revenue-squeezed government, highly dependent on foreign capital creates major security vulnerabilities." This is a serious warning to heed. Defense spending as a percent of GDP still remains at a generational low, the lowest since the 1960s, but there has been an increase in the national debt and in foreign holding of U.S. government debt. Should events such as another major terrorist attack or a shift in international defense alliances and threat result in a significant realignment in global financial markets, the result could be a dramatic reduction in capital inflows from foreign investment in US Treasury bonds, a spike in US interest rates and a further collapse in value of the dollar, in some ways a more profound effect than any terrorist could hope to achieve.

In his historical review, we may laud three Secretaries of Treasury, Hamilton for his handling of the post-Revolutionary way crisis; Salmon Chase who guided the Union through its fiscal and monetary efforts in the Civil War, and William Gibbs McAdoo who helped the U.S. withstand the global economic storm set off by WWI. In addition, it is important to note that President Dwight D. Eisenhower steered a middle course between a military-industrial complex and a sometimes irresponsible tax-cutting Congress. In his farewell address, mostly known for its statement about the military-industrial complex President Eisenhower also warned, "We cannot mortgage the material assets of our grandchildren without asking the loss also of their political and spiritual heritage. We want democracy to survive for all generations, not become the insolvent phantom of tomorrow." (Eisenhower, 1961: 3)

Since the collapse of financial markets in 2008 and the rise in US government debt to the present critics have observed that the US is at another turning point in its history where it may be risking its future by imprudent fiscal behavior. This notwithstanding, the critics' main message is that the US has found appropriate fiscal strategies and competent leaders to implement them in times of danger and war. We may note, however, that the legacies of at least two former Presidents, Johnson and George W. Bush, were marred by the inability to either win or withdraw from publicly contentious and expensive wars. In retrospect, had the Union not prevailed over the Confederacy in 1865 perhaps Lincoln would have been similarly criticized.

The long sweep of history teaches us that techniques of budgeting are not as important as the purposes for which the money is spent and the ideas that drive those purposes and the decision makers who pursue them. While the budget is in a constant state of change, the budget process provides a resilient and flexible procedure able to accommodate to changing conditions. However, such adaptability also raises other problems.

By 2011, most observers had agreed that using supplemental appropriations to fund ongoing war operations in Afghanistan and Iraq was a flawed process, yet in separating out and highlighting those costs, the supplementals served a purpose. Part of the genius of the American character resides in its impulse to find a better way, but not to depart radically from tried and tested methods. Social engineers, rationalists, and autocrats might have done better, but experience with the budget process leaves us with a demonstration that representative government works, although it often appears to do so awkwardly and inefficiently. Thus it appears that some degree of inefficiency in policy and budgetary decision making in a democracy is the price to be paid for operating under this form of governance. As Otto von Bismarck, the renowned German Chancellor and statesman who guided the unification of his nation in 1871, is reputed to have observed about democratic government: if you like sausage you don't want to see how it is made.

NOTES

1. There were fifteen appropriations bills passed between the founding of the Federal government and May 8, 1792. A listing of these may be found at 3 Annals 1258-1259. From 1789 to 1792, the United States had a surplus of $21,762 on revenues of $11,017,460. (3 Annals: 1,259) During this period, general appropriations grew steadily from $639,000 in 1789 to $1,059,222 in 1792, and ranged from a high of $2,849,194 for payment of interest on the national debt in 1790 for 1792 to a low of $548 for sundry objects in 1790. Protection of the frontier was a growing expense: $643,500 in 1792, not included in the general appropriation. In 1791, Congress appropriated $10,000 for a lighthouse; in 1792, $2,553 was appropriated for a grammar school. Of the $11 million raised to the end of 1792, over $6.3 million was applied to the debt, either interest or principal. It is clear that these were still transitional years for the federal budget.

2. Fisher notes that lump-sum appropriations are especially noticeable during periods of war and national depression, when the crisis is great and requirements uncertain. At these times, the Congress tends to delegate power (Fisher, 1975: 61).

3. This position was supported by both academics and public administrators. Notable among the former were Arthur Smithies in The Budgetary Process in the United States (1955: 53), and Leonard D. White in his four-volume history of the federal government, notably, The Jeffersonians, (1951, p: 68–69);

The Jacksonians, (1954, p: 77-78); and The Republican Era, (1958: 97). The bureaucrats who argued this included two budget bureau directors, Maurice Stans and Percival Brundage; see Fisher (1975: 269-270). Fisher describes the growth of executive budget power as a steady accretion manifest in numerous statutes, financial panics, wars, "... a splintering of congressional controls," and demands from the private sector for economy and efficiency.

4. Burkhead observed that the interest of the business community in reform was the crucial element in this mixture. Business expected lower taxes. (Burkhead, 1959: 15) To understand other variables at work, see Schick, 1966: 243–258.

5. Burkhead (1955: 17) suggests the primary motive of Congress in passing the Budget Act was to reduce taxes, not to improve executive leadership.

6. One of the most comprehensive analyses of the Congressional Budget and Impoundment Control Act is found in Schick, (1980).

7. From 1789 to 1849, accounts for the U.S. government show a surplus of $70 million. Notwithstanding any individual year deficits, they practiced what they preached. OMB, 2008.

8. As the first Secretary of the Treasury in 1790, Alexander Hamilton explained that the huge debt run up during the Revolutionary War was the 'price of liberty' and had to be repaid. (1790; 1791)

CHAPTER 3

BUDGETING FOR NATIONAL DEFENSE

INTRODUCTION

Defense Budgeting is Different

Defense budgeting presents unique challenges not faced elsewhere in the federal budget process, nor in budgeting for state and local governments. The research and writing of budget guru Aaron Wildavsky shaped how students and scholars viewed budgeting for more than three decades, but most of his work explicitly ignored defense budgeting. The original *Politics of the Budgetary Process* in which Wildavsky's theory of budgeting as incremental behavior was unveiled (Wildavsky, 1964: 13–16) had no chapter on budgeting for national defense. Wildavsky is not alone in this omission. Most of the literature on public sector budgeting ignores defense budgeting, despite the fact that spending in this area typically represents a large part of the discretionary budget of the federal government. For example, far more is written and published in analysis of the roughly $50 billion the budget allocates annually to social welfare programs versus the more than $500 billion spent on national defense. Recognizing the deficiency, Wildavsky added an insightful chapter on defense budgeting to his classic 1964 work when he revised it as *The New Politics of the Budgetary Process* (Wildavsky, 1988: 348–395). Wildavsky recognized that for the reasons we explore in this chapter, budgeting for na-

Financing National Defense: Policy and Process, pages 71–89
Copyright © 2012 by Information Age Publishing
71

tional defense might be differentiated from budgeting elsewhere in the federal government.

Total Spending Matters

Defense and non-defense budgeting differ in important ways. The defense budget is an instrument of foreign policy and other nations react to changes in funding levels and priorities. U.S. defense budgets also respond to spending in nations that pose threats to our interests and those that we rely on as allies. When other nations change their defense allocations, e.g., as a percentage of Gross Domestic Product (GDP), U.S. decision makers must determine how such changes should affect our defense spending. Will we have enough trained personnel, ships, and aircraft, spare parts, maintenance capacity, and all the other capabilities that establish and sustain military force readiness to counter changes in levels of threat? Money buys capability. More capability means an increased ability to deter threat or to inflict damage to others who then must counter that increased ability with increases of their own, or find strategic alliances to negate the threat. Other nations do not monitor the total amount the U.S. spends on education or health, or if they do, it does not have the same salience. Defense budgeting is about deterring or countering threats that exist, or that may emerge in the future. Shifts in funding are early warnings that the threat scenario is changing and responses to it must also change. Thus, Wildavsky observed that one difference between defense and non-defense budgeting is that total spending means something to other nations and the amount of change from one year to the next is carefully watched for hints about future behavior.

Recognition of this fact was perhaps the most significant insight of the Presidency of Ronald Reagan. Reagan and the Congress appropriated funding levels for national defense that literally drove the Soviet Union to bankruptcy and contributed to the end of the Cold War. In this respect, U.S. spending on defense during the entire Cold War era represented willingness on the part of American people and their leaders to spend what was necessary to counter what was perceived as an ever-expanding threat from the USSR. After the end of the Cold War, Presidents Bush and Clinton and Congress struggled with the question of how much to spend on defense. The terrorist attack of September 11, 2001 forced the U.S. into a new posture to counter a type of threat never faced before. The result is increased defense budgets and spending plans for the first decade of the 21st century and perhaps longer. Former President George W. Bush articulated the resolve of the United States of America to do, "... whatever is necessary for as long as it takes," to prevail in the war on terrorism. To a great extent, the

American people have supported the view of spending whatever is necessary to counter a threat that is far less easily identified and attacked than any other the nation has faced in it history.

The Defense Budget Provides Opportunity to Reward Constituents

National defense comprised 20% of discretionary spending in the President's FY2012 budget request. Because so much of total federal spending has been placed into entitlement accounts (e.g., Social Security, Medicare, Medicaid) that are permanently authorized rather than appropriated annually, the defense budget is one of the few discretionary pots of money available where a Congressman can seek to get money to fund local projects. Moreover, since defense appropriations tend to be veto proof (Reagan vetoed one because the out-year funding provided by Congress was too low), Congress is often tempted to attach unrelated items to defense appropriation bills. For example, the fiscal year (FY) 1994 federal budget carried a sum for breast cancer research, as well as dollars for various museums and memorials. The Senate version of the supplemental appropriation

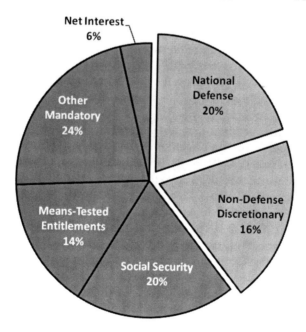

Figure 3.1 Distribution of Federal Outlays, FY2012. Source: Office of Management and Budget, 2011

to cover the cost of the war in Iraq in 2003 included funding for fisheries management in Lake Champlain, subsidies for catfish farmers, research facilities in Iowa and the South Pole, repair of a dam in Vermont, money to designate Alaskan salmon as "organic" and an increase in postage allowances for Senators to communicate with their constituents. The defense appropriation bill is much larger than any other appropriation bill and presents Congress with a continuing temptation to add unrelated items to it. Thus, the defense appropriation sometimes is referred to as a "Christmas tree" bill because it provides an opportunity for everyone to get something.

A Different Relationship with OMB

The Office of Management and Budget (OMB), the President's budget office, has a different relationship with the Department of Defense (DOD) than it has with other federal departments and agencies. OMB generally does not play the same adversarial role toward defense that it plays in examining the budgets for agriculture or education and other domestic programs. Rather, OMB and the Office of the Secretary of Defense (OSD) team up in review of the defense budget, a process done in-house in the Pentagon. Other domestic departments submit their budgets for review to OMB. When OMB cuts them, these agencies may appeal the actions to the President; the pattern is reversed with Defense. If OMB wants to cut the defense budget, this first must be negotiated with officials of the Office of the Secretary of Defense (OSD). Then, on large spending issues, OMB must appeal decisions not approved by the Secretary of Defense directly to the President, a tradition that Wildavsky and Caiden (1997: 234) date to the Kennedy administration.

Appeals to the President may not support the OMB position versus defense. This depends on the priorities of the President. For example, David Stockman, Director of OMB under President Reagan, related (Stockman, 1986) that he would tell the President defense spending increases could not be accommodated without increasing the annual budget deficit to unacceptable levels ("... deficits as far as the eye can see"). Invariably, Stockman was argued down by Defense Secretary Caspar Weinberger in front of the President, and Stockman would be sent back to OMB headquarters, grumbling to himself, after having been told to "rework the numbers" so that defense increases would be "deficit neutral" in the President's budget. The point is that Presidential priorities for defense and national security dominate the relationship between OMB and the Defense Department. With national defense, it is clear that policy drives the budget—not the other way around. This cannot be said for the rest of the domestic portion of the budget. The prominence of policy is highly evident in the post-

September 11, 2001 era where defense budgets have increased to fight the war on terrorism.

The types of decisions made solely by OMB with respect to other agencies are made in concert with the Department of Defense at the Office of Secretary of Defense level. The partnership that exists between OMB and DOD is different from that which exits between OMB and other cabinet level agencies. OMB staff work at the Pentagon and are involved not only in budget review, but also in development and review of program structure in the POM phase of the PPBES process where resources planners decide what program structure needs to be maintained, improved, created or de-emphasized to meet changes in the threat. Tyszkiewicz and Daggett state, "The defense budget is unique in the extent to which OMB is directly involved throughout the budgeting process." (1998: 28) To some extent, this results from the fact that defense is different from other federal departments. Its appropriation is much larger than that of other federal departments; it employs many more people than the others do; and its input mix of people and capital equipment is very different as it provides trained people and equips them to defend the country against various threat scenarios decades into the future. Its focus is also somewhat different as it prepares to meet what might happen as opposed to reacting to what did happen last year. Most importantly, the defense budget is simply too large for OMB to review on its own. It is not staffed to perform this tremendously time-consuming task.

No Ceilings for Potential Defense Spending

Another unique aspect of the defense budget is that the process for defining requirements begins with no ceiling in the Planning, Programming, and Budgeting Execution System (PPBES) used by DOD to define the threat, and to plan and budget to counter it. However, annually ceilings are introduced in the budget cycle, initially by the President and DOD, and then by Congress. Historically, this ceiling has been depicted in the President's budget in billions of dollars and as a percentage of GDP, a percent of the federal budget, and a real dollar (inflation adjusted) percentage amount over a five year period. OMB represents the President's priorities as it helps negotiate overall defense spending in the defense budget process prior to presentation of the budget to Congress.

Cycles versus Straight Line Growth
Defense spending has exhibited a cyclical long-term pattern different from that of other budget accounts. For example, human resource spending has soared since the 1960s in a relatively straight line, but defense has

experienced a feast or famine profile, consisting of roughly seven lean years and seven rich years. This pattern is revealed in the build up and down for the Korean and Vietnam wars in the period 1950 through 1980, the increase during the Reagan years, the decline during the Presidencies of George H. W. Bush and Bill Clinton and the rapid post 9-11 rise under George W. Bush and then Barack Obama. With the fall of the Berlin wall and the end of the Cold war, defense budgets declined by about a third in dollars and personnel to the mid-1990s and stayed roughly flat to almost to the end of the decade. Increases in appropriations began to flow into the personnel accounts to keep salaries and benefits equal with inflation and to solve retention problems within the defense community in 2000. Then, the onset of the war on terrorism initiated a new spike upward in of defense spending.

Absence of Consensus on How Much to Spend for Defense

Economists have depicted budgeting for national defense as a competition between funding "guns or butter." There are few milestones to indicate what is sufficient spending for defense. During the 1960s and again in the 1980s the media often portrayed defense spending as taking from the "poor" (human resource spending) to give to the "wasteful" (defense). One trend identified by Wildavsky is the disappearance of consensus over national defense policy and budget goals. Defense was budgeted under bipartisan consent from the beginning of WWII to the early 1960s. However, what was determined by consensus in the 1950s became subject for conflict in the 1960s. During this period, considerable consensus existed in other areas of budget policy. However, constrained resources, an increasing national debt, the end of the Cold war, an aging population, and increasing health care costs have driven great fissures into the previous state of budgetary consensus, so that policy that was consensual has now become embroiled in the politics of dissensus. The result of this dissensus is that budgets have tended to be bitterly contested and generally passed late each year. Moreover, part of the price of passage involved spending on projects and programs that benefited certain areas or regions more than other areas. In the 1990s, for example, numerous projects were forced on DOD to maintain local employment. This was sometimes called 'maintaining the defense industrial base.' In other instances, it was simply called 'pork.' Observers have also suggested that DOD responds to political pressures by placing programs in as many states and districts as possible to strengthen its political base in Congress (Smith, 1988). The terrorist attack on the U.S. on September 11, 2001 and the subsequent war on terrorism produced a new consensus on the need for increased defense spending. Defense spending rose significantly in 2001 and thereafter and large supplemental bills dominated by defense and homeland security interests were also quickly passed.

Increased Participation in Defense Budgeting and Policy Making

Lindsay, Wildavsky, Jones and Bixler and others have suggested that since the 1960s budgeting for defense has gone from an "insiders" game to an "outsiders" game. (Lindsay, 1987; Wildavsky, 1988; Jones and Bixler, 1992) In the 1950s, it was possible for just a few members of Congress serving on appropriation sub-committees to dictate decision-making. This was true for non-defense budgets as well. However, since the end of the 1960s there has been a loss of power by the powerful committee chairmen who rose to power based on seniority and safe electoral districts and could bestow rewards and punish almost with impunity. To some extent, stability in budget decision making was purchased with the coin of secretive, elitist decision-making.

The seniority system for selecting committee chairs and members remains important, but committee positions may also be gained through caucus elections. Powerful committee chairman can be upset and disciplined through this election process. The traditional power of political parties to enforce voting blocks has been eroded by the formation of shifting coalitions of legislators seeking to unite with others to vote for specific interest that cross party line boundaries. Thus, the power of committee and subcommittee chairs has been reduced and can be moderated. Moreover, when the Democrat or Republican caucus picks a committee member by vote, the vote of the newest freshman is equal to that of the most senior member. Furthermore, as opportunities for funding projects for local constituencies in other parts of the budget have declined, more effort has been exerted to gain access to and influence over the defense budget.

In the 1950s, the defense budget was composed and reviewed in approximately six committees. By the mid 1960s, the number had expanded to ten (Wildavsky, 1988). Since then there has been a proliferation of committees involved in defense policy making and budgeting. Now, defense policy and budget hearings are held in approximately 28 different committees and subcommittees. This includes the full appropriations committees and five subcommittees (military construction, defense, energy and water, HUD and the Independent agencies, and Commerce, Justice and State), the budget committees and the committees on Armed Services, Commerce, Energy, Government Affairs, Select Intelligence, Small Business and Veterans Affairs or their analogs in each chamber. GAO estimated that in the period from 1982 to 1986, 1,306 DOD witnesses testified before 84 committees and subcommittees, presenting 11,246 pages of testimony in 1420 hours (Wildavsky and Caiden, 1997: 243). Jones and Bixler found that the size of defense authorization bills increased from one page in 1963 to 371 pages by 1991 (Jones and Bixler, 1992: 49). Appropriation bills increased in the same period from 18 pages to 59. Similar increases were found in the length of committee reports accompanying the budget. They also found that the number of days the House of Representatives devoted to debate on defense authorization bills rose by a factor of 10 from 1961 to 1986, and the number of proposed amendments increased from 1 in 1961 to 140 in 1986 (Jones and Bixler, 1992: 68–69). The number of directives related to the budget made by Congress to DOD increased from 100 in 1970 to 1084 in 1991 (Jones and Bixler, 1992: 78). DOD made fewer than ten presentations on the budget to Congress annually in the 1950s, but by the early 2000s, this number was over 100.

It should also be noted that DOD has a sizeable "black" or secret budget, once inadvertently leaked to the press to be in excess of $30 billion. Wildavsky and Caiden (1997: 234) assert that the black budget had increased from about $5.5 billion in fiscal 1981 to $28 billion in 1994. In 2007, it was estimated that DOD was asking for $31.9 billion for classified acquisition programs, about 18% of the acquisition total of $176.8 billion, as estimated by Steven Kosiak of the Center for Strategic and Budgetary Assessments. About $14 billion was for procurement and about $17 billion was for RDT&E. Estimates are that about half of the testimony on defense is classified secret and takes place behind closed doors. Thus, even if the public does not have access to this testimony, members of Congress do. Experts assume that 'black' programs typically get fewer restrictions on their funding streams, and less oversight by lawmakers and Pentagon leaders and that this leads to performance problems and undue cost growth. Of the services, Air Force gets the largest share of the black budget, perhaps 80%, because it handles so much of DOD's command, control, communications and intelligence tasks, including satellite and space launches. (Bennet, 2007: 22)

More committees mean more places where decisions about defense are made, more opportunities for outsiders to hold congressional committee seats from which influence can be leveraged, and more necessity to coordinate final decisions on the floor where each member has one vote. These committees demand lots of testimony from DOD, for budget making and for oversight. In addition, turf wars between committees, for example the authorization and appropriation committees, must be negotiated among a larger number of participants and sometimes the dialogue spills over to the floors of the House and Senate where outsiders have voice in voting on bills such as the defense authorization, appropriation, and military construction bills and amendments to them.

In summary, the number of players in the defense policy and budget arena has expanded from a handful of members to a cast of hundreds, all supported by personal and committee staff. Thus, budgeting for defense has gone from an insider game to an outsider game, where considerable dissensus exists about defense policy priorities. While the terrorist attacks of 2001 seemed to provide a renewed public appreciation for national defense, anti-war demonstrations took place even as Operation Iraqi Freedom went forward and American and Coalition troops were under fire. How much should be spent for defense, and on what, remains a highly debatable and political question.

THE POWER OF THE BUDGET BASE AND THREAT ASSESSMENT

Each year the defense budget debate focuses more on the changes to the base—on new plans, programs, and activities. Here is where benefits may be sought for constituents. Thus, the dialogue in most congressional committee venues is on the increment of change; the base will be questioned in most cases only to the extent that proposed changes expose its strengths or weaknesses. In the main, this process is incremental, historical, and reactive. The defense budget is large and bewilderingly complex, and so is the process that produces it. Size and complexity make complete review of the annual defense budget impossible.

The main drivers of the defense budget are the policy priorities of the President combined with assessments made in the Planning Programming and Budgeting and Execution (PPBES) process about the nature of future threat, the direction and demands placed on U.S. foreign policy, the nature of existing defense alliances worldwide, and the disposition and behavior of allies and other nations and their military capabilities. From these assessments, a scenario is created, much like a large mural, describing the world of the near term future. Not all the details are filled-in, nor are all the details clearly seen. This scenario exists some five years into the future and the iterative, cyclical nature of the defense PPBES process keeps this scenario alive and moves it forward year by year. The scenario depicts threats to U.S. interests at home and abroad. Each annual budget fleshes out details, but only for the most important parts of the picture at that moment. There is a constant evaluation of the scenario viewed from different levels within the Department of Defense and other parts of the government concerned with national security. A continuous budget process attempts to clarify the near term aspects of the scenario.

The strength of PPBES is asserted to be the supply of long-term stability to defense planning and budgeting. Conflicting political priorities, complicated congressional budget procedures and annual delays in passing defense appropriations make it more difficult for DOD to link the multiple budget years it administers at any one time with proposed funding for the fiscal year sent to Congress by the President. This situation has perplexed a succession of defense leaders. Former Secretary of Defense (SECDEF) Donald Rumsfeld viewed the PPBES cycle as too long and repetitive, and in the period 2001–2003 took steps to shorten it to provide decisions in a more timely manner. Other changes in this system have been made as we describe in a subsequent chapter.

In sum, the threat and the capabilities needed to respond to threats drives the budget, but antecedent to this is US foreign policy and threat identification. In some cases foreign policy initiatives can reduce threats or cause them to disappear, through treaties and alliances, but diplomats know that diplomacy without the threat of force and the capability to project power is useless. DOD supplies the threat of force, but foreign policy defines where force may be needed and risk assessments dictates how much force capability needs to be purchased through the budget.

When defense leaders deploy the armed forces, they are using the tools and personnel that their predecessors built. They literally fight with a force and force structure someone else built. Their challenge is to build a force that will meet the demands that will be placed on their successors some five to ten years in the future. In terms of what the budget buys, the military must fight with yesterday's force as they create tomorrow's force. Tomorrow in this sense may be five years away, or it may be just over the horizon where a potential opponent introduces new technology, e.g., North Korean nuclear capability. In budgeting then, the crucial questions for DOD leaders are "What is the threat?" followed by "How has it changed from last year?" and then "Given this change, how do we need to change force structure and how much will this cost?" The further imperative is that only so many dollars may be allocated to defense. Defense leaders always feel that whatever is allocated is not enough, thus they continually fight to let no dollar go to waste. DOD always budgets in an environment of scarcity. Other departments do not have quite the same conceptual burden of reading the future and interpreting the past that DOD annually faces in the budget process.

Notwithstanding these complexities, some simplifying routines do exist in defense budgeting. The threat scenario is fine-tuned from one year to the next. Fundamentally, this is an incremental process. Indeed, in defense some patterns are even more stable than for non-defense programs, e.g., treaty commitments are long-lasting, binding agreements that shape the defense configuration. Acquisition of defense hardware, weapons, aircraft, ships, tanks, and similar purchases involve long-term commitments. Once a decision has been made to purchase an aircraft carrier, a type of aircraft, or an advanced submarine, the posture of defense capability has been set for the next two to three decades or longer. For example, in 2007 the average age of the Air Force KC-135 Stratotanker was 48 years, this for an airplane that was in daily use around the globe.

What happens when a fundamental change takes place, e.g., at the end of the Cold war or the scaling back of US middle eastern operations? At such points, the entire threat and response capability must be reevaluated, and this takes time that runs beyond the annual budget cycle. It is arguable that long-term planning systems such as PPBES do not handle revolutionary change very well. A major portion of defense workload disappeared with the end of the Cold war. The 1991 Gulf war hastened the transition from Cold war thinking and planning. Still, it has taken two decades and the emergence of the new threat of terrorism to transform U.S. defense forces planning away from a Cold war mentality. Moreover, even through the late 2000s, some of the defense budget remains Cold war influenced and many of the assets in the military force structure are those designed to counter the Cold war threat. Replacing these aging assets has become a major budgetary problem for DOD.

The defense resource allocation process is inextricably entwined with foreign policy as we have noted, and foreign policy itself is constantly evolving. New threat and policy responses can redefine the workload demand placed on the military forces. By choosing to deal with the threat in a variety of ways, the demand placed on defense capability varies. Examples are decisions to go to war in Iraq and Afghanistan, but to negotiate the threat posed by Iran and North Korea. The point is that the defense resource allocation process may have huge burdens thrust upon it or taken from it in ways that are not characteristic for other parts of the federal government. The PPBE system and DOD as an organization have problems in quickly accommodating substantial changes in direction, not only in deciding how to meet new threats but because its asset base (trained personnel, aircraft, ships, weapon systems) once purchased and deployed take time to redirect and rebuild to meet newly emergent threats.

DIFFERENCES IN DEFENSE BUDGETING RELATED TO FEDERAL BUDGET PROCESS DYNAMICS

The federal budget process contributes to the difficulty of enacting and changing the direction of defense policy and the budget. That the budget process has no end is a lament heard in the Pentagon and Congress about the resource allocation process. The PPBE process is an iterative process, so it purposely recreates and adjusts the threat and response scenario on an annual basis. Additionally, because much of the defense budget is for acquisition of military hardware this means that programs are almost always at risk for elimination, for modification, delay, "stretch-outs" (buying weapons over a longer period of time than intended), and other changes deemed politically attractive in Congress. For acquisition program managers, their programs are always under scrutiny and decision closure is hard to get—what is agreed to in one year may be undone or changed later. This is primarily a defense phenomenon due to the high percentage of the non-defense federal budget that is mandated. Thus, defense budget players know that each budget process has an end game in which decisions are finally taken to produce a President's budget or to get an appropriation bill passed. However, they are also aware that as soon as decisions have been made, the next budget cycle will crowd in with its deadlines, crises, problem programs and issues that seemed to have been settled but must be visited anew. Hence, the adage, "There is no ninth inning" for the defense budget is accurate.

For many domestic programs, the annual budget process has fairly stable rhythms. In addition, for many domestic agencies funding increments since the early 1980s have been small or in budgets that have been in steady state. In contrast, the defense budget is not really an annual budget, although much of it is annually authorized and the majority of it is appropriated and obligated annually. Rather, it may be viewed as a stream of decisions and resources whose intent is to sustain a force to safeguard the future by being ready for use if called upon.

Since the early 1960s, defense funding has gone through cycles of increase, major decline, significant growth, another major decline, and most recently, an increase. These changes in funding levels resulted from the Vietnam war spending boost, post-Vietnam war cuts that left what has been referred to as a "hollow force," the Reagan defense build-up, the Cold war "peace dividend" reduction, and new growth to fight the war on terrorism. In the late 1980s, the Senate Armed Services Committee required a biennial defense budget intended to provide stability and to extend the decision time horizons, but the appropriating committees never adopted it. To a considerable extent, by virtue of the design of the PPBE system, defense resource planners have operated beyond a two-year time horizon since the 1960s, notwithstanding that most defense dollars are provided in annual appropriation bills.

Short vs. Long Spending Accounts

In the defense budget, plans, programs, and activities (PPAs) are created and provide budget authority (BA) through passage of appropriations legislation that is eventually signed into law by the President. Budget authority is a promise to pay over the life of a program but not all BA to fund acquisition and other programs is provided in any one fiscal year. The figure below indi-

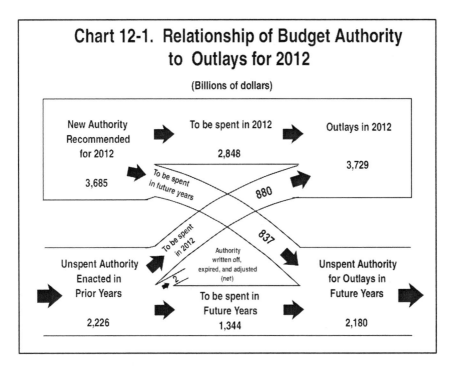

Figure 3.2 Budget Authority for Current and Future Years (Chart 12-1) *Source*: Office of Management and Budget, 2011, Budget of the United States Government, FY2011. Analytical Perspectives, Chart 12.1: Relationship of Budget Authority to Outlays

cates the amount of new budget authority recommended for FY2012 as well as how much of FY2012's outlays would result from budget authority created in prior years ($880 billion). Thus in total, the President proposed to spend $2,848 billion in 2012 and added another $837 billion to the stock of budget authority to be spent in future years, totaling $2,180 billion. to be spent in future years, even if no further appropriation actions were taken in the future. Thus, although the U.S. budget process is accurately described as an annual budget system, the budget process regularly makes legally appropriated future year budget authority commitments whose cumulative size proposed for FY2012 was approximately 58% of total proposed budget year outlays.

Total Obligational Authority

In DOD, budget authority for the current year is merged with budget authority created in past years, but available in the current year. This is called

Total Obligational Authority (TOA) and consists of all funds available for obligation in a single year. At the end of the fiscal year, unexpended funding will be rolled forward as long as the authority for it has not expired. Figure 3.3 shows the relationship of TOA, outlays, and budget authority provided for defense. The years after FY2011 are estimates. TOA may exceed outlays and budget authority when the stock of unspent authority from previous years falls in any particular year, e.g., FY2009. Budget authority may exceed TOA for any particular year, because budget authority is created for this and future years, thus budget authority is more than TOA in 1990, less in 1991 and about the same as TOA in 1998. Outlays are made as goods and services are delivered to execute obligations. Outlays can be more or less than TOA and BA. From figure 3.3 it appears that obligations created in 1991 were satisfied and paid for in 1992 through 1995, thus making outlays greater than either TOA or BA. The one relationship not shown on this exhibit is obligations; as we will discuss later, obligations should never be more than total obligational authority. TOA is the stock of authority DOD has to use to create obligations in any one year.

Obligation and Outlay Rates

Defense budget accounts have different outlay or "spend out" rates, e.g., military personnel spends out at about 96% in a single fiscal year, Operations and Maintenance (O&M) accounts spend out about 82% in their first year and have five additional years of expenditure availability, although by the end of the third year 99.8% of the funds were outlayed. Acquisition spending for major programs may be spread over a 5 to 10 year time horizon. For example, aircraft procurement has a three-year obligational window and five years beyond that for outlay. APN accounts obligate 78% of their funds in the budget year and outlay 16%; ship construction obligates 63% in the first year and outlays 7%. Military construction appropriations may take 20 years to be spent. Thus, the promise is on the books, but as successive budgets are reviewed and approved in Congress, changes in programs may be made and budget authority may be adjusted accordingly. This is a normal part of the turbulence in the defense budget process.

Managing Multiple vs. Single Years

Defense budget managers have multiple year budgets to monitor and control in execution, and execution issues and problems differ by type of appropriation account. A manpower analyst in the Pentagon may work only with current fiscal year accounts, e.g., FY 2004, using money from the

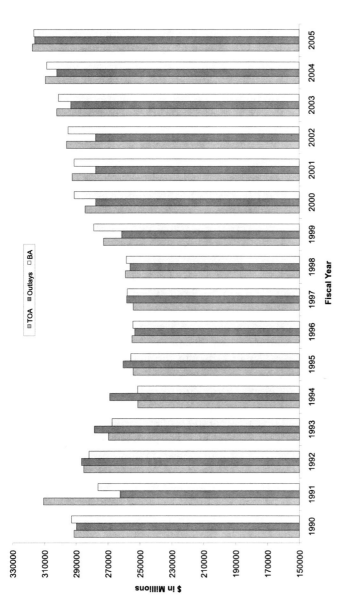

Figure 3.3 Defense Total Obligational Authority, Outlays and Budget Authority. Source: Duma, 2001: extracted from National Defense Budget Estimates for Fiscal Year 2001 Budget, the Green Book: 62-145. TOA equals budget authority created for the current year and budget authority created in previous years to be spent in the current year. (See figure 3.2) TOA tells DOD administrators how much they have to spend in the current year. BA is the amount of budget authority created by Congress in the current year; some of this will be for the current year and some will be for future years. Outlays consist of dollars actually spent in a particular year; past years are actual amounts while future years are estimates based on historical outlay to TOA relationships.

Military Personnel (MILPERS) account. On the other hand, a Navy fleet comptroller will work with current fiscal year accounts and accounts for the two previous fiscal years to fund Operations and Maintenance (O&M) where money to pay for fuel, ship repairs, steaming and flying hours is expended. On the other hand, an acquisition program manager managing the construction of a weapons system may have three past years and 5 to 7 future years to manage to complete a purchase from Aircraft Procurement Navy (APN) or Ship Building and Construction Navy (SCN) accounts. In contrast, non-defense federal budgeteers basically operate on a single year time horizon as budget authority or its equivalent is created, obligated, and outlayed during the fiscal year.

Managing Outlays vs. Obligations

Attempts to control the federal deficit using outlay limits in deficit score-keeping from 1986 through the late 1990s further complicated defense budgeting by making outlays more important than budget authority. Prior to this period, defense budget analysts were not as concerned about outlays; budget authority was what counted. During the late 1980s, outlays became critical targets in attempting to manage the annual operating budget deficit. This presented difficulty in managing accounts with greater than annual spending patterns. Currently this would amount to about 40% of the defense budget.

When spending reductions have to be made because of outlay targets, it also means that the faster spending accounts are cut first (e.g., Military Personnel, Operations and Maintenance), because a dollar in budget authority for them is closer to a dollar in outlay in the budget year. For defense budgeting, time displaced merit in an outlay driven context to an extent not applicable in the non-defense arena. For example, merit might dictate that a ship purchase be cut from the budget, but its first year spend out rate (outlay) may be only 5%. Thus, to meet the outlay savings target might take a cut of ten ships; hence, the cut would be made in personnel, because 98% of outlay in this account is realized immediately. Whenever quick cuts are sought in defense, the temptation is to put at risk those accounts that have a high annual spend-out ratio, irrespective of what this does to total force capability. Because it appears that the federal budget will be in deficit for most of the coming decade, the outlay scoring game may once again become more important for defense budgeting in the future than during the brief four year period when annual federal budgets were in surplus.

Maintaining Balance in Managing Budget Accounts

Balance is an important concept within defense due to systems critical-ity. When one budget category, program, or element is adjusted, typically several others have to be re-adjusted so that all the elements of a defense program arrive at a predetermined time in the future, e.g., aircraft, pilots, trainers, repair and maintenance manuals, weapons systems, avionics, spare parts. Constant attention has to be paid to balance. Non-defense systems generally do not have the same systems criticality and do not need to continuously scan budget lines for balance.

The procurement/acquisition and Research, Development, Test, and Evaluation (RDT&E) budget accounts make up approximately 40% of the defense budget and are heavily scrutinized by Congress. Balance in managing these accounts is a critical for defending them from reduction by Congress in annual budget review. In addition, within DOD, balance has to be maintained between the Military Construction account provided as a separate appropriation bill (MILCON) and the main DOD appropriation legislation; a change in one may well affect a program in another. The task of budget analysts is to make the corresponding changes to keep the program in balance.

OVERVIEW: POLICY AND BUDGET NEGOTIATION WITH CONGRESS

The stakes are high when DOD budget growth or decline is negotiated with Congress because defense spending is such a sizeable component of the federal budget. Successful lobbying by constituent interests produces a significant payoff in terms of jobs, income, and preservation of a stable defense industrial base. Even when the climate exists to substantially change the defense budget, for example to reduce it and harvest a "peace dividend" in the early 1990s or to increase it substantially to respond to the terrorist attacks of 2001, the demands of electoral politics may be expected to compete with the national interest in providing for a common defense.

Congressional review and enactment of the defense program and budget requires action by six core committees and several others, assisted by their subcommittees. In addition, floor, conference committee, and reconciliation votes are required to pass enabling legislation authored by authorization (Armed Services) and appropriation committees. Given the decentralization and, at times, disorganization of congressional resource decision making, it is little wonder that the DOD, like many other federal departments and agencies, engages in some degree of strategic representation in justifying its program and budget. Both Congress and the DOD attempt to satisfy constituent demands in the budget. After all, pork barrel politics is nothing more or less than democracy in action, albeit at the loss of military effectiveness.

The defense plan and budget are prepared by the OSD, the military departments, and uniformed services on a programmatic basis with requests to Congress divided into 11 program elements. This program structure also is cross-walked into appropriation, function, sub-function, object of expenditure, and other budget formats by DOD for presentation to and review by Congress. Although DOD prepares the program budget, Congress does not review or enact the budget on the DOD programmatic base. Instead, separate authorization and appropriation processes are employed for policy, program, and budget decision-making by the six major committees that negotiate and enact the defense budget (House and Senate budget, armed services, and appropriations). This is also the case for the rest of the federal budget. Congress reviews, negotiates, and executes much of the DOD budget proposal on a project and object-of-expenditure basis. Also similar to budget review for other federal departments and agencies, the most detailed analysis of the defense budget outside the DOD occurs in Congress at the subcommittee level.

Congress requires the DOD to submit its budget in a variety of forms and at a highly disaggregated level of detail. Appropriation committees, for example, receive voluminous computerized R-1 and P-1 program exhibits that show proposed spending on a line-item basis for every research and development or procurement program in the defense budget. DOD reports required by Congress typically indicate the item, quantity of purchase, and costs proposed by the DOD for each procurement by each service branch in thousands or millions of dollars and also contain information on previous purchases and product vendors.

With this level of detail, subcommittee members and staff can attempt to surgically manipulate the DOD budget request to satisfy national security needs as well as the various constituent interests represented effectively by lobbyists in the highly decentralized Congressional decision process. So-called "add-ons" and "plus-ups" that provide funds for programs not requested by the DOD or that increase spending over DOD proposals are a normal element of congressional budgeting. However, in spite of the degree of budgetary power and influence wielded by congressional subcommittees, decisions reached in subcommittee are not final. Many opportunities exist to add or cut programs and money in full committee or on the floor of either house of Congress, in conference-committee negotiation, through reconciliation legislation, in ad hoc budget "summit" conferences, as well as through reprogramming and transfers in oversight of the budget once enacted. Review of supplemental budget requests from the DOD provides additional opportunity for congressional program, budget, and policy direction.

As we have noted, there is *never* a final defense budget—the budget is constantly up for negotiation. Despite the appearance of programmatic vitality or morbidity in the short-term perspective of one-year budget negotiation, programs that seem dead rise again as advocates find new windows of opportunity. Conversely, programs that appear to be blessed with everlasting life may be put at risk of reduction or elimination at almost any time in the nine-to-eleven month decision cycle that is typical for annual congressional action on the budget, and in the total budget cycle including preparation, negotiation, enactment, and execution that lasts approximately two and one-half years. If audit and evaluation phases of the process are included, the total cycle takes three, four or more years, i.e., the length of a single Presidential administration and double the length of the two-year terms of members of the House of Representatives.

In resource negotiation with Congress, the DOD advocates its position assertively in a myriad of public and not-so-public forums. Since the 1970s the number of formal committee and subcommittee hearings held on the defense budget has increased substantially. Despite DOD claims that it is "micromanaged" to incredible excess by Congress, DOD legislative strategy is developed and executed to play in the congressional budgetary system as it is rather than the way many critics believe that it ought to be. This is done through political issue positioning and a considerable degree of strategic representation that is a necessary response to congressional tendencies to micromanage, as with other federal departments and agencies. The details of congressional budgeting for national defense are provided in chapter 5.

CONCLUSIONS

The purpose of this chapter is to provide a basic understanding of some of the characteristics that make defense different from other players in the federal government budget process and cycle. This chapter has reviewed how budgeting for national defense differs from budgeting for other entities in the public sector in the U. S, and virtually anywhere else in the world, including how Congress makes decisions on defense spending. In the chapters that follow we provide detailed analyses of the various components of defense budgeting and resource management. To understand defense budgeting and financial management thoroughly, we delve into the operation of the PPBES process, the intricacies of congressional defense budgeting and supplemental appropriation, DOD budget execution, the roles played by some of the top level participants in the defense budget process in the Pentagon, and financial management issues that confront defense resource managers and decision makers. To these topics we now turn.

CHAPTER 4

THE PLANNING, PROGRAMMING, BUDGETING AND EXECUTION SYSTEM

INTRODUCTION

How the United States determines how much it will spend on defense is a function of many influences inside and outside of government. At all times how much to spend is first a policy question related to what amounts of resources are needed to counter the actions of opponents during time of war, or to deter war, the threat of war or to prevent the need for engaging in other types of armed conflict in the present and foreseeable future. Thus, threat and the resource capabilities needed to counter it drives military spending. In the majority of years, the most influential budgetary factor influencing the amount of funding to be provided to the Department of Defense and the military departments and services is the annual defense budget request and its associated justification which is imbedded in the President's budget delivered to Congress annually. While Congress has the sole power of the purse and the Constitutional power to "raise and support armies" and "provide and maintain a navy" the majority of programs in the annual defense budget, with some high profile exceptions, are authorized and appropriated by Congress roughly as proposed by the Secretary of Defense and the President. Thus, a key factor influencing the size and shape of the defense program is the budget proposal developed by the Depart-

Financing National Defense: Policy and Process, pages 91–151
Copyright © 2012 by Information Age Publishing
91

ment of Defense (DOD) and the military departments and services. The process internal to the DOD for developing the budget is the Planning, Programming, Budgeting and Execution system (PPBES).

More specifically, PPBE is not a single process, but a system of processes. Each phase of PPBE is also a set of subordinate processes that are dedicated to strategic planning, resource allocation, budgeting and program execution. Further, PPBE is embedded among systems of strategic and operational planning processes for the military. PPBE is dedicated to the task of determining budgetary allocations for the manning, training and equipping of the military and the operation and support of defense systems that support national security objectives. The resourcing processes intend to create program and budget plans that align with, and support, the military's operational plans.

To best understand the PPBE system, this chapter will first provide a theoretical foundation of public budgeting in general, leading to an understanding of program budgeting. There are several budgeting systems any government may use and program budgeting is one specific form. We will draw the distinction between theories of public budgeting, systems of budgeting and techniques used by budgeters within those systems. Second, given that theoretical and conceptual view, this chapter will provide an historical view of the PPBE system. We examine the rationale behind the original implementation of PPB in the DOD over fifty years ago and briefly describe the political and managerial changes that brought DOD to the current version of the system.

Third, this chapter addresses the critical topic of structures. All program budgets depend upon a program structure: that set of programs for which resources are identified and the relationships between them that must be analyzed during the several phases of PPBE. In addition to the program structure, it is important to consider other structural factors that influence the size and shape of the defense budget. Budget formats, budget rules, and appropriation rules all influence budget outcomes and, thus, will be briefly explained.

Fourth, the four phases of the PPBE system will be described. We strive to do so in sufficient detail to give the reader an appreciation for the depth of analysis, complexity, and interconnectedness of the processes. At the same time, we strive to do so in fairly general terms. The titles for specific outputs of the system, and the offices responsible for them, change frequently and vary across the military services. The intent here is not to describe the process exactly as it occurs as this book goes to print, but to explain the enduring rationales behind the transient specifics. Many descriptions of PPBE have been written over the years, most are correct on the essentials and wrong on the details. Our goal is to stress those essentials and avoid the confusion that would result from documenting the current details.

PPB-type systems have had their share of criticism, and DOD's PPBE is no exception. PPB-type practices were attempted in domestic federal agencies, in state and municipal governments, and many of those attempts failed. The system has been charged with being incomplete, overly optimistic, responsible for undesirable secondary effects, it is said to be inefficient and ineffective. The final section of this chapter will analyze PPBE and address some of that criticism.

The objectives of this chapter are to enable comprehension of PPBES as a system as follows:

1. A common theoretical foundations for public budgeting
2. Common budgeting systems and techniques
3. The history of PPBE and recent reforms
4. To understand key defense budget structures and the structural influences on resource allocation decision-making for defense
5. To understand the PPBE system: the goals of each stage, the processes used to achieve those goals, critical outputs, and the transitions between stages
6. To describe how PPBE relates to other DOD and government processes
7. To analyze PPBE along critical dimensions
8. To assess some of the strengths and criticisms of the PPBE system

THEORETICAL FOUNDATIONS

Nearly all discussions of budget theory begin with V.O. Key's classic observation that the *basic budgeting problem* is, "on what basis shall it be decided to allocate *x* dollars to activity A instead of activity B?" (1940: 1137). Key suggests that while this looks like a rudimentary applied economics question, it is really a political question:

> It is not to be concluded that by excogitation a set of principles may be formulated on the basis of which the harassed budget official may devise an automatic technique for the allocation of financial resources. [...] The doctrine of marginal utility, developed most finely in the analysis of the market economy, has a ring of unreality when applied to public expenditures. The most advantageous utilization of public funds resolves itself into a matter of value preferences between ends lacking a common denominator. As such, the question is a problem in political philosophy. (1940: 1143)

In search of an answer to Key's fundamental question, scholars have drawn from the theoretical foundations of many disciplines: organizational behavior, economics, political science, decision science, sociology, and behavioral finance to name a few. As Bartle notes, "there is no single theory, there are several" (2001). Budgeting, at its core, is a group decision-making process and decision-making can be observed and understood through many lenses. This section outlines a few of the more significant theories to have captured the attention of economists, political scientists, and public administration scholars. Each, the reader will note, has an element of truth and can explain certain phenomena in certain circumstances, but none are robust enough to constitute an overarching theory of public budgeting. Given the stakes involved and the nature of humans, it is likely that none ever will.

The theories shown fall into different categories. There are normative theories (how the world should work) and descriptive theories (how the world does work). The descriptive theories either describe budgeting processes or budgeting outcomes or both. The theories described below, arguably, can be categorized as shown in Figure 4.1.

The Theory of Public Finance

Musgrave's Theory of Public Finance is rooted in macroeconomic theory and describes three separate, but interrelated, functions of public budgeting: adjusting the allocation of resources, adjusting the distribution of income and wealth, and securing economic stabilization (Musgrave, 1959). First, the allocative function of budgeting seeks to correct in some ways for economic market inefficiencies (e.g., monopolies, excess rents, undesirable social costs, in some cases unjustified public goods) through the allocation

Normative Theories	
Theory of Public Finance	
Descriptive Theories (process)	Descriptive Theories (outcomes)
Incrementalism	
"Muddling Through" and Bounded Rationalities	Punctuated Equilibrium
Multiple Rationalities	
Institutionalism	
Garbage Can Model	

Figure 4.1   Public Budgeting Theories

of goods and services to achieve "social benefits." The determination of so-cial benefits or wants occurs through political, not economic processes. The provision of goods and services to satisfy social wants comes from general revenue, but may or may not be supplied directly by the government. The design of both expenditures and taxes are tools of the allocative function. Second, whereas the allocative function seeks to move the nation's resources from the satisfaction of private wants to public wants, the distribution func-tion seeks to correct market inefficiencies or achieve social aims in the dis-tribution of income and wealth from one individual to another. Political pro-cesses tend to design that redistribution. The tools of redistribution—taxes, transfer payments, and the provision of goods and services—will overlap the allocative function and may even undermine allocation goals. Finally, the stabilization function seeks to ensure a high level of resource utilization (e.g., high employment) and stable prices (low inflation). This can be done by restricting or expanding economic activity which can be affected through taxation, lending, money supply, and transfers. Once again, the budgetary functions overlap. Actions in one dimension have effects in the others. The challenge, of course, is meeting all three objectives as optimally as possible in simultaneous fashion. Musgrave suggests a normative theory of budgeting (how it should be done) more than a descriptive one (how it is really done).

Incrementalism

Most closely associated with the work of Aaron Wildavsky, incremental-ism suggests that the "largest determining factor of this year's budget is last year's" (Wildavsky and Caiden, 2001: 47). That is, incrementalists sug-gest that budgets do not result from a comprehensive rational economic process, but rather from the consideration of a narrow set of increases and decreases that are "in play" at a given moment. Incrementalism suggests that most of the budget is fixed from year to year because of factors such as established ideas on perceived fairness, core requirements, historical prec-edent, and sunk costs from previous decisions. Thus, only a small fraction of the budget on the margin is available for reallocation at any time. In-cremental decisions result from a fairly simple, repetitive, and fragmented process. Wildavsky's theories are based in political science and address the process of budgeting as well as the outcome of budgets.

Muddling Through and Bounded Rationalities

Most closely associated with the work of Charles Lindblom (1959) and Herbert Simon (1997), the idea of bounded rationality arose from behavioral observation. People simply do not act as fully rational decision-makers: subjects who seek to maximize expected utility in the presence of perfect knowledge of preferences, utilities, beliefs, probabilities, payoffs, etc. The *rational-comprehensive* method (as you should learn in an economics or decision science class) follows the basic path of: clarify values and objectives prior to considering means, perform a comprehensive means-ends analysis, select a "good" policy because it is the one with means that are most appropriate to the desired ends. Lindblom asserts that while the comprehensive-rational decision making model is ideal, actual decision-making takes the form of *successive limited comparisons.* That is, goals and analysis are intertwined so means and ends are not distinct; therefore a comprehensive means-ends analysis is inappropriate or even impossible.[1] Analysis is constrained rather than comprehensive; therefore some outcomes, policies, and values are neglected in the search for an acceptable policy. A "good" policy is frequently defined as the one that analysts and policymakers come to agree to implement, without them necessarily agreeing it is most appropriate to the objective (1959: 81).

Similarly, Simon (1997) argues against the "economic man" in favor of "administrative man." According to Simon, people are only boundedly rational: they pursue self-interests but often do not know what those are; they are aware of multiple possible decisions, but limit their search; they do not analyze comprehensively due to limitations of data, attention span, time, and cognitive abilities; and emotions and norms of behavior impact decision-making. People tend to stop searching for the perfect solution once they find an adequate one. This is sometimes referred to as *satisficing* behavior. Lindblom and Simon, like Wildavsky, approach budgeting from a process perspective; Lindblom as a political scientist and Simon as a psychologist (although he earned a Nobel Prize in Economics).

Punctuated Equilibrium

Bryan Jones (2001) noted that incrementalism was a useful explanation for budget changes if one looked at rather narrow time horizons or slices of the budget, but when one expands the time frame or breadth of the analysis, significant and sudden changes can be observed. An evolutionary or incremental theory does not adequately explain those changes. He hypothesized that the bureaucracy, politicians, and interest groups operating near equilibrium will produce incremental budgets. The belief is that opposing forces cancel each other, mistakes are corrected in subsequent budget cycles, and participants desire stability. In short, conflicting forces for change are largely offset and the system remains balanced with only marginal changes. But there are times when the attention span of those involved in public budgeting align in a particular direction resulting in sudden and large swings. This is often precipitated by environmental considerations—a change that begins outside the organization. Examples affecting U.S. federal budgets include the civil rights movement, the end of the Cold War, the terrorist attacks of 2001, and the passage of the Congressional Budget Act in 1974. This is a descriptive theory of budgetary outcomes.

Multiple Rationalities

The multiple rationalities theory is also referred to as the "stages heuristics" theory, and is most closely associated with the work of Thurmaier and Willoughby, and Rubin who recognized the social-political complexity of the budgeting process. Thurmaier and Willoughby assume that budget actors behave rationally, even if that rationality is bounded, but that public budgeting consists of various actors at various stages of the process, each of whom possesses his own rationality. What is rational for a systems analyst inside an agency (the data used, the methodologies employed, the problems considered, the objectives sought) will likely differ from the rationality of an official in the central budget office or in the appropriations committee in the legislature. The stages heuristics model suggests that in each stage in the budgeting process—agenda setting, policy formulation and legitimization (or enactment), policy implementation, and policy evaluation–problems come in several forms. They may be technical, economic, social, legal, or political problems. Each type of problem is best served by a differently rational way of approaching a solution (Thurmaier and Willoughby, 2001).

Rubin asserts that public budgeting consists of five separate but linked "decision clusters." Revenue decisions involve estimates of how much will be available; process decisions involve the manner in which allocation decisions will be made and the distribution of decision rights; expenditure decisions involve estimates of spending; balance decisions involve the relationship between revenue and expenditures; and implementation decisions are concerned with compliance or deviation from the budget plan (Rubin, 1993: 21–23). Decision clusters occur separately in different parts of government but are interrelated. She theorizes that budgets are the result of non-linear, real-time decisions influenced by whichever one or more of the five decision streams are converging at the moment. These theories describe budgeting processes more than they do outcomes.

Institutional Theory

Similar to the multiple rationalities approach, institutionalists note that group decisions result from the interplay of actors with differing goals.

Institutions can exist within organizations or can include multiple organizations (e.g, line officers in the Navy comprise an institution, the Navy as a whole is an institution as is the Army, likewise public service is one). Institutions are guided by both formal and informal rules. With respect to budgeting, formal rules include the structure and timing of the budget documentation, approvals within a chain of command, budget laws, etc. Informal rules are based in the group's norms and beliefs which create expectations for individuals serving in particular roles. Individuals who serve in those roles bring their own biases and beliefs to bear on decisions. The rules can result in a decision logic based more in appropriateness than in economic rationality (March, 1994). Actors meet to make decisions in "action arenas" that are affected by the formal and informal rules, by technology, by environmental considerations, and outcome expectations. The individual actors strategize based on their individual goals and the net result of their meeting is a decision. Such action arenas can occur in layers (such as up or down a chain of command) and across entities (such as in a process flow), and each actor's strategies are continuously revised in response to the differences between realized and desired outcomes. Understanding outcomes is a function of understanding the strategies of the actors, thus this is more of a process theory than an outcome theory. (Ostrom, Gardner and Walker, 1994)

The Garbage Can Model

Developed in studies of organizational decision-making under extreme uncertainty, this theory suggests that decisions are not hierarchically organized, but result from the interaction of streams of problems, solutions, participants, and opportunities (Cohen, March, and Olsen, 1972). The theory applies to organizations that have problematic preferences, unclear technology, and frequent turnover of personnel. [To a limited degree, budgeting in DOD conforms to those criteria.] The theory suggests that the interplay of the four streams is not deliberate or standardized. Solutions may precede problems (think of the personal computer's impact on organizations) and problems may have no solutions. Opportunities can be created or are imposed; decisions may be serendipitous or demanded by external considerations. The overall pattern of organizational decisions appears random and occurs at those moments when solutions and problems meet with the attention of participants. The theory seems to hold best in high-tech situations (e.g, Silicon Valley) or specialized industries (e.g., movie production). This theory describes both outcome and process.

BUDGETING SYSTEMS AND TECHNIQUES

Before describing the DOD process for program budgeting we should first define what program budgeting is and how it compares to other forms of public budgeting. In short, *program budgeting* considers the goals of a government organization and identifies a set of programs that can achieve those goals. Programs are individually priced and the optimum "market basket" of programs is selected that is most able to produce the desired goals. Program budgeting relies on analytical techniques like systems analysis and seeks allocative efficiency. Program budgeting answers the question: what combination of activities, things and human resources must be purchased to achieve the agency's goals and how much do they cost? (Hitch and McKean, 1960; Schick, 1966)

Program budgeting is distinct from other systems of budgeting such as object of expense budgeting. One should note that given the complexity of the DOD budget, although it uses a program budgeting system, other forms may be—and often are—used within that overarching program structure. That is, a new program may develop its first budget using a zero-based approach and then shift to an incremental approach. All operation and maintenance budgets in the DOD include a budget exhibit formatted by object of expense, and the cost breakdowns in procurement budgets are another form of object of expense budgeting. Let's look very briefly at these other budgeting systems:

Object of expense budgets consist of a simple listing of what the funds will procure without regard to the output of the agency. Thus, object of expense budgets can be structured the same for all agencies of a government and for all activities within an agency. Objects of expense include thing such as salaries, rents, utilities, consumable materials, capital purchases, and service contracts. Object of expense budgeting provides detailed information about the cost of inputs (what is to be bought), but no information about the nature or value of the program or activity it supports. It answers the question: on what items will the money be spent?

Incremental budgets take the extant level of funding for an organization, program, or activity and addresses whether that level is insufficient, adequate, or excessive. Reallocations occur on the margin with little investigation of the core amount. It presumes the function of government (or agency) itself is valuable, but questions the relative amount of it. It answers the question: how much more or less should be spent on this function next period compared to the current period?

Zero-based budgets are the functional opposite of incremental budgets in that they directly confront the core amount and in so doing address whether the organization, program, or activity should be funded at all. It presumes the value of nothing and requires every aspect of the budget to be (re)examined. It answers the question: if we started this function anew, how much would we need?

Performance-based budgets apply a data-based incremental approach to (usually) a program budget structure. They require the presence of clear organizational goals and an understanding of the causal relationships between activities and goals. They require financial and non-financial data regarding inputs, outputs and outcomes and a measurement system to assess the efficacy (or efficiency) of activities in achieving the goals. Performance-based budgeting answers the question: to what extent should activities see an increase or decrease of funding to best affect the desired outcomes? (Robinson, 2007; Miller, Hildreth and Rabin, 2001; Posner and Fantone, 2008)

Subordinate to all the budgeting systems are specific techniques that may be employed. That is, one may use any or all of these techniques within any of the budget systems. Open-ended (or unconstrained) budgeting considers all the needs of an agency without regard to an upper limit. At the opposite extreme are fixed ceiling budgets that set an upper limit and require the agency to construct a program within that amount. Cost-benefit analysis examines the relationship between the cost of programs and their impact on agency goals; those items with the highest benefit per unit cost are funded. Priority listings compile all the needs of an agency by descending order of significance and a line is drawn where the resources expire; things above the line are funded, below the line are not. A prioritized, relative-benefit approach is a tool for combining multiple priority listings (usually in a complex organization with dissimilar functions or goals) by considering the relative benefit of items on different lists. A relative priority list can also be used to rank alternative "bundles" of items from a single priority list. Unit costing breaks the activities or outputs of an agency into component costs so that one can state how much of an activity (or number of units of output) can be provided for a given amount of funding. Across-the-board methods will increment or decrement all programs proportionally as ceilings raise and lower. As we will discuss a little later, these techniques are primarily used in the programming phase of PPBE, but may also be used in budgeting.

When done effectively, public budgeting systems display characteristics that uphold democratic ideals of accountability and transparency and which reveal professional management practices. In their Public Expenditure Management Handbook, the World Bank (1998) provides a list of such characteristics. At the top of their list are comprehensiveness and discipline. An effective public budget takes a holistic approach to societal problems and that the processes for collecting revenue and allocating those resources across capital and operating expenses should be appropriately linked to those goals. A disciplined process is one that restrains government. A disciplined process is also predictable; decision-makers, managers and the public can rely upon a standard process and can have faith that decisions, once made, will be implemented. Budgeting systems should be legitimate in that those who decide budget allocations consider and accept input from those who implement or are directly served by those policies. With such input to the process, decisions are contestable. Those with a stake in the policies can voice objections and objective program evaluation informs the process. An effective budget is flexible enough to allow operational managers the freedom to adapt to contingencies while being restrained by a clear strategy. "Too often in the public sector, implementation is tight but strategy is loose." (World Bank, 1998: 2. On budget strategy see also Meyers, 1994). Flexibility is balanced by accountability and transparency—decision makers and public managers owe an accounting to the wider public and should be held responsible for policy outcomes. Finally, the budget must rely upon honest information in the form of unbiased projections and technically sound analysis of revenues and costs. As this chapter describes the PPBE process (and other chapters describe other aspects of national and defense budgeting), think about how well the U.S.A. and DOD uphold these characteristics of ideal budgeting.

While DOD manages its internal resource management systems, this is done under the watchful eyes of Congress. A number of issues related to planning and budgeting for national defense confound DOD and congressional decision makers annually. Among these are how to perform effective and competent threat assessment and the consequences of doing this job well or poorly. Another issue is how much is made available to spend on national defense, the so-called "top line." This is determined in part by the perceived threat, but is also influenced by national fiscal policy and the dynamics of the politics of budgeting for defense. Defense spending is also known to vary in response to public opinion about perceived threats and spending priorities (Hartley and Russett, 1992; Higgs and Kilduff, 1993; Brook and Candreva, 2007). Debate and consensus building for national defense budgets is part of our democratic political tradition. Budgeting for national defense is always complicated by conflicting political opinion and information, but also the need for selective degrees of secrecy with respect to identifying and evaluating the threat and budgetary responses to it. These conditions make marketing the need for national defense spending an inevitable task and part of the obligation of defense advocates working in an open political system.

Because so much of the policy framework and budget of the Department of Defense is determined by Congress—which under the U.S. Constitution has sole power to tax, spend, authorize and regulate the military—analysis of resource allocation for defense cannot ignore the political context within which decisions are made and executed. Policy development and resource planning for defense is inextricably linked to constituent politics in defense budgeting. National security policy choice and implementation is made more difficult by the highly pluralistic nature of the resource allocation decision environment (Wildavsky, 1988: 191–193; Adelman and Augustine, 1990; Augustine, 1982). In the final section of this chapter we will discuss whether PPBE does this adequately. Still, disagreements over policy and resource allocation should be anticipated and, indeed, welcomed in a democracy.

HISTORICAL FOUNDATION OF PPBES

Policy development, planning, and resource allocation decision making for the U.S. Department of Defense is a task of enormous complexity due to the nature and size of the Defense Department and the highly differentiated nature of its mission and activities. The Department of Defense plans, prepares, negotiates, and makes decisions on policy, programs, and resource allocation using the Planning, Programming, and Budgeting System. The Department's approach to this task has evolved over time.

The post-WWII sequence of budget reforms that led to PPBS in the 1960s started with performance budgeting in the 1950s, as having been strongly recommended by the Hoover Commission in 1949. Such techniques, in fact, were seen earlier in the recommendations of the Taft Commission of 1912 and its implementation in the Department of Agriculture in 1934 and the Tennessee Valley Authority in the later 1930s (McCaffery and Jones, 2001: 69). In 1949, Congress required that the budget estimates of the Department of Defense be presented in performance categories. In this first wave of performance budgeting (the second wave would hit in the 1990s) great effort was exerted to develop measures of performance and relate these to appropriations and spending. In fact, many of the measures developed in this era did not measure performance. Instead, because it was easier (and perhaps the only approach possible), workload and input cost data were used in place of real measures of performance. Still, budgeting in this era moved far from the simple objects of expense formats of the past. Formulae and ratios between proposed spending and actions were integrated into the Executive budget along with explanations of what the measures demonstrated and how they related to justifications for additional resources. (McCaffery and Jones, 2001: 69)

The emphasis of budget reform shifted in the early 1960s to what was termed "program budgeting." The Planning, Programming and Budgeting System (PPBS) was implemented in DOD during the Kennedy and Johnson administrations by Defense Secretary Robert McNamara and by comptrollers Charles Hitch, Robert Anthony and others (Thompson and Jones, 1994). Prior to 1962, the DOD did not have a top-down coordinated approach for planning and budgeting (Puritano, 1981; Korb, 1977; Korb, 1979; Joint DOD/GAO Working Group on PPBS, 1983). Until this time, the Secretary of Defense (SECDEF) had played a limited role in budget review as each military service developed and defended its own budget. McNamara had used PPBS when he was the President of the Ford Motors Corporation and with Hitch, his Comptroller, had confidence that the system would be valuable for long-range resource planning and allocation in DOD. McNamara wanted PPBS to become the primary resource decision and allocation mechanism used by the DOD and implemented the system after President John F. Kennedy tasked him to establish tighter civilian control over the military services. As a former member of Congress, Kennedy was highly distrustful of the military service planning and budgeting. Consequently, the initial motivation for establishing PPBS had as much to do with control and politics as it did with rational resource planning and budgeting (Thompson and Jones, 1994; Feltes, 1976; Korb, 1977; Korb, 1979).

Hitch implemented PPBS and the tools of systems analysis throughout DOD, but most of the program analysis was done by his "whiz kids" in the Office of the Secretary of Defense (OSD) under the auspices of the Comptroller and the office of Program Analysis and Evaluation. The military departments were not anxious to implement PPBS, but had to do so eventually to play in the new planning and budgeting game run and orchestrated by Hitch and his staff. PPBS was not just budget reform—it was a new approach to analysis and competition between alternative programs, weapons systems and, ultimately, multi-year programmatic objectives.

The Programming, Planning, Budgeting System was intended to be a thorough analysis and planning system that incorporated multiple sets of plans and programs. PPBS drew upon methods from various disciplines, including economics, systems analysis, strategic planning, cybernetics, and public administration to array and analyze alternative means and goals by program and then derive benefit/cost ratios intended to indicate which means and ends to choose. Budgeting under this system was to become a simple matter of costing out alternative means to achieve the goal chosen. (Lee and Johnson, 1983; Hinricks and Taylor, 1969; Merewitz and Sosnick, 1972; Schick, 1966; Schick, 1973; McCaffery and Jones, 2001: 70)

In theory, the program budgets that resulted from PPBS were supposed to provide the Executive and Congress information on what the federal government was spending for particular categories, e.g., health, education, public safety, etc. across all departments and agencies. Program budgets may best be understood as matrices with program categories on one axis and departments on the other. (The DOD matrix has since evolved to four dimensions, adding appropriation and joint mission capability area.) Thus, in the fully articulated program budget Congress could determine how much was spent on health or education in total in all departments and agencies and this would promote deliberation over whether this was enough, too much or too little.

President Lyndon Johnson thought that PPBS was so successful in DOD that in 1966 he issued an executive order to have it implemented throughout the federal government. Regrettably, although Executive branch departments prepared their program budgets and related spending to objectives, Congress largely ignored what it was presented, preferring to stick with the traditional appropriations framework for analysis and enactment of the budget. Although the government-wide experiment with PPBS was suspended by President Richard Nixon in 1969, this was done more for political than efficiency reasons (Schick, 1973). Still, the system continued to be used in the Department of Defense, in part because DOD purchases substantial long-lived capital assets and since PPB requires long-range planning as its first component, it suited the needs of the Defense Department, an agency whose planning demands are more volatile than other government agencies.

While the manner in which PPBS operates has varied under different Presidents and Secretaries of Defense, the basic characteristics of the system have remained in place for more than 40 years. During this period, three significant reform initiatives have influenced the PPB system: the Laird reforms, the Goldwater-Nichols Act, and the Rumsfeld transformation in 2001–2003 that renamed the process adding the word execution, i.e., the system is now referred to as PPBE.

Laird Reforms

In 1969, Melvin Laird was appointed Secretary of Defense by President-elect Richard Nixon to succeed McNamara. Laird brought a different management orientation to the Defense Department, one more in keeping with its historical predilections, emphasizing decentralization and military service primacy. If McNamara increased scientific decision making in the Pentagon, he also installed a centralized management approach. Systems analysis, top-down planning, and benefit/cost analysis supported this centralized focus. One of the key bureaucratic players was the Office of Policy Analysis, which made use of the tools cited above to help McNamara centralize decisions in the Office of Secretary of Defense (Thompson and Jones, 1994: 68–73). Laird's methods ran counter to this approach, emphasizing participatory management and decentralization of power. Beginning in 1969, Laird shifted decision making power away from the DOD staff agencies to the Military Department Secretaries, because there were, "...many decisions that should be made by the Services Secretaries and they should have the responsibility for running their own programs. I have no business being involved in how many 20mm guns should go on a destroyer. That is the Secretary of the Navy's business. I must let the Services take a greater role." (Feltes, 1976) Laird also pursued a process of participatory management, in which he hoped to gain the cooperation of the military leadership in reducing the defense budget and the size of the forces.

Laird was preoccupied with disengaging from Vietnam, but not to the exclusion of other issues, such as burden-sharing costs with other nations, maintaining technological superiority (e.g., B-1 bomber, Trident submarine), improved procurement, enhanced operational readiness, and strategic sufficiency and limitations on the nuclear build-up (Feltes, 1976; Armed Forces Management, 1969). On the management side, Laird gave the military department secretaries and the JCS a more influential role in the development of budgets and force levels, but he also returned to the use of service program and budget ceilings (fixed shares) and required services to program within these ceilings. This concept of ceilings or "top-line" still influences DOD budget requests today, as services are expected to balance their program and budget against the total obligational authority they are given at various stages in the planning and budget process.

The Goldwater–Nichols Act of 1986

It may be argued that the creation of the defense department in 1947–49 never really took hold in that, by and large, the military departments continued to go their separate ways within the envelope of the Department of Defense until the reforms of the 1960s and, to some extent, until implementation of the Goldwater–Nichols Act of 1986 (Thompson and Jones, 1994: 78–79, 246). By 1981, the sitting JCS Chairman, General David Jones was writing that the system was broken and asking Congress to fix it (Jones, 1982). The fact that General Jones as CJCS was voicing such criticisms was in itself very significant (Chiarelli, 1993:71). Jones (1982) suggested that because of the decentralized and fragmented resource allocation process driven by parochial service loyalties, there was always more program than budget to buy it; that the focus was always on service programs; that changes were always marginal when perhaps better analysis would have led to more sweeping changes; that it was impossible to focus on critical cross-service needs; and the result was that an amalgamation of service needs prevailed at the Joint Chiefs of Staff level. General Jones argued that staff to the Chairman of the JCS was so small that the Chairman could focus only on a few issues and that all of this undercut the authority of the unified command structure established in the Defense Reorganization Act of 1958 (Thompson and Jones, 1994: 51–53).

In 1986 Congress passed a sweeping reform plan, commonly referred to as the Goldwater–Nichols Act (for its congressional sponsors), over the ardent objections of many in the Pentagon, including Secretary of Defense Caspar Weinberger (Locher, 1996: 10; Locher, 2002) who thought it would break apart the DOD management system. The legislation is too complex to detail here, but among other things it strengthened the hand of the Chairman of the Joint Chiefs of Staff as chief military advisor and spokesman to the Secretary of Defense and to the President, provided the CJCS with a larger staff and identified important phases in the PPBS process where the JCS would play in setting requirements and reviewing the plans of other players. It established the national command authority to run from the President to the Secretary of Defense to the unified commanders in chief (CINCs, now called Combatant Commanders). This increased their formal authority so that rather than using whatever forces the military services would allow them to use in their geographical area, the unified CINCs had war fighting and command responsibilities and the military service roles were to provide them with the wherewithal to do so (Thompson and Jones, 1994: 51–53, 79, 223–224). This distinction clearly put the military services in the role of manning, training and equipping forces on behalf of the warfighting missions of the geographically based unified commanders. The military department secretaries hold most of the DOD budget authority, while the combatant commanders identify requirements in the planning and programming phases. While a provision was made for combatant commander review of service budgets, such reviews tended to be pro forma until the late 1990s.

Goldwater–Nichols also created the position of Vice-Chairman of the Joint Chiefs of Staff. Generally, the officers who have served in this spot have been strong innovators and, through various committee structures, have had a substantial impact on the resource planning process within DOD. The VCJCS chairs several influential committees that govern the business of DOD including the validation of capabilities and requirements.

The Rumsfeld/Gates Reforms

In 2003, the DOD announced significant changes to the PPB system, renaming it the Planning, Programming, Budgeting and Execution System (PPBES), in Management Initiative Decision (MID) 913 (Secretary of Defense, 2003a; Jones and McCaffery, 2005). While the basic structure and approach of PPBS remained, it was changed in three important ways. First, the reform merged what had been separate programming and budget reviews into a single review cycle, and merged what had been separate databases of program and budget information. Logically, the budget is written to describe and justify the mix of programs suggested by the services in their Program Objective Memoranda (POM) and programs in the POM must be funded. Reviewing the POM for compliance with planning and programming guidance separately from the budget occasionally led to inconsistencies and, "often failed to integrate strategic decisions into a coherent defense program" (Secretary of Defense, 2003a: 2). The decision also directed the use of more performance-based analysis, consistent with the Bush Administration's management agenda. OSD facilitated the reviews by shortening the time allotted for formulation and extending the time allotted for review. This also served to put the budget in the hands of the OSD for a longer period of time at the expense of time spent by the individual services. This was widely interpreted as a shift to greater centralized control of decision-making by the Secretary of Defense.

Second, the changes shifted emphasis from service-level, threat-based planning to joint-level, capabilities based planning. In the decade or so since the Cold War ended, strategic planning in the Pentagon was difficult. Planning processes and systems were designed around a peer competition and struggled with the new global order that left the US as the sole superpower. Defense budgets fell in pursuit of a "peace dividend" and planners struggled to identify the military threats that would justify retaining or changing force structure. This was a period of time in which there were many vocal change advocates, a "revolution in military affairs" was a topic of speculation, and advances in information technology changed the nature of security and threats to it. It was decided that planning would not be done in response to *specific* threats, but rather based on a portfolio of capabilities designed to counter *possible* threats. Not only was budget guidance distributed on a capability basis, but programs were assigned to capability portfolios to be reviewed by OSD staff and Combatant Commanders, and readiness reporting shifted to a new system (Defense Readiness Reporting System) that was organized by joint mission capability areas comprised of mission essential tasks.

Third, the MID-913 reforms defined a biennial budget process and changed the cycle for OSD provision of the top level planning information to the military departments and services. It intended two distinct two-year cycles within a four-year presidential administration. "The Department will formulate 2-year budgets and use the off year to focus on budget execution and program performance" (Secretary of Defense, 2003a: 2). Section 922 of the FY2003 National Defense Authorization Act aligned Quadrennial Defense Review submission dates to the second year of a presidential administration, allowing the first year to conduct the review. During even numbered, or "on", years, the Secretary of Defense would issue detailed planning and budgeting guidance and a full cycle of POM development and biennial budgets. The addition of "E" to PPBS signaled the emphasis on execution review in the odd, or "off", years. The four year schedule is portrayed in Figure 4.2.

Portions of the third element were short-lived. Only the CY 2004/2005 cycle could be said to conform to this new model. By CY2006 and later, budgeting returned to a more traditional process and in 2010 the Deputy Secretary of Defense removed all pretenses of biennial budgeting and announced that there would be annual POMs and annual budgets. (Insid-

4 Years in the 2-Year Cycle	
Year 1: Review and Refinement	**Year 3: Execution of Guidance**
Early National Security Strategy	- - -
Restricted fiscal guidance	Restricted fiscal guidance
Off-year Defense Planning Guidance, as required, incorporating new Administration's priorities, fact-of-life acquisition program changes, and congressional changes.	Off-year Defense Planning Guidance, as required, incorporating fact-of-life acquisition program changes, and congressional changes.
Limited changes to baseline program and budget	Limited changes to baseline program and budget
Program, budget and execution review initializes the on-year Defense Planning Guidance	Program, budget and execution review initializes the on-year Defense Planning Guidance
President's Budget and Congressional Justification	President's Budget and Congressional Justification
Year 2: Full PPBE Cycle— Formalizing the Agenda	**Year 4: Full PPBE Cycle— Ensuring the Legacy**
Quadrennial Defense Review (QDR)	- - -
Full fiscal guidance	Full fiscal guidance
On-year Defense Planning Guidance, implementing the QDR	On-year Defense Planning Guidance, refining alignment of strategy and programs
POM / BES submission	POM / BES submission
Program, Budget and Execution review	Program, Budget and Execution review
President's Budget and Congressional Justification	President's Budget and Congressional Justification

Figure 4.2 MID-913 Change to PPBE. *Source*: Secretary of Defense 2003a: 3

eDefense.com, 2010: 1–2) The remaining reforms, however, continued to be institutionalized throughout Robert Gates' tenure as Secretary of Defense and continued under his successor Leon Panetta.

DEFENSE BUDGET STRUCTURES

Earlier in this chapter we said program budgeting considers the goals of a government organization and identifies a set of programs that can achieve those goals; programs are individually priced and the optimum "market basket" of programs is selected that is most able to produce the desired goals. Let us now expand on that idea and discuss how the DOD structures it program. The basic building block of the defense program is a program element (PE). According to the DOD's Future Years Defense Program (FYDP) Structure Handbook (DOD, 2004), a PE "generally represents aggregations of organizational entities and resources related thereto" (8) for a unique defense program. PE's "are both mutually exclusive and exhaustive" (5) so that every DOD resource is assigned to one and only one PE. One cannot fund a program without first having a PE assigned. There are thousands of PEs and they include things as diverse as Space-Based Interceptors, Base Operations for Naval Air Stations, Sensor Fused Weapons, Pollution Prevention, Army Reserve Division Forces, Dental Care, and Special Operations Technology. Informally, PEs can be aggregated in any number of ways: to display total resources for a specific program, to display weapons systems with their support systems, to display families of systems (e.g., a ship with all its associated weapons systems or an entire battle group), or to compile logical groupings for analysis (e.g., all surface ships or all tactical aviation or all health care). Formally, each PE is associated with one of eleven major force programs (MFP). According to the FYDP Structure Handbook, an MFP is "an aggregation of PEs that contain the resources needed to achieve an objective or plan" (6), but that is not really true if one looks at how broadly defined each MFP is. While specific objects and organizations are assigned PEs, operational plans and objectives normally require a combination of many programs from several of the MFPs. Each MFP contains programs that support countless objectives or plans. MFPs are displayed in Table 4.1.

One of the lessons that emerged from the Johnson/Nixon era attempts to use PPBS broadly was that program structures greatly influenced the manner in which budget decisions were made and most agencies struggled to define a program structure that accurately and adequately captured the work of the agency. In fact, to facilitate a smooth implementation and to preserve accountability, program structures normally mirrored existing organizational structures, thereby reducing the potential value of the program approach (Black, 1971, Schick, 1985). DOD's major force programs

---

TABLE 4.1   Major Force Programs

MFP-1 Strategic Forces	Those organizations and associated weapons systems whose force missions encompass intercontinental or transoceanic inter-theater responsibilities. Includes strategic offensive and defensive forces including associated headquarters, support, and logistics elements.
MFP-2 General Purpose Forces	Those organizations and associated weapons systems whose force mission responsibilities are, at a given point in time, limited to one theater of operation. Includes associated headquarters, support, and logistics elements.
MFP-3 Command, Control, Communications, and Intelligence	Intelligence, security, communications and functions such as charting, mapping, and geodesy activities, weather service, oceanography, special activities, nuclear weapons operations, space boosters, satellite control and aerial targets. Intelligence and communications functions that are specifically identifiable to a mission in other major programs shall be included with the appropriate program.
MFP-4 Mobility Forces	Sealift, airlift, traffic management, and water terminal activities. Includes associated headquarters, support, and logistics elements.
MFP-5 Guard and Reserve Forces	Guard and Reserve training units in support of other MFPs.
MFP-6 Research and Development	All R&D that has not been approved for operational use. Includes basic and applied projects and the development, test, and evaluation of new weapons systems.
MFP-7 Central Supply and Maintenance	Supply, maintenance and service activities including transportation, overseas port units, industrial preparedness, commissaries, and depot maintenance.
MFP-8 Training, Medical, and other General Personnel Activities	Training and education, health care, permanent change of station travel, family housing. Similar costs directly related to another major program are included in that program.
MFP-9 Administration and Associated Activities	Administrative support of departmental and major administrative headquarters, field commands and administration activities not accounted for elsewhere. Includes construction planning, public affairs, claims, and criminal investigations.
MFP-10 Support of Other Nations	Includes support of Military Assistance Program, foreign military sales, NATO infrastructure, and humanitarian assistance.
MFP-11 Special Operations Forces	Special operations forces (active, guard, and reserve), including command organizations and direct support.

Note: MFP-1, 2, 3, 4, 5 and 11 are combat force programs; the others are combat support
Source: DOD 7045.7H, Future Years Defense Plan (FYDP) Structure, November 2004: 6–8

---

are extremely broad, particularly **MFP-2 General Purpose Forces** which includes nearly everything used in conventional warfare, whether conducted at sea, on the ground or in the air, plus a good deal of the infrastructure supporting it, including such diverse things as Base Realignment and Clo-

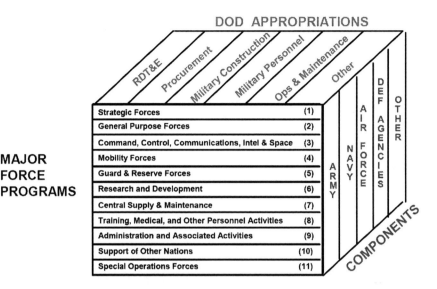

Figure 4.3 Future Years Defense Program (FYDP) Structure.

sure (BRAC) funds, Child Development Centers, Foreign Currency Fluctuations, and Bachelor's Quarters.

Similar to the fact the federal budget of the USA is structured by budget function, federal agency and object of expense, the defense budget is arranged by Major Force Program, by organization (military services and defense agencies) and by appropriation title. The cube-shaped diagram in Figure 4.3 illustrates that every program falls into a particular MFP and PE, it must be funded by the proper appropriation(s) and is administered by one of the military services or defense agencies. In other words, every program that receives funding, has an "address" in the cube. Secondly, the cube illustrates that one of the primary goals of defense budgeting is to achieve balance on each dimension. How that is done is described in the next section of this chapter. Data for Fiscal Year 2010 actual defense outlays is provided for all three dimensions in Tables 4.2 to 4.4.

In the previous section we noted that one of the recent reforms to PPBE is an emphasis on planning for capabilities over threats. We also mentioned that defense activities have been defined as falling within taxonomy of Joint Capability Areas (JCA). There are nine JCAs: force support, battlespace awareness, force application, logistics, command and control, net-centric, protection, building partnerships, and corporate management and support. Some of these parallel the MFPs, but others do not. Within the nine JCAs, there are dozens of specific joint tasks. Each PE is linked to a JCA. As this system develops, it is unclear whether it will eventually replace the MFP

TABLE 4.2   Defense Budget Allocations by MFP, 1962–2011: Highest Proportion, Lowest, Average, and Most Recent

	Strategic Forces	General Purpose Forces	C3, Intel & Space	Mobility Forces	Guard & Reserve Forces	Research & Development	Central Supply & Maintenance	Training Medical & Other	Admin & Assoc	Support of Other Nations	Special Ops Forces
High	21%	44%	15%	4%	9%	10%	11%	25%	13%	3%	3%
Low	1%	32%	6%	1%	3%	5%	4%	10%	2%	1%	1%
Average	6%	37%	10%	3%	6%	8%	8%	18%	3%	1%	1%
FY2010	1%	38%	13%	3%	5%	8%	4%	10%	13%	2%	2%

Source: FY2011 and FY2012 National Defense Budget Estimates (DoD Greenbook), Tables 6-5.

TABLE 4.3   Budget Authority: Summary by Component

Component	FY 2010	FY 2011 CR	FY 2012
Army	138.8[a]	136.8	144.9
Navy	155.3	155.6	161.4
Air Force	142.4	143.2	150.0
Defense-wide	91.5	90.5	96.8
Total	**527.9**	**526.1**	**553.1**

[a] ($ in billions)
Note: Numbers may not add due to rounding.
Source: Fiscal Year 2012 Defense Department Budget Request Roll Out Briefing Slides, February 2011

TABLE 4.4   Budget Authority: Summary by Appropriation Title

Appropriation Title	FY 2010	FY 2011 CR	FY 2012
Military Personnel	135.7[a]	135.2	142.8
Operation and Maintenance	183.9	184.5	204.4
Procurement	103.2	104.8	113.0
RDT&E	79.3	80.4	75.3
Military Construction	20.5	15.9	13.1
Family Housing	2.3	2.3	1.7
Revolving and Management Funds	3.1	3.1	2.7
Total	**527.9**	**526.1**	**553.1**

[a] ($ in billions)
Note: Numbers may not add due to rounding.
Source: Fiscal Year 2012 Defense Department Budget Request Roll Out Briefing Slides, 2011

structure or will become a fourth dimension on which the defense budget is built and balanced.

The programming phase of PPBE operates within the MFP and JCA structures, organizing personnel, hardware, and financial data for each PE into operational and support categories. The budgeting phase primarily operates within appropriation structures: the division of appropriations into the various budget activities and budget line items. Since budget structure impacts and constrains decision-making, it is important to note that the legal structures that define appropriations affect how programs are priced and budgeted. These legal structures are examined in detail in the Budget Execution chapter. Perhaps the most striking example comes from the full and incremental funding policies. Programs that are funded in procurement appropriations normally must follow the full funding policy that mandates the full cost of acquiring a complete, usable system must be budgeted and available before contract can be let. Thus, the full procurement cost of an aircraft or ship is budgeted in the year in which the order is placed, even if that item takes several years to build. Conversely, programs that are funded in research and development or operation and maintenance appropriations are budgeted on an incremental basis. Incremental funding means that one year's worth of an ongoing activity is requested in the budget.

HOW DOES THE PPBE SYSTEM OPERATE?

Now that we have considered the theoretical foundation, history and structures that govern the process of building the defense budget, let us now turn to a description of the actual PPBE system. By way of introduction, we offer a few reminders and additional process background. First, the reader is reminded that the PPBE system only defines those processes that occur *inside* the DOD. Thus, the system tends to consider the military environment when making resource allocations as that military environment drives considerations about force structure, but the system tends to ignore the domestic political environment and the international diplomatic environment, both of which serve to constrain it. As will be discussed in section 7 of this chapter, some have argued that PPBE relies too much on systems analysis and too little on policy analysis (Dror, 1967; Wildavsky, 1969). Readers should keep this in mind and at appropriate points in the descriptions of the processes recognize the external actors and processes that either influence PPBE or are influenced by it.

Second, PPBE is a decision-support system concerning resource allocations that determine force structure. Those decisions should comport with decisions related to other aspects of staffing, operating, equipping and maintaining the military. The linkages to those processes will be discussed.

Third, readers should understand the temporal aspect of PPBE. While PPBE consists of four sequential phases, those phases overlap. Each phase begins before the preceding phase has ended. This is the result of several factors: the federal budget calendar, the complexity of the process, the intrinsic linkage between programming and budgeting, and the interdependencies between the phases themselves. Thus, each phase begins with only a partial understanding of the results of the phase that precedes it. This means that as each phase progresses those engaged must cope with increasing constraints. Only the middle two portions of PPBE are purely calendar-driven with discrete start and stop points. Planning tends to be perpetual and so is execution. PPBE is a calendar-driven process inside an ongoing and event-driven organization. Further complicating the calendar is that while PPBE is often described as an annual process, in reality there are inter-annual portions (such as supplemental appropriations, reprogramming funds and transfers of functions between organizations), annual portions (the POM and budget itself), biennial portions (much strategic planning occurs in approximately two-year cycles, such as the National Military Strategy), quadrennial portions (the Quadrennial Defense Review or QDR) and ad hoc components. Given that phases overlap, operate on different cycles, or are continuous, any significant budgetary event confronts the question of: what impacts to which years? Is this something that can wait for the next budget cycle, does it require an amendment to the budget currently on Capitol Hill, should funds be transferred immediately, or some combination of all three? The case of Mine Resistant Ambush Protected Vehicles in response to the threat of improvised explosive devices in Iraq and Afghanistan clearly demonstrates the complexity—and use—of all these choices (Blakeman, et. al., 2010).

Fourth, each military department is organized a little differently with respect to who is responsible for what portion of PPBE (Hill, 2008). What we try to do in this section is to describe the system generally so that it assists one in understanding PPBE, regardless of the particular version of it you are in. When differences are particularly acute, we will make note of them.

As we head into a description of each phase of PPBE, let us pause first to look at the forest, then the trees. PPBE is a budgeting system designed to link agency strategy setting and plans to a set of programs that will most effectively achieve that strategy, within fiscal limitations. Corresponding budget materials are prepared for the legislature who will consider and amend the request and provide authorization and funding for the programs. Using that funding, the agency executes the budget, implements the programs and seeks to achieve the strategic goals. Thus,

- The goal of *planning* is to identify gaps or overmatches between strategy and capabilities and produce objectives for programming to address them.

- The goal of *programming* is to allocate resources among programs across a mid-range time horizon that best achieves the planning objectives.
- The goal of *budgeting* is to justify the programming decisions in a format that serves the process of legitimization (enactment).
- The goal of *execution* is to implement the policy direction and to create the desired capabilities.

The Planning Phase

In assessment of policy and planning for defense, it is important to understand that the goals and mission of the DOD are not deliberated and set exclusively within the PPBES system. Policy direction comes from the President, the National Security Council, and other executive branch agencies, and from Congress. Under the separation of powers constitutional political system in the U.S., Congress always has the authority to assert policy priorities independent of the executive branch.

Another set of factors that drives the policy and planning phase of PPBES are the treaties, international commitments, agreements, and understandings of U.S. defense obligations negotiated by policy makers over the past century and particularly since World War II. Critical to policy development and planning is the assessment of worldwide threats to U.S. and allied interests that are monitored constantly by a variety of intelligence agencies. In addition, a distinguishing set of factors that drive PPBES are the broad defense policy and programmatic objectives set by the President and his advisors, for example, in the areas of readiness, sustainability, force structure, and modernization.

Not only is defense planning partially out of the control of the department; the strategic planning considerations inside the department are complex. DOD policy development and planning may be differentiated into semi-autonomous systems: one for macro-international security planning, a second for warfighting planning, and a third for defense management and resource planning. OSD conceives of force planning as a four dimensional construct. The first three dimensions—*employing the force, managing the force, and developing the force*—occur in the immediate, short-, and medium-term time horizons, respectively. The fourth dimension is continuous and consists of *corporate support*—the administrative and support activities that enable the other three. In addition, the DOD may be viewed to operate organizationally using two or, arguably, three interdependent management control systems: one for military operations, a second for general administration, and a third for financial management and budgeting.

Organizationally, we have already discussed the individual military services, departments, defense agencies, and combatant commands, each of which performs some degree of individual strategic planning and control. Alignment of these semi-autonomous systems of planning and control has been a problem for DOD. OSD and the military departments have attempted to address alignment problems through governance structures that forge improved linkage, networking and coordination, and through the use of better business practices, such as enterprise-wide business and IT architectures and the adoption of more joint systems and business rules.

Essentially, the planning stage of PPBE is about the intersection of strategy, operating concepts, and capabilities. There is a relationship, but a distinction, between the strategic planning processes that support war fighting and those that drive resource allocation decision-making. Considering first the support to war fighting and operational planning, we see in Figure 4.4 a series of documents that become less policy oriented and more task oriented; they become less politically and economically oriented and more militarily and operationally oriented. The *National Security Strategy* is viewed as the Grand Strategy that is superior to, and guides, the strategic planning that occurs inside the DOD. The NSS is developed by the National Security Council with input from all those agencies involved in the security of the state: defense, state, homeland security, FBI, CIA, etc. It is promulgated by the President. The DOD input to the NSS is coordinated by the Joint Staff and is embodied in a Comprehensive Joint Assessment of the security environment, health of the force, military risk assessments, results of joint experimentation, and proposals for new concepts of operations. The *National Defense Strategy* describes the defense department's portion of the NSS and is promulgated by the Secretary of Defense. It sets objectives to guide the department. It may be included as part of the Quadrennial Defense Review or a separate document. Whereas the NSS and NDS are issued by civilian authorities, the *National Military Strategy* is issued by the Chairman of the Joint Chiefs of Staff and focuses on specific military objectives and capabilities. The *family of joint operating concepts* describes how the joint forces will operate in the medium to long term (8–20 year time horizon) and provides the conceptual framework for future operations which drives the development of military capabilities. It is about this point in the process where strategic direction diverges. The capabilities and force structure demanded by the joint operating concepts feed into the requirements process that eventually feeds into the defense acquisition process. Likewise, the capabilities and force structure demanded by the joint operating concepts also defines the objectives that need to be resourced by the military services in the PPBE process. As noted by the Naval War College,

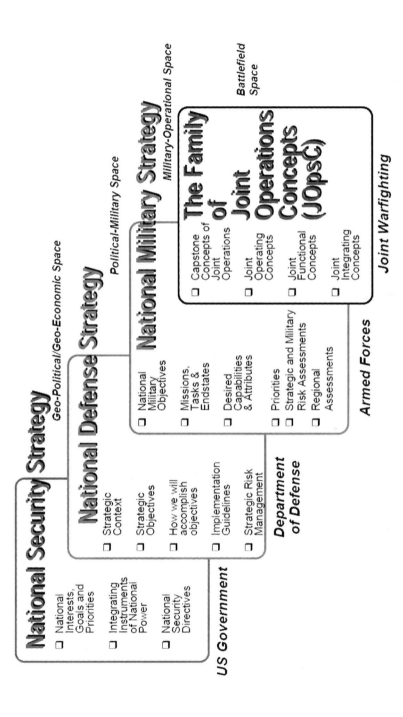

Figure 4.4 Linking Strategy to Operational Capabilities. *Source:* Sullivan, Sean C. (2010) The Four Year Integrated Defense Planning Cycle: The Formal Processes in U.S. Defense Planning, A Desk Reference, Newport, RI: Naval War College.

The process of using strategy to develop operating concepts, capabilities, and force structure is extremely challenging. The link between strategy and force structure is complicated by many challenges. Strategy is developed based on a projection of the future security environment. Strategic objectives are assigned to the military and the military is tasked to develop the operating concepts and capabilities that accomplish these objectives. This task requires operating concepts, capabilities and forces that are relevant and effective in the security environment. Complicating this is the lead time required to create force structure. Defense programs often take several years to develop, procure and field. Also, uncertainty in the security environment can manifest itself by unexpected events, surprise, or miscalculation. This can increase risk and potentially impact the effectiveness of force structure in the future. (Sullivan, 2010: 27)

In addition to the *Comprehensive Joint Assessment*, the Joint Staff conducts a series of *Joint Strategy Reviews* of existing strategic guidance and estimates of the force's capability of achieving those strategies. Reviews are done on such areas as intelligence, capability gaps, operating concept development and experimentation, logistics, and personnel. These assessments are done annually and form the basis for three products from the Joint Staff: the *Chairman's Risk Assessment* which includes his advice regarding the strategic environment, the *Chairman's Program Assessment* of how well the military department and service POMs support the Combatant Commanders, and the *Chairman's Programming Recommendation* on acquisition program priorities for the joint force. The latter two directly impact resource decision-making within the services' programming phases.

The goal of planning, with respect to resource allocation decision-making, is to identify any gaps or overmatches between the national military strategy and the extant and pending capabilities, and to produce objectives for the programming phase to address them. Front end assessments of the fit between capabilities and strategy are a critical component of this part of the process. Each year the Secretary of Defense assigns a set of topics for specific study. The result of these assessments, coupled with the strategic processes already described culminates in guidance issued by the Secretary of Defense to the military departments to direct their programming priorities. This guidance has had a number of names over the years (Defense Planning Guidance, Joint Programming Guidance, Guidance for the Development of the Force, Defense Planning and Programming Guidance). Regardless of the name, this document signifies the end of the planning phase and the beginning of programming. For the purposes of this discussion, we will refer to this as the *Defense Programming Guidance* (DPG).

It is important to note that these strategy documents are written at the national level with a joint force and DOD-wide emphasis. Similarly, each military department undertakes a parallel process within its own ranks, setting strategy and investment criteria for its roles and missions within that NDS and NMS. The Department of the Navy, for example, issues a Maritime Strategy and the Department of the Army has The Army Plan. Consistent with the Maritime Strategy, the Chief of Naval Operations issues a Navy Capability Plan and the CNO Investment Strategy to draw the strategy-to-capability-to-investment link at the navy level that the DPG performs at the defense level.

Who does planning? At the OSD level, it is guided by the Undersecretary of Defense for Policy, USD(P). The work of writing strategy often is done within the "5" codes (J5, N5) on the military staffs. The work of assessing current capabilities is dispersed to the relevant functional codes: the "1" code reviews manpower, the "4" code reviews logistics and readiness. Often the "8" code is the one that knits the two together and provides oversight to the programming role. The title given the N8 on the Navy staff is Deputy Chief of Naval Operations for the Integration of Capabilities and Resources, a rather apt description.

Accompanying the DPG is the *Fiscal Guidance* (FG). Until now, most of the planning and analysis has been relatively unconstrained by resources. Planners will assume a funding profile in the outyears that is consistent with last year's budget plus or minus an incremental amount. The level of precision is low because the task is not to allocate or request budget authority, but to set strategy, assess current capabilities, define desired capabilities, and identify the mismatches between them. As the work transitions to the programming phase, hard resource allocation decisions will be made and a budget constraint is imposed in the FG. The FG originates in OMB with the White House sending early signals about the budget year fiscal policy. These figures may change through the summer and into the fall in response to real world situations, economic factors, and the actions of Congress on the current budget. The DOD comptroller receives the guidance and issues controls (top line targets) to the military departments and defense agencies.

One criticism of the DPG has been that it sets a fairly broad set of goals, broader than the FG can afford. Another criticism is that it comes out too late (April or May) and most of the programming work in the services is nearly complete by then. The service budget review in the summer months is often spent re-doing the early programming work to fit the DPG's direction; because of that, early rounds of programming are often over-resourced so that there is a sufficient body of programmatic analysis to consider all the options when it comes time to allocate the constrained resources in accordance with the ultimate guidance. From the SECDEFs perspective, the broad goals yield options and allow him to see the funding risk in the budget. It forces the services to prioritize consistently with the strategic di-

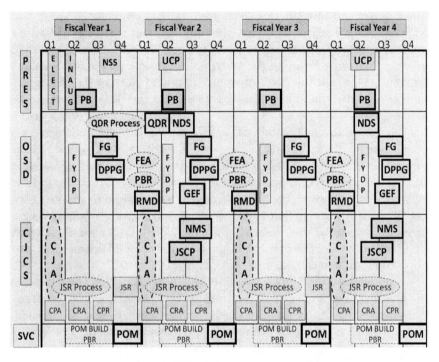

Definition Key

CJA – Comprehensive Joint Assessment	NDS - National Defense Strategy
CPA – Chairman's Program Assessment	NMS – National Military Strategy
CPR – Chairman's Program Recommendation	NSS – National Security Strategy
CRA – Chairman's Risk Assessment	PB – President's Budget
DPPG – Defense Planning and Programming Guidance	PBR – Program Budget Review
	POM – Program Objective Memorandum
ELECT – Presidential Election	POM BUILD – Service POM Building Period
FEA – Front End Assessments	QDR – Quadrennial Defense Review
FYDP – Future Years Defense Program	RMD – Resource Management Decisions
INAUG – President's Inauguration	SVC PLANS – Services Planning and Programming Period
JSR - Joint Strategy Review	
JSCP – Joint Strategic Capabilities Plan	

Figure 4.5 Four Year Integrated Defense Planning Cycle. *Source*: Sullivan, Sean C. (2010) The Four Year Integrated Defense Planning Cycle: The Formal Processes in U.S. Defense Planning, A Desk Reference, Newport, RI: Naval War College.

rection. For the most part, the services would prefer to have clearer direction earlier.

The Programming Phase

When one thinks of budgeting in the generic sense, one tends to think of creating work plans, evaluating alternative courses of action and choosing from among them, and defining boundaries on revenue and expenditure. One might envision a number of decision analytics in use, depending on the nature of the resource allocation decision: capital budgeting techniques for big purchases, incremental approaches for running costs. Best practices for public sector budgeting are to consider a medium term time horizon (3–7 years) and to annually update that medium-term plan. One may also think of the group dynamics involved—the politics of decision making—framing arguments, taking positions, negotiation, persuasion, coalition building, strategizing and game playing. But at the same time, there are pressures to make a budget contestable and legitimate so stakeholders and others with a vested interest have a voice in the process. All these things are descriptive of the programming phase of PPBE more so than the budgeting phase. The Pentagon's core resource allocation decision-making occurs in programming. Whereas the planning phase of PPBE determines the goals, objectives, operating concepts, and desired capabilities of the military, the programming phase allocates resources across a set of programs that are believed to best achieve those goals and objectives.

Simply put, *the goal of programming is to allocate resources, constrained by the fiscal guidance (FG) and appropriations rules, among programs across a midrange time horizon that best achieves the objectives defined in the Defense Programming Guidance (DPG).* As noted in the section on structure, allocations of resources are made among the thousands of program elements (PEs) and consist of obligation authority, personnel, and quantities of items. The midrange time horizon is set by the terms of the future years defense plan (FYDP) and, at the time this is written, is defined as the budget year plus four additional years. The principal output of the programming phase is the Program Objectives Memorandum, or POM.

What does the work of programming look like? It is important to note that of all the phases of PPBE, programming is the least documented. Key outputs from the strategy phase are publicly released (e.g, NSS, NDS, NMS, QDR) and it is not difficult to find descriptions of the process of strategizing and planning (Sullivan, 2010). After programming, thousands of pages of budget documentation are publicly released and there are detailed budget guidance manuals one can read (DOD, 2011 (FMR); DON, 2011). In contrast, the POM is an internal DOD working document; it is not shared publicly. Correspondingly, the process for creating it is not widely shared. Even if it was documented, the documentation would be quite voluminous because each military service takes a different approach to the task and different cells within each service will employ different analytical tools to do their work. As leaders of specific functional areas of the service staffs rotate in and out of those assignments, the manner in which decisions are made might change with them. Despite all that diversity, there are a number of common procedures and analytical techniques we will describe, and we will also identify key differences between the military services.

Programming is an iterative process in which tentative allocations of resources are alternatively disaggregated to be evaluated at the program level and consolidated to be evaluated on increasingly broad integrated levels. Take naval aviation, for example. Program estimates for costs and quantities of F/A-18s would be developed by the sponsor on the navy staff with input from the program management office. That proposal would be evaluated on its individual merits and then all the proposals for navy tactical aircraft would be evaluated as a group to ensure a proper balance across aircraft types. Then the distribution of resources across tactical aircraft would be evaluated in the context of resources for support aircraft, product improvement, depot maintenance, and operating costs for aircraft. Then naval aviation would be evaluated against naval surface forces, submarine forces, education, base support and all else the navy does. At each level, different evaluative criteria are used, feedback is provided, incremental decisions are made, and the process repeats.

The starting point for programming is the most recent version of the FYDP, generally the one that corresponds to the President's Budget submission for the prior year. Programming mainly occurs in the late winter and early spring, so the POM-14 process would occur just after the 2013 budget is submitted to Congress. Unless there is a significant deviation in the Fiscal Guidance from the last FYDP, there is no new money to be allocated. The database rolls forward one year, another year is added to the back (POM-13 covered 2013 through 2017, POM-14 covers 2014 through 2018), and figures are adjusted. Thus, nearly all changes to the POM are give-and-take. Someone loses in order for someone else to gain. When there is a difference between the desired plan and existing plan, a programmer has four options: (a) fund the difference, but this requires resources from somewhere; (b) restructure the program so that it can still be executed for the given level of funding, this entails adjusting the program's results which more often than not means fewer items or slower progress; (c) reengineer the program to improve the efficiency of the process, but that is largely out of the control of the programmer and may become a problem for the program, and (d) accept the risk and simply give the program fewer resources and hope they can execute effectively.

Consider the pair of tables that follow. Let us assume that the first table represents the 2013 President's Budget. We see eight approved programs across the FYDP and three programs for which there were insufficient funds. This forms the baseline for the 2014 budget, in the second table. Here,

TABLE 4.5 Programming Example

	Programs	President's Budget - FY2013				
		2013	**2014**	**2015**	**2016**	**2017**
Funded	A	100	150	200	250	300
	B	500	515	530	545	560
	C	1250	1000	600	200	0
	C'	100	500	1300	2000	2500
	D	700	720	740	760	780
	E	600	700	600	500	400
	F	800	600	300	200	0
	G	0	0	100	400	500
		4050	4185	4370	4855	5040
Unfunded	H	200	205	210	216	222
	I	0	0	300	900	1500
	J	100	300	500	550	600

	Programs	President's Budget - FY2014					
		2013	**2014**	**2015**	**2016**	**2017**	**2018**
Funded	A	100	150	200	250	300	350
	B	500	515	530	545	560	577
	C	1250	1000	600	200	0	0
	C'	100	500	1300	2000	2500	2750
	D	700	720	740	760	780	800
	E	600	700	600	500	400	300
	F	800	600	300	200	0	0
	G	0	0	100	400	500	800
		4050	~~4185~~	~~4370~~	~~4855~~	~~5040~~	~~5577~~
			4100	4200	4600	4800	5000
Unfunded	H	200	205	220	235	240	
	I	0	0	300	900	1500	
	J	100	300	550	600	650	
New	X		300	500	200	100	0
	Y		0	100	300	500	900
	Z		200	210	220	235	250

Source: Authors, 2011.

the programmer's task is to adjust to the new top line allocation resulting from the fiscal guidance (note the sums of the columns are crossed out and new, lower, figures inserted), but he not only will need to determine how to cut back on programs A through G, he needs to consider funding the unfunded requirements, and needs to address the new requirements. Various options for the FY2014 budget will be considered including adding one or more unfunded programs (H, I, J) and new requirements (X, Y, Z), obviously at the expense of existing programs (A–G). Note there are two related programs (C, and C'); it is often the case that programs are interrelated and changes to one necessitates changes to the other. Examples include ships and their crew, ships and their weapons systems, building construction and facilities maintenance, and new product development transitioning to production. How does the programmer approach this task?

There are several analytical methods a program sponsor might employ to craft a proposal. Programs that consist of well understood cost elements in predictable processes may be modeled. *Models* have been created for a large number of mostly expense-type programs. There are models for ship operations, flying hours, building maintenance, and military manpower to name a few. Consider military manpower. A command will have a manning requirement for a certain number of enlisted and officers, of varying specialties, qualifications and ranks. We could take that requirement and add the costs of those people: basic pay, housing allowances, special and incentive pays, retirement account accrual. We can also look at the employment of that command to determine if conditional costs will be incurred, such as hazardous duty pay, combat pay, or sea pay. Adding or deleting commands, or changing their missions, will affect the manpower requirement and may result in change of station moves of personnel. Since personnel represent both a stock and a flow, one needs to consider the flow and the associated costs: recruitments, reenlistments, promotions, new qualifications, longevity pay raises, transfers, retirements, deaths. All of these factors can be built into a model that is used to analyze alternatives: what if reenlistment rates are higher/lower by 1%, 2%, 3% how much more/less does that cost? What if fewer/more retire than planned, how much more/less does that cost? What if we accelerate the commissioning of a new ship, how much more do we need in the manpower account for the cost of the crew? What if we substitute civilians for these 100 airmen, how much less is the manpower cost? Many models can also be run backward so that instead of determining the cost of a force structure, we determine feasible structures for a given cost: if we only have $41 billion and not $41.4 billion for manpower, how many fewer soldiers is that?

If a model is not appropriate, a sponsor may conduct a *cost-benefit analysis* (CBA) or a *cost-effectiveness analysis* (CEA). In a CBA, the objective is to weigh the total expected costs against the total expected benefits of one or more alternatives in order to choose the one that provides the greatest return. Generally, the costs and benefits are expressed in dollar terms. A CEA is similar but tends to be used more often when it is difficult, or improper, to express the benefits in dollar terms. In some applications, such as health care options and military capabilities, it is distasteful or too arbitrary to express benefits in dollar terms and so proxies are used, such as years of life or levels of readiness. A CBA or CEA analysis might be used to evaluate options for a vehicle service life extension program: if we overhaul the vehicle after 5 years or 50,000 miles what is the cost/benefit versus waiting until 7 years or 70,000 miles? Many of the arguments for and against an alternative engine for the Joint Strike Fighter were based in CBA analyses. One weakness of a CBA/CEA is that the assumptions used in the calculation or the proxies chosen for benefits may be points of contention among consumers of that analysis, depending on their level of support for the program.

Some programs do not lend themselves to a clear-cut model or economic analysis. One example is basic research. One cannot input the variables and turn a crank and get an incremental cost per research output. Nor can one quantify—sometimes one cannot even identify—the benefits of basic research. In this case one would use a fairly primitive *level of effort* analysis. The sponsor would ask, "The Army invested $422 million in basic research in FY09, are we satisfied with the results of that investment in light of our strategic goals? Should we increase or decrease that level of effort in the budget year?" Level of effort programs are highly susceptible to "salami slice" budget cuts since the impacts are difficult to quantify; they are inherently difficult line items to defend. Allocations are often based on the judgment of experts and a quest for equity in the allocation.

In some cases, a sponsor may be evaluating alternatives for a large number of programs and has model outputs, CBA/CEA figures, and proposed levels of effort. In a consolidated and integrated decision process, how might these dissimilar items be evaluated when confronted with a budget constraint? The sponsor might employ a *priority list* or *relative priority list*. The Marine Corps uses a version of this method. Initial evaluations of programs occur in several functionally organized Program Evaluation Groups (PEG). The PEGs focus only on the benefit of each program, using professional judgment and consensus building to assess the relative value of each program. They deliver a prioritized list of initiatives, normalized on a 0–100 scale. The chairs of the various PEGs form a POM Working Group (PWG) that focuses on the resource requirements to support each initiative to "buy" the benefit identified by the PEGs. The cost is multiplied by the value and new cost-benefit value is assigned. The various PEG lists are merged into a single list that is re-ordered in cost-effectiveness sequence and constitutes an order-of-buy. The most cost-effective program is bought first and others in turn until resources are exhausted. This list is forwarded to a Program Review Board (PRB). The PRB is comprised of senior officers and civilians of higher rank than the PWG and they conduct further analysis, look holistically and strategically across the proposed program, and recommend changes to the Marine Corps Requirements Oversight Council who makes the final POM recommendation to the Commandant.

Who does the work of programming? It was mentioned that programming is done by the service staffs. Programming is a headquarters function since it is fundamentally a policy-setting process. But the service staffs do not necessarily have all the knowledge they need in order to program effectively. While others, organizationally nearby, conducted the planning phase and articulated strategies and capabilities, gaps and overmatches, the comparatively more detailed work of programming requires data be gathered from outside the Pentagon. Each service does that differently. The Air Force POM is built through a set of Force Mission and Mission Support Panels. These panels cover the broad missions and equipment of the Air Force and consider the inputs from Program Element Monitors (PEM), Integrated Process Teams (IPT), and major commands. The major commands are those who execute the Air Force mission, be it operational, support or acquisition. The major commands possess much of the data regarding the current status of forces, budgets and acquisition programs. They also do the feasibility analysis of various program proposals. The PEM is the link between the major command and the Air Staff. They represent one, or a logical group of several, program elements and are responsible for the POM for these PEs. The work of building the POM occurs in the IPT consisting of PEMs, major command representative and other functional area experts. The Panels review the tentative POMs on an integrated basis with an eye on Air Force capabilities; they suggest adjustments to the POM and the IPT continues its work. As the POM is refined, it works its way up the Air Staff chain of command through an Air Force Group (1-stars), an Air Force Board, to the Air Force Council. (Hill, 2008)

In the Army, the major commands use a Management Decision Package (MDEP)—a justification format for a group of logically connected PEs—to build a POM submission. Program Evaluation Groups at the Army staff review the MDEPs. There is a Council of Colonels who begin to review the work of the PEGs and integrate the Army POM before it is staffed by a Planning, Programming, Budget Committee that ensures the entire PPBE process flows smoothly. An Army Resources Board makes the final decisions. (Hill, 2008)

Whereas the Army major commands propose POMs to the service staff, and the Air Force uses an IPT structure to collaboratively build a POM, in the Navy most of the POM work is done by the staff with ad hoc input by the major commands. Some program managers and major commands submit formal POM requests, others respond to data calls, depending on the needs of the Navy staff. On the Navy staff are resource sponsors (who perform as do PEMs and PEGs in the other services). These resource sponsors are responsible for one or several PEs. Resource sponsors are organized hierarchically; as one rises in the hierarchy, a broader set of navy force structure and operations is considered. Individual Sponsor Program Proposals (SPPs) are reviewed and reworked. Twice during the process a fully Integrated Sponsor Program Proposal (ISPP) is compiled and reviewed by senior leadership for comment and direction. The ISPPs are also reviewed by those who conducted the planning phase to ensure the POM comports with the plan. A complicating factor for the Navy is that near the end of budgeting, the Marine Corps POM is integrated with the Navy POM. The two are crafted independently after negotiating a "blue-green split" of the Fiscal Guidance and DPG. For the Navy, the Deputy Chief of Naval Operations (Integration of Capabilities and Resources, N8) takes charge of the PPBE process and N80 directs the Navy POM process. USMC (Programs and Resources) builds the USMC POM. The Secretary of the Navy has a "large group" and a "small group" with both USN and USMC representation to review the POM and make final decisions.

What criteria are used when the POM is reviewed? As the POM is built and reviewed by the chain of command, what do they look at? Most significantly, the service staffs are looking to ensure there is balance among the various program elements. Does the POM represent a balance between current readiness and future capabilities; balance between combat arms and combat support; balance among sea, air, space, ground, and C4I capabilities; balance among the major force programs; balance among the joint capability areas; and the like. The Navy has used an Integrated Warfare Assessment (IWAR) model to achieve this balance. The Air Force uses Concept of Operations (CONOPS) champions. At the OSD level, Joint Capability Portfolios are used. Secondly, the POMs are reviewed to ensure they respond to the guidance issued in the DPG. The DPG may contain specific guidance for individual programs (e.g., the Joint Strike Fighter) or more general guidance for types of programs (e.g., unmanned systems). The DPG may suggest how the scale should tip when balancing the POM along the various dimensions. Third, the POMs are reviewed to ensure they also serve the interests and aspirations of the decision-maker at each level. This is complicated because there is not complete alignment between the officials involved. A particular program advocate is attempting to optimize his area of responsibility, even if that means a suboptimal service. Conversely, the service chief is trying to advocate for his service and may sub-optimize individual programs or the OSD balance in the process. In turn, OSD is seeking a broad balance that must serve political as well as military aims.

There are also technical factors that are reviewed in programming. (a) *Programming models* go through a validation, verification and accreditation process to ensure they are logically sound, statistically sound, and are accepted by the stakeholders. Data sources must be approved and reliable. (b) *Acquisition cost estimates* that originate in program offices are independently validated by service cost estimators and validated by the OSD Office of Cost Assessment and Program Evaluation (CAPE). (c) *Program schedules* are verified to ensure they comply with annual full-funding rules, are not excessively lengthy, and comport with acquisition and production lead time estimates. (d) *Program pricing* is validated against proposed pay increases for labor, inflation estimates for non-labor, learning curve, promises of improved efficiency, and staffing strategies (e.g., outsourcing or substitution civilians for military). (e) *Current execution* of programs are examined for evidence that funds may be excessive or inadequate, technical progress is being made, schedule slips or advances, readiness rates are achieved, level of effort is reasonable, as all these factors affect the level of future funding. (f) *Congressional actions* on the intermediate year budget may affect the POM. Is Congress proposing the addition or deletion of particular programs? Are they inserting restrictions on the size or scope of programs (e.g., the 2009-2011 debates concerning an alternate engine for the Joint Strike Fighter).

Remember that Congress is enacting one budget as the prior budget is being executed and the next budget is being formulated. In Figure 4.6 the vertical dotted line reminds us that at any point in time, multiple budgets

Resource Allocation Process

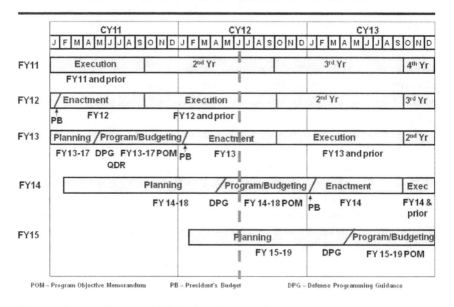

Figure 4.6 Overlapping PPBE Cycles. *Source*: Authors, 2011.

are "in play." This particular line shows us that (going from bottom to top) while starting to plan the FY2015–2019 period, programming of FY2014–2018 is occurring, congress is enacting the FY2013 budget and FY2012 (and prior) year budgets are being executed. Thus, any major programmatic decision will not only be incorporated in the POM, but may require action in other phases. When the announcement was made to cancel the Expeditionary Fighting Vehicle in POM-12, the DOD still had not received its FY2011 appropriation. The department was still under Continuing Resolution Authority. This necessitated Secretary of Defense Gates' testimony before Congress regarding the 2012 budget include a request to modify the year-old 2011 budget request for the program, and he directed the program office to alter its course of action.

Below, in Figure 4.7, is a depiction of the POM-2012 process for the Navy. The bulk of the POM process occurs in the late winter and early spring months of the year, just after the last budget is submitted to Congress. One can see the early influence of the planning phase in the NSP (Naval Strategic Plan) and FEA (Front End Assessment). Note that the N81 office that did the FEA is also providing the warfighting capability plan to the resource sponsors as input to their sponsor proposals. The numbers surrounding the box entitled Resource Sponsor are the various codes on the Navy staff that have a role in programming: N1 programs military manpower, N4 programs operations and logistics support and construction, N8 programs acquisition and modernization. N80 directs the process of programming and integrates the individual Sponsor Program Proposals into the ISPP. N81, having done the planning work, performs a program assessment. And one can see the pricing validation teams and program assessment steps as the POM is reviewed.

The Budgeting Phase

In the history of PPBE section of this chapter it was noted that a recent change to the system was the merging of databases containing programming and budgeting data, and the associated merging of program and budget reviews. The logic behind that change is that the budget is written to describe and justify the mix of programs suggested by the services in their Program Objective Memoranda (POM), and programs in the POM must be funded; the two are inextricably linked. However, whereas programming is about deciding between alternative allocations of resources, the budgeting phase is about justifying that decision in a manner that supports the enactment phase of the budgeting process. Considering the four phases of budgeting in the public sector (Figure 8), all we have discussed so far in this section—planning and programming—occurs in the executive bud-

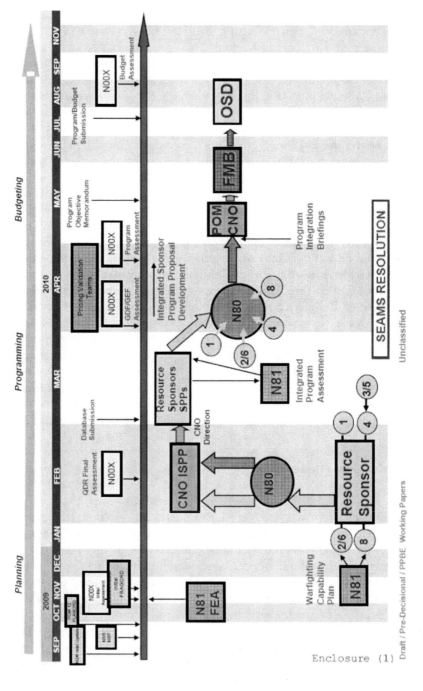

Figure 4.7 POM-2012 Navy Process. *Source:* Flavell, 2010.

get formulation phase. The budget justification books are the documents that serve as the transition between formulation and legislative enactment. Thus, *the primary aims of the budgeting phase are to ensure those justification books accurately describe the decisions made in the POM, are aligned with and reflect the plans to accomplish the NSS, NDS, NMS and other planning documents, and are formatted in a fashion that serves the legislative function.*

The Budget and Accounting Act of 1921 requires that the President submit to Congress an annual budget proposal for the consolidated functions of the federal government. The Act also created the Office of Management and Budget (OMB, formerly the Bureau of the Budget) and the Government Accountability Office (GAO, formerly the General Accounting Office). This law established the modern budgeting process for the executive branch. The Congressional Budget Act of 1974 established the modern legislative process. The budget—the formal document submitted by the President—is due to Congress not later than the first Monday in February of each year. In order to meet that deadline, work begins many months in advance. We saw in the last section that much of the task of service-level programming occurs in the spring to allow for an early summer service budget submission and review. The OSD level budget and program submission occurs in late summer to early fall, and final decisions are made in December. January is spent drafting the final budget proposals.

The budgeting phase for the federal government formally begins with the annual issuance of OMB Circular A-11, *Preparation, Submission and Execution of the Budget.* Often referred to as The Budget Call, the circular does more than that. It provides an overview of the budget process and includes reference material such as the basic laws that regulate budgeting and the terms and concepts. It covers the development of the President's Budget and tells agencies how to prepare and submit materials required for OMB and Presidential review of agency requests. It discusses supplementals and amendments, deferrals and Presidential proposals to rescind or cancel funds, and investments. It provides instructions on budget execution, including guidance on the apportionment and reapportionment process (SF 132), a report on budget execution and budgetary resources (SF 133), and a checklist for fund control regulations. Since OMB is responsible for both budget and management, the circular describes the Administration's approach to performance management and requirements for strategic plans and annual program performance reports. Readers should consider using the most recent A-11 as a supplement to this text.

Figure 4.8 displays the more significant steps in the process of building the annual President's Budget. This section will elaborate on the actions taken inside DOD (the far right column), but it is helpful to put those actions into context. Note that it begins with budget and fiscal policy from the President and consists of a flow of information between the executive agen-

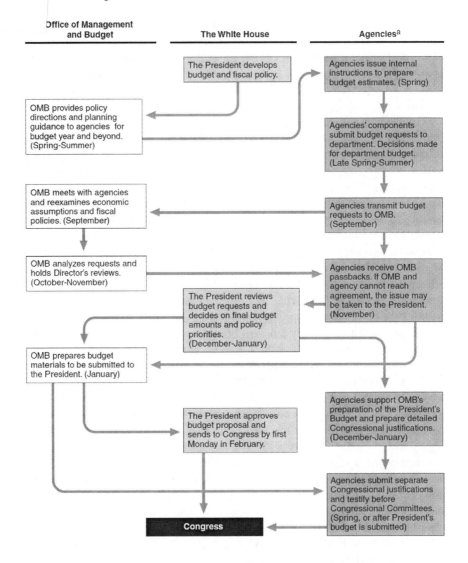

Figure 4.8 Executive Branch Budget Formulation. *Source*: Senate Budget Committee, December 1998, adapted by GAO.

cies and the White House, including the OMB as the budget is refined. Although not depicted in the figure, there are contextual analyses that impact the guidance given to the agencies: for example, the Treasury is projecting receipts and the Council of Economic Advisors is projecting the state of the economy. The A-11 is represented by the box in the upper left corner. Based on the guidance in the A-11, the agencies will draft budget proposals during the summer months which are forwarded to OMB for review. The OMB review considers both the fit with the President's agenda and will consider revisions to the economic assumptions based on fact-of-life changes since they were first promulgated. The budget proposals are passed back to the agencies for revision and final review by the President. The agencies will then prepare detailed budget justification books for Congress and will provide data to OMB to construct the various tables and exhibits in the *Analytical Perspectives*, the *Historical Tables*, the *Appendices* and other documentation that comprise the President's Budget.

The Roles of Budget Offices

OMB Circular A-11 is of greatest concern to the staff at OSD (Comptroller) that prepare the formal Appendix to the President's Budget and maintain the formal budget database at OMB. For the military departments, services, and major commands within the services that draft the detailed exhibits, more detailed budget calls flow down the chain of command with specific instructions. Included in that guidance are things such as: which budget justification exhibits to prepare, what the threshold value is to report individual line items (i.e., the guidance may say to list all contracts valued at $10 million or more and to consolidate all the rest into "miscellaneous"), inflation factors to use for various objects of expense, due dates, the website where templates can be found (or the agency may be told to use last year's budget as the template for the current year). The budget call that arrives at the budget submitting office is often accompanied by a set of advance review questions. Some of these questions are routine and are asked each year to provide the military department budget analyst who will review the budget background information or information about current year execution. Other questions may be directed to a specific program and may involve issues identified during programming. These questions are not part of the formal budget justification material that would be given to Congress, rather they are part of the internal service-level and OSD-level review process.

Before going further into detail regarding the budget formulation and review processes, a word about organizational structure. In most of the military departments, the detailed budget justification books are written at the source of the information: in the major commands and in the acquisition program offices. The Army and Air Force refer these as major commands (MajCom or MACOM) and the Navy refers to them as Budget Submitting Offices (BSOs). A BSO is a large, single-purpose command that usually has several subordinate commands. In the Navy, there are the systems commands (e.g., NAVAIR, NAVSEA, SPAWAR), the major support commands (e.g., BUPERS, CNIC, BUMED), and the operating commands (e.g., CFFC, PacFlt). Budget materials may be drafted by a subordinate unit of the BSO (such as a program management office at a systems command), but are consolidated, reviewed, and submitted to the service budget office via the BSO comptroller. Each military department has an Assistant Secretary for Financial Management and Comptroller (FM&C) and within those offices is a Deputy Assistant Secretary for Budget, who is usually a two- or three-star admiral or general. The DON refers to the office of budget as FMB (Financial Management and Budget). This is the office that directs the budgeting phase of PPBE for each service. They issue service-specific budget calls, they collect the budget drafts, they review the budget, make modifications, and then issue guidance for the second round of the budgeting phase, the OSD/OMB budget. At the DOD level, the office of the Undersecretary of Defense (Comptroller) assumes the same role as the Assistant Secretary (FM&C) did for each service. Because the DOD is such a large department, rather than have separate OSD and OMB reviews, the two are done together. Every other agency of the federal government reviews its budget at the department level first, then submits it to OMB. The participants in the DOD budgeting process are described and analyzed in more detail in chapter 8.

BUDGET FORMULATION AND REVIEW

Analyzing how the budget is formulated then how it is reviewed runs the risk of seeming repetitive. We believe the reader can infer the formulation process by looking at the review criteria and the review process. The intent of the budget review is to ensure the budget was formulated accurately and properly, that the program is executable, and that it follows the budget call guidance. Before the military department budget office is comfortable defending the budget at the OSD review, they will first scrutinize the data and methods behind the formulation of the budget. Thus, we will structure this by looking primarily at the budget review criteria and will amend that as necessary to cover the process of formulation.

Let us remember what a budget is. A budget is document that expresses in financial terms the plan for accomplishing an organization's objectives. It describes the objectives of the program, the specific objects of expense to achieve that goal, a time horizon and the associated financial cost to achieve the goal. So formulation is about determining the plan, including cost estimates, and review is about validating and approving that plan. There are three key components of a budget review: the program, the schedule and the funding. The *programmatic review* is about measuring technical performance against the plan. The *schedule review* is about validating the timing of events (production schedules, deliveries, contract awards, fielding of a new system, rate of activity for an operating account). The *funding review* validates cost estimates and assesses the adequacy of financial resources. Those conducting a budget review are performing financial due diligence;

whether the right amount of money is in the right year for each program. They focus primarily on the budget year with some consideration of the out-years of the FYDP. But the review also considers current and prior year program and financial execution. For example, if it appears the program has excess funding in the current year, the budget year may be adjusted under the assumption the program will have progressed further than originally planned. Program managers and resource sponsors tend to be optimistic in their projections, but budget analysts tend to be skeptical and risk-averse. The budget review is expected to identify programmatic and financial risk to decision-makers and, where appropriate, provide funding alternatives that mitigate the risk. The alternatives are similar to those programmers have: (a) add funds to mitigate risk and maintain schedule, (b) adjust the program and re-price it accordingly, (c) adjust the quantity to be acquired or reduce the level of activity, (d) terminate or truncate existing programs or defer future programs, or (e) deliberately assume the risk.

There are two broad categories of budgets in the DOD, expense-type budgets and investment-type budgets. When we discussed budget structures, we alluded to this difference. Let's elaborate on that idea. Expense-type budgets are used when the program primarily supports current operations, are consumed during or shortly after use, have nominal value, or represent a continuous activity. Examples of expenses include military and civilian labor, education and training, facilities maintenance, base operations, and the operation and maintenance expenses of vehicles, aircraft and ships. Expense-type budgets are incrementally funded which means they request sufficient funding for one year's worth of expenses: one year of salary, one year of flying hours, one year of education. Expense-type appropriations include Operation and Maintenance (O&M), Military Personnel (MilPers), and Research, Development, Test and Evaluation (RDT&E).

Investment-type budgets are used when the program primarily supports future operations, durable goods are acquired, at relatively high costs. Examples of investments include armored vehicles, helicopters, buildings, weapons, and large-scale IT systems. Investment-type budgets are fully funded which means they request sufficient funding for one year's worth of production of complete, usable systems. The full cost of acquiring a complete, usable system must be budgeted and available before a contract can be let. In other words, if a future year appropriation is required to complete construction of an item, it is not fully funded. Thus, the full procurement cost of the item is budgeted in the year in which the order is placed, even if that item takes several years to build. Investment-type appropriations include Shipbuilding and Conversion, Navy (SCN), Military Construction (MilCon), and Procurement.

Table 4.6 lists the most common appropriations in the defense budget, basic characteristics, cost drivers that affect the formulation of the budget

TABLE 4.6 Budget Drivers and Review Criteria by Budget Type

Expense-Type Appropriations

	Military Personnel	Operation & Maintenance	Research, Development, Test & Evaluation
Obligation Period	1 year	1 year	2 years
What Does it Buy?	Officer and enlisted personnel salaries, bonuses, and special pays. Permanent change of station moves. Health care accrual.	Administrative expenses, civilian labor, temporary travel, spare parts, equipment maintenance, fuel, minor purchases and minor construction, education & training, recruiting, base support.	Basic and applied research, advanced technology development, prototypes, system development and demonstration, RDT&E management support, operational system improvements.
Budget Content	Narrative description and justification, cost data, explanation of year-to-year changes, inventory data (end-strength and average-strength of categories of personnel), actuarial data	Narrative description and justification, cost data by object of expense, explanation of year-to-year changes, performance data, inventory data (number of civilian personnel, number of operational units supported)	Narrative description and justification, cost data by project, contract data, explanation of year-to-year changes, development schedules
Budget Drivers	Force structure (overall size of force, mix of force), promotion rates, retention rates, policies (pay scale, entitlement rules for bonuses), employment of the force (sea pay, combat pay, family separation allowances)	Force structure (overall size of force and mix of force), deployment rates, inter-deployment maintenance plans, contracting strategies, in/outsourcing strategies, commodity prices, rate or pace of activity.	Depending on the type of R&D: development schedules, technology maturity, availability of laboratory or range facilities, engineering issues and test results, changing requirements, pace of level of effort.
Budget Review Criteria	Appropriation cognizance (active vs. reserve), links to changes in structure and employment of operating forces, costs are annualized, active/reserve mix, technical factors (inflation rates, control numbers, arithmetic, completeness), consistency with prior and related submissions	Appropriation cognizance, Expense/Investment criteria, price and program growth, annualization of costs, links to operating tempo, maintenance plans and staffing strategies, link to delivery of new operational units, technical factors (inflation rates, control numbers, arithmetic, completeness), consistency with prior and related submissions	Appropriation cognizance, incremental annualized funding with minimal carryover, past and current performance impact on future schedules and funding needs, level of effort analysis, concurrency with procurement, technical factors (inflation rates, control numbers, arithmetic, completeness), consistency with prior and related submissions

Investment-Type Appropriations

	Procurement	Construction
Obligation Period	3 years	5 years
What Does it Buy?	Durable equipment such as vehicles, weapons, aircraft, industrial equipment, and major modifications to the same. Initial outfitting costs for the same.	Shipbuilding, conversion, and construction, including initial outfitting, and nuclear refueling (SCN). Construction of buildings, acquisition of real property, improvement of real property (MilCon)
Budget Content	Narrative description and justification, variable cost breakdown by unit of purchase, one-time and fixed costs, contract and contractor data, production and delivery schedules, inventory data.	Narrative description and justification, cost data by project and by project phase, projects by location, contract and contractor data, schedule data, inventory data, special considerations (historical sites, environmental protection, floodplain management, etc.)
Budget Drivers	Unit cost, hardware and acceptance costs are fully funded, support and program management costs are incrementally funded, rates of production, pricing dynamics (inflation, economic price adjustments, economic order quantities, foreign currency fluctuation, learning curve, input/output schedules), acquisition strategies (joint programs, competition, multi-year procurement, dual sourcing, buy vs. lease, block vs. spiral development), production initiation or termination.	Size and use of building, environmental considerations, climate at building site, construction method and materials, supporting facilities, site preparation, utilities, features that are inherent to the building versus those that are considered furnishings, contingency costs, supervision and overhead, studies and permissions, contracted versus in-house portions of work.
Budget Review Criteria	Appropriation cognizance, concurrency with RDT&E, Expense/Investment criteria, appropriate funding (full or incremental) for types of expenses, past performance impact on future amounts and schedules, program management and support costs in relation to production costs, cost estimation methodology, modification schedules compared to operating schedules of same, quantities and schedules (minimum sustaining rate or economic ordering quantity), advance procurement and multi-year procurement aligned properly, funded delivery period (a single year budget buys a single year of production), technical factors (inflation rates, control numbers, arithmetic, completeness), consistency with prior and related submissions	Appropriation cognizance (especially construction costs versus construction oversight costs versus building equipping and furnishing costs), past performance (especially for multi-stage projects), status of design, impact on operating budgets, basis for cost estimate, acquisition strategy, compliance with applicable environmental and safety regulations, construction schedule, technical factors (inflation rates, control numbers, arithmetic, completeness), consistency with prior and related submissions

Sources: Flavel, 2010; U.S. Navy budget guidance manual; DOD Financial Management Regulations; FY2012 Budget for the Department of Defense.

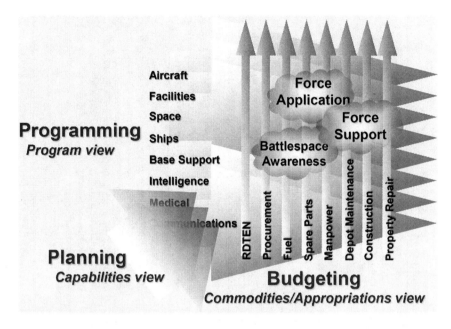

Figure 4.9 Resource Allocation Views. *Source*: Flavell, 2010.

and the key criteria used by the service budget offices to review the justification books.

Thus far we have discussed the development of the defense budget from the planning phase through the programming and budgeting phases. Figure 4.9 depicts how each phase takes a different view of the defense program. Remember, the goal of planning is to identify gaps or overmatches between strategy and capabilities and produce objectives for programming to address them. Planning views DOD through a capability lens. The goal of programming is to allocate resources among programs across a mid-range time horizon that best achieves the planning objectives. Programming views DOD through a program lens. And the goal of budgeting is to justify the programming decisions in a format that serves the process of legitimization (enactment). Budgeting views DOD through an appropriations and budget line item lens.

Schick (1966) asserts that a budget is a "process for systematically relating the expenditure of funds to the accomplishment of planned objectives" and that all budgeting systems support three administrative processes: strategic planning, management control and operational control. PPBS was the first budgeting system designed to fully integrate all three. Thus, it produces a budget that is consistent with the strategy and operational planning processes in the DOD (planning phase). It produces a budget that allocates resources to activities and objects that are best able to implement that strategy (programming, management control). And it produces detailed plans that are used to request appropriations and to guide the actions of the department during execution (budgeting, operational control). Let us briefly turn our attention to the enactment and execution phases of defense budgeting. There are separate chapters dedicated to both, so this discussion will be rather limited.

Budget Enactment

We have noted that the PPBE system only describes those activities that occur within the DOD. But a very significant series of events occur after the B(udget) is submitted and before E(xecution) begins. As each phase of PPBE defines and constrains the subsequent phase, the budget is a basis for congressional authorization and appropriation, and those acts prescribe, constrain and enable the execution of the budget. The DOD does not execute the budget that was drafted and submitted, they execute the budget that was authorized and appropriated by Congress. For much of the routine and recurring business of the department, congress tends to appropriate what was requested and the enactment phase may seem like a formality. For other programs, significant changes occur.

Assuming the DOD wants to execute the budget it drafted, it has some responsibility to draft a budget that will be acceptable to lawmakers. Operating above the highly analytical and rational analysis layer of PEMs and resource sponsors, is a policy layer that considers the political viability of the defense program. Recent issues about which congresses have been closely engaged include health care coverage, staffing levels, management reform, specific weapons programs (e.g., C-17, Joint Strike Fighter, V-22, MRAP), aircraft carrier homeport decisions, base closures, support contractors, and emerging threats (e.g., cyberwarfare, counterinsurgency). Recognizing the overlap in PPBE phases, it is imperative that defense officials understand the nature of congressional interest as they authorize and appropriate fiscal year 201X so that appropriate guidance can be given to the services as they build the POM for fiscal year 201X+1.

Analogous to the distinction between the POM and budget, congress both authorizes programs to exist and appropriates budget authority to pay for them. The execution phase is about exercising that budget authority in a fashion that is both compliant with the law and achieves the programmatic objectives defined in the planning phase. Chapter 5 addresses the enactment process and congressional—defense relations in detail.

The Budget Execution Phase

Where budgeting is creating a plan and enactment is approving the plan, execution is working the plan. Where budgeting is a request for authority and enactment is a grant of authority, execution is the exercise of authority. Where planning, programming and budgeting are about policy choices, execution is about policy outputs and outcomes. Because executing the budget is fundamentally different from formulating a budget, it is covered in detail in Chapter 7. In this section, we briefly introduce the core issues of execution with specific emphasis on how execution is part of the system of PPBE.

The goal of execution is to implement the programs and policies that were described in the budget—as approved or modified in the authorization and appropriation process—in order to deliver the desired military capabilities, and to feed information into subsequent rounds of the PPB process. Execution can be thought of in two related, but distinct ways: program execution and financial execution. In the first, programs are executed: training occurs, operations are conducted, equipment is purchased or modified or maintained, and people perform tasks. In this view, military capabilities are developed or maintained and the security of the nation is assured. In the financial view, budget authority is allocated, cited, transferred, expended, recorded, and reported. The financial resources of the department are managed and the actions of the department are controlled to ensure compliance with appropriations law. Generally speaking, line managers are concerned about program execution and the finances are merely a means to an end; staff managers (e.g., comptrollers) are concerned about the department's fiduciary responsibilities and the means are the end. Ideally, the two work collaboratively and cooperatively to ensure programmatic goals are achieved while complying with the regulations.

Chapter 7 deals with the key issues of budget execution: the legal framework, the flow of funds, monitoring and measuring financial execution, the annual cycle of events, and the tension between flexibility and control. With respect to the role of execution in PPBE, it is accurate to say that the "E" is still not an institutionalized part of the process. PPBE is overwhelmingly forward-looking and most of the participants are concerned about the FYDP. That said, execution influences the process in two key ways that parallel the programmatic and financial views of execution in practice. First, at the OSD level, the Office of Cost Assessment and Program Evaluation (CAPE, formerly Program Analysis and Evaluation) is responsible for reviewing, analyzing, and evaluating programs for executing approved strategies and policies and to prepare programmatic guidance on which the FYDP is based (DODD 5141.01). Similar functions occur with each military service. These are programmatic reviews: whether the outputs and outcomes of the programs are consistent with the plans laid out in the budget; whether they provide the intended military capability; and whether the technical specifications, quantities and schedules should be modified. Thus, analyses of programmatic execution feed into future rounds of programming and budgeting.

In addition to the reviews of programmatic execution, financial execution is reviewed and may influence future budgets. Such reviews are done by the service and DOD comptroller community and budget offices. They consider the sufficiency of funding for a program, whether all programmatic goals are being achieved and to what extent the level of funding is a contributor. They monitor the rate at which funds are obligated and expended and deviations from plans or standard rates of consumption invite scrutiny. A program that appears to have excess funding is at risk of having future year's budgets reduced.

One should note that even before the E was formally added to the PPBS system, current year execution played a similar role in program and budget formulation. This activity is not new, but rather has a renewed emphasis. Since the Rumsfeld/Gates reforms, there has been an increase in the authority of CAPE and a wider division between program analysis and budget analysis. Curiously, as program and budget reviews were combined during formulation, execution reviews have become more organizationally distinct. The USD(Comptroller) noted that he was unable to speak about the details of individual programs as well as he did when the Program Analysis and Evaluation (PA&E) group was part of his office; the CAPE is a separate office and conducts those reviews. However, he added that while he lacked first-hand knowledge, the two offices had a good working relationship and he believed budget outcomes were better (Hale, 2011).

ANALYSIS AND CRITICISM OF PPBES

Over the past 50 years, there have been many critics of program budgeting, far more it seems than there have been supporters. Some have discussed the challenges associated with establishing a program budget system (Frank, 1973, Churchman and Schainblatt, 1969). Some of criticized the quality and completeness of the analysis (Dror, 1967, Schick, 1973, Wildavsky, 1969; Jones and McCaffery, 2008), including the perverse incentives the systems creates (Chwastiak, 2001). Whereas others have criticized the recommendations the system produces or their secondary effects or unintended consequences (Spinney, 1985, Kadish, et al, 2006). One can also criticize whether the design and operation of the processes are efficient and effective independent of the outputs of the system (Frank, 1973). Much of that analysis occurred in the 1960s and 1970s as PPBS was attempted throughout the federal government. Recent practices are not as well documented or analyzed in the literature with the notable exceptions of Porten, et al. (2003) and Jones and McCaffery (2005) and we raise additional contemporary issues here and pursue reform options further in chapter 10.

Certainly there are things about PPBES that work well or the DOD would have probably abandoned it years ago. A comprehensive program structure permits the analysis of programs on an individual level and in various combinations. The essence of resource allocation involves analyses among alternatives which program budgeting facilitates well. Program budgeting is also a useful tool when formulating a budget for an agency that operates in a contingent environment that demands frequently shifting strategies. The deliberate links between strategy, allocation, and control make it a particularly effective budgeting technique compared to other techniques. For example, DOD could not incorporate recent changes like counter-insurgency or humanitarian assistance as newly recognized strategic missions using only an object-of-expense budgeting system. A comprehensive program structure ensures that all things are considered in each round (or phase within a round) of budgeting and ensures that emergent requirements are validated and approved before receiving funding. The PPBE process also has a logical division of labor: planning matters (what should we be capable of doing) are distinct from programming matters (how should that be done), budgeting matters (how much does that cost), and the actual program execution. Program analysis is distinct from economic or cost analysis. Each task demands a different discipline and set of analytical tools; DOD is large enough that such analysis can be done in specialized organizational units. Specialization is also possible among types of programs: weapons systems, logistics, installations management, and manpower.

Prior to analysis and criticism of any system as complex as PPBES we should first ascertain the attributes that are relevant for assessment. Further, we must determine on what basis and which criteria should be used to evaluate such a system. James E. Frank provided a framework for analyzing PPBS success (1973). He noted that there were various forms of PPB systems and began with a typology for analysis using the four critical characteristics of PPB systems as the basis: identifying fundamental objectives of the agency, explicit consideration of future year implications, the consideration of all pertinent costs (capital, non-capital, and indirect), and systematic analysis of alternatives. He noted there were two categories of PPB components, data configuration aspects that include program categories, multiyear presentation, and inclusion of indirect costs. The second category are the analytic aspects: measurement of outputs and effects, examination of alternative programs, and examination of goals and objectives. These categories are not mutually exclusive and fall along a continuum; every instance of PPB does both to varying degrees. Figures 4.10 and 4.11

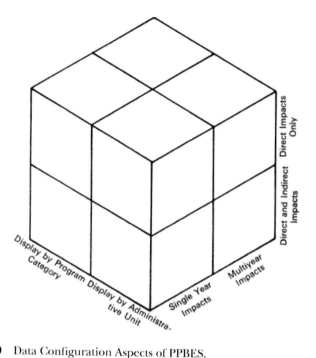

Figure 4.10 Data Configuration Aspects of PPBES.

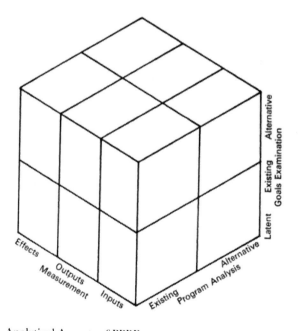

Figure 4.11 Analytical Aspects of PPBE.

illustrate the multiple dimensions of both the data manipulation aspects and the analytical aspects.

With $2 \times 2 \times 2$ variables for data manipulation and $3 \times 2 \times 3$ for analytical aspects, there are 144 possible combinations that could describe a particular PPB system. The interesting thing about the DOD's instantiation of PPBE is that it attempts to be all things. Data are displayed by both program and by the organizational unit that manages the program. Both single year and multi-year impacts are assessed, but the accuracy of the latter years is highly suspect as we note later in this section. Absent even a reasonably good cost accounting system, only direct costs are considered, but analysts work tirelessly to identify impacts to related programs when a given program is adjusted. On the analytical side, the department clearly measures inputs and does fairly well with outputs; measuring effects, as is the case with most public functions, is extremely challenging (Robinson 2007). Program analysis is conducted on both existing and alternative programs, but varies in quality depending on the financial or strategic materiality of the program and its variability (a level-of-effort program like basic research receives considerably less attention than a new technology that is transitioning to production, for good reasons). Goals are examined in the planning phase, but whether goals drive budgets or budgets drive goals is a longstanding issue and most often is determined by the fiscal policy and national security contexts. During wartime and when there is less concern over deficits, the strategy tends to drive budgets (e.g., fiscal year 2004) ; during periods of peace or heightened concern over deficits, the budget tends to drive strategy (e.g., fiscal years 1993 or 2012).

DOD attempts to be all things as a function of the sheer size and complexity of the defense program. Given thousands of program elements, there are several that properly situate in each of the 144 possible mappings on the chart. As a system of budgeting, though, such a complex mixture of analytical approaches creates problems. Those problems are described in the paragraphs that follow and include incomplete decision-support, reinforcement of the status quo and the hoarding of resources, and a mismatch between plans and reality.

Frank (1973) discusses variables that determine success in effectively implementing a PPB system, but those same variables seem to also suggest success in effectively using an entrenched PPB system. They include: the actor(s) in the policy environment who advocate for success, the mechanism(s) that actor uses to ensure success, and the responses of the system to those influences. Each of those three can be applied to both the data configuration aspects and the analytical aspects. With respect to data configuration, the salient variables are the quality of information about programs and the flows of that information through the organization. The quality of budgetary information can be affected by numerous factors: adequacy of accounting systems, cost estimating methodology, maturity of knowledge and designs of systems, non-financial data and the depth of knowledge regarding causal factors (when assessing performance), and underlying assumptions. The flow is impacted by formal processes, number of levels in the hierarchy, methods of aggregating data, competition for resources, trust among actors, and budgeting gamesmanship.

With respect to the analytical aspects, the key issues are the identification of goals and objectives, identification of alternative programs, the measurement of outputs and effects, and integrating processes with conflicting cycle times (Frank, 1973). As we saw in the description of the planning phase, there is no shortage of documentation of goals and objectives. Perhaps there is too much. With a cascading series of strategy documents at the national, DOD, military department and often the major command and subordinate command levels, they are never completely aligned. Operating concepts evolve with technology and world events, strategies and resource goals shift accordingly. Given the magnitude of the analysis of existing programs, the system does not have a large appetite for analyzing alternative programs. Often, once a program is approved, analyses of alternatives are done within the parameters of the program not about the program. That is, the alternatives considered are about accelerating or slowing schedules, increasing or decreasing quantities, not whether the program should continue. Strongly vested interests inside and outside the Pentagon may limit the nature of the alternatives considered. The prevalence of multi-mission systems further confounds comparison of alternatives as the balance among many missions and capabilities is intertwined. Measurement is often difficult, particularly as programs are aggregated and concepts become more abstract. Counting training events and assessing inspection results is straightforward, but measuring something like "readiness" is a perpetual struggle and the extent to which changes in funding affect readiness is not well understood (CBO, 2011, Webb and Candreva, 2010). Lastly, integrating multiple years of program alternatives is challenging as the department does not well integrate capital purchase plans with operating and support costs. Trading off changes to fully funded investment accounts with impacts to annually funded operating and support accounts is not only empirically difficult, the two forms of analysis are done in different organizational units. Chapter 10 of this book is devoted to issues related to reform or replacement of PPBES in DOD with a different type of financial resource decision and management system.

Process, Outcome, and Design Issues

The PPBE system is probably the most influential and poorly disciplined process in the Pentagon. There are a number of likely reasons for this. Given the materiality of the process—nothing occurs in government without resources appropriated for its purpose—and the lead time involved (2 year cycle) compared to the typical tenure of a flag/general officer or political appointee (also about 2 years), changes to the system are made frequently and on a rather ad hoc basis. At the time this is written in the summer of 2011, the DOD instruction and directive governing PPBE are 14 and 11 years old, respectively. In comparison, the instructions governing the requirements process (JCIDS) is updated once per year on average and the instructions governing the acquisition system are updated every 18 months to two years. From the OSD level, the PPBE system has been modified in recent years through Directive Type Memoranda, Management Initiative Decisions, and Resource Management Decisions. At the service level, the processes are updated frequently—sometimes more than once per cycle— through serials, orders, memoranda, notices, and other semi-formal pronouncements. Perhaps because it is not really a process at all, but a family of processes, each owned by a different subunit in the department, and many of which are tailored at lower levels of the hierarchy, writing a definitive instruction is futile. Over time, responsibility for all or part of PPBE has belonged to USD(C), ASD (PA&E), USD(AT&L), and CAPE. To support action officers in the Pentagon, informal PPBE guides exist in the Air Force and Army, but not the Navy (Hill, 2008).

PPBE received its share of criticism in the early years from the segment of public budgeting scholars that come to that field from political science backgrounds. PPBE, it was argued, was too analytical, too rational, too much reliant on economic analysis and not reliant enough on policy analysis. Wildavsky (1969:193) quipped, "I have been told that in a better world, without the vulgar intrusion of political factors (such as the consent of the governed), PPBS would perform its wonders as advertised." Frank (1973: 533) expounded on the political and behavioral aspects of budgeting, "Emphasis must be given to the interplay between the technical and behavioral phases of the system" and "Treating PPB as a purely technical problem overlooks the behavioral dimensions of the political-bureaucratic decision system in which it is [being] installed." Dror (1967: 198) argued that "The invasion of public decision-making by economics is both unavoidable and beneficial, but fraught with danger. [. . .] It is fraught with dangers because of the inability to deal adequately with many critical elements of public policy-making and the possible distortion in decision-making resulting there from." Schick (1973:147) elaborated on this point by noting "PPB's effectiveness was impaired by the failure of many analysts to comprehend the connection between their work and budgeting" although he said that DOD did better than the civilian agencies through the direct involvement of the Secretary of Defense. Wildavsky (1969: 190) asserts that "policy analysis aims at providing information that contributes to making an agency politically and socially relevant. Policies are goals, objectives, and missions that guide the agency. Analysis evaluates and sifts alternative means and ends in the elusive pursuit of policy recommendations." Then he assesses "the damage that the planning-programming-budgeting system has done to the prospects of encouraging policy analysis" concluding that while "measuring effectiveness, estimating costs, and comparing alternatives" should be done, it "is a far cry from being able to take the creative leap of formulating a better policy." (193).

What Wildavsky noticed is that while PPBE purports to offer alternative policy choices, in practice it is only a tool for defining alternative means to those ends. In that regard, PPBE tends to reinforce the status quo rather than question it or improve upon it. Programs, once started, are rarely questioned. Courses of action, once initiated, are refined and economized, but not second-thought. Thus, program sponsors tend to hoard resources rather than offer them to the next best use. Chwastiak (2001) went so far as to assert that the mechanistic, value-neutral use of PPBS normalizes war and took critical decision-making out of the hands of the professional soldier and put it in the hands of technocrats and "elevated rationality and analysis over military expertise" (508). And despite focusing on alternative means to ends, those means do not always provide the desired ends.

Franklin Spinney's classic book, *Defense Facts of Life: The Plans/Reality Mismatch* (1985) highlights flaws in the department's requirements and resourcing processes. The persistent demand for the latest technology, more readiness, and additional capabilities assures that the cost of the desires of the military always exceed available resources. To tame the appetite, processes have been created to validate what are requirements from what are merely "desirements," but those processes have been criticized for not adequately considering resource constraints and the result is an army of program sponsors marching through the Pentagon waving "validated requirements" presuming a stream of funding exists. Programmers, under pressure to fund these requirements, use the tools at their disposal: resource, restructure, reengineer or assume risk. More than a modest amount of risk is assumed or programs are directed to reengineer in order to achieve "efficiencies," usually in the outyears. The result is a budget that is oversubscribed: more program than funding levels can reasonably pay for. The further out in the FYDP one looks, the more risky the program. The Defense Acquisition Program Assessment (DAPA) of 2006 described a "conspiracy of hope" in which the department understates cost, risk and technical readiness of acquisition programs which renders them unexecutable. (Kadish, et al., 2006: 29)

Inevitably, schedules slip and costs rise and fewer items are purchased in the outyears than originally planned. Congress gets upset because of the missed expectation and unrealized plan. In tables 4.7 and 4.8 we see the ship and aircraft procurement plans for the U.S. Navy as described in the

TABLE 4.7   Naval Ship Construction, Plans versus Actual

	2003	2004	2005	2006	2007	2008	2009	2010
FY11 Presidents Budget								7
FY10 Presidents Budget						3	10	8
FY09 Presidents Budget						4	7	8
FY08 Presidents Budget					7	7	7	12
FY07 Presidents Budget				6	7	7	11	12
FY06 Presidents Budget			8	4	7	7	9	10
FY05 Presidents Budget		7	9	6	8	8	17	
FY04 Presidents Budget	5	7	8	7	7	9	14	
FY03 Presidents Budget	5	5	7	7	11			
FY02 Presidents Budget								
FY01 Presidents Budget	8	8	7					
FY00 Presidents Budget	8	8	9					
FY99 Presidents Budget	7							
High to Last Difference	**−3**	**−1**	**−1**	**−1**	**−4**	**−6**	**−7**	**−5**

Source: DOD, 2011

TABLE 4.8   Naval Aircraft Production, Plans versus Actual

	2003	2004	2005	2006	2007	2008	2009	2010
FY11 Presidents Budget								207
FY10 Presidents Budget						182	200	203
FY09 Presidents Budget						183	206	225
FY08 Presidents Budget					162	225	213	228
FY07 Presidents Budget				134	165	203	257	269
FY06 Presidents Budget			115	138	174	212	249	261
FY05 Presidents Budget		103	104	127	184	246	285	
FY04 Presidents Budget	95	100	100	133	191	258		
FY03 Presidents Budget	83	85	105	147	193			
FY02 Presidents Budget								
FY01 Presidents Budget	173	177	187					
FY00 Presidents Budget	183	187	201					
FY99 Presidents Budget	164							
High to Last Difference	**−88**	**−84**	**−86**	**−13**	**−31**	**−76**	**−85**	**−62**

Source: DOD, 2011

annual budget highlights books for budget years 1999 to 2011 describing plans and actuals for production years 2003 through 2010. Looking at production year 2007, for example, we see in the first budget that projected that far (FY 2003 Presidents Budget), the plan was to build 11 ships and

193 aircraft. By the time that year was complete and reported in the FY2008 budget, the Navy actually built 7 ships and 162 aircraft, 4 fewer ships and 31 fewer aircraft than originally planned. For every production year, one can see the outyear plans were almost always more optimistic than the eventual reality, on average 3.5 fewer ships were built each year than the most optimistic forecast for that year, and 65 fewer aircraft were procured than earlier projections indicated.

Another criticism of PPBE relates to its use—or more precisely, nonuse—in the early 21st century, post 9/11 period. (Jones and McCaffery, 2008) Given that PPBE is a system deliberately designed to integrate goals and objectives with the means of accomplishing those goals and using that as the basis for a budget request, it is significant that PPBE has been used only for the "base budget" requirements and not the war on terrorism. Throughout most of the modern budgeting era—that period after the passage of the Congressional Budget Act of 1974—supplemental appropriations have primarily been used in two circumstances, response and recovery from national disasters and military contingencies (McCaffery and Godek, 2003). Both circumstances represent costs that are unpredictable and urgent. Historically, extended military operations have used supplemental appropriations only for the first two years until the lead-time for budgeting permitted the cost of the war to be included in the annual defense budget. Such was the practice for funding the Korean conflict, Vietnam conflict, and military contingency operations in the 1990s (Daggett, 2006). In similar fashion, supplemental appropriations were used immediately after the terrorist attacks of September 11, 2001 (Candreva and Jones 2005), but their use has continued longer than the customary two years. Between September 2001 and July 2010, the war on terrorism had been funded in 34 different acts (26 of which provided funding to DOD, the other 8 funded other federal agencies) and numerous transfers of funds from other appropriations (Belasco, 2010).

Not only were the operations funded in *ad hoc* supplemental appropriations, the budget request was formulated in a process distinct from the routine PPBE cycle. In some military services, one budget staff formulated the base budget while another formulated the cost of war budget. The line dividing the two was inconsistently defined and the military was accused of buying base budget requirements with cost of war funding. For example, the Air Force included two Joint Strike Fighters in their supplemental request in 2007 despite the fact the JSF would not see action for many years and the Navy included the purchase of V-22 tilt-rotor aircraft that had not been deployed to the theater of operations (Karp 2007). Senior leaders understood this practice distorted the base budget and feared that eventually the end to the conflicts would result in cuts to the base. Others worried about the link between the two. For example, equipment used heavily in the war requires more maintenance than equipment used in training, but those costs and depot maintenance workloads may not have been properly planned and the costs estimated in the base budget. Why ad hoc cost of war supplementals were used rather than combining those costs with the base request is a broader subject than the goals of this chapter. Briefly, there were political advantages to both the White House and the Congress to keep costs and processes separate. From a military perspective, the lead time to formulate a cost of war request and process it in Congress is much shorter than the base budget, so they were likely to be timelier and more accurate.

Lastly, PPBES can be criticized on design factors. The division of specialization discussed earlier can have negative consequences if the various organizational units are too disjointed. Separation of duties and specialization of analytical tasks work well if there is ample communication, well-designed interdependencies, and common understanding as the output from one organization (or process) becomes an input to another organization (or process). Frequent changes to the processes and reorganizations of service staffs hinder the ability to develop tightly coupled processes. There always seems to be role ambiguity or process questions that are negotiated at the expense of executive attention to the content of the system. Managing the seams between the organizations and phases is a perpetual challenge.

The other design matter is that the evolution of PPBES from PPBS is yet incomplete. While the E represents execution, it probably should be called evaluation as that has a broader connotation. Execution normally refers to spending the money, evaluation is a richer concept that also connotes an assessment of whether the objectives were achieved. The department has increased its program evaluation capability, but at the time this is written the establishment of the CAPE is too new to say whether it will be able to institutionalize a feedback mechanism that considers current and past performance in program and budget decisions. As noted, more on possible ways to reform PPBES is provided in chapter 10.

CONCLUSIONS

Winston Churchill is credited with saying, "Democracy is the worst form of government except for all those others that have been tried." One might observe that PPBE is the worst budgeting system for DOD except for all the others. Despite its flaws, the system still does an admirable job of supporting the fundamental mission of the military departments: organizing, supplying, equipping, manning, training, and maintaining a first-class fighting force. It includes all the essential elements any defense planning and public budgeting system should entail. It ensures the budget process considers and supports the strategic plans and operating constructs of the force. And it produces an annual budget request the Congress and American public generally find acceptable.

PPBES is a less effective system the further out one looks at the medium term spending time horizon; plans are simply not always realized and in most cases they are changed relative to changes in the threat environment. The system also is less effective during periods of high operational tempo—in operational terms, i.e., immediate military needs and contingencies related to fighting wars may get stuck inside the resource allocation OODA loop.[2] PPBES probably is overly analytical and oft-times insensitive to political realities. The ponderous way it functions may be responsible for negative side effects when important military funding contingencies are not addressed. Not only does the content of data and the decisions to be made change from year to year, the processes, organizational units responsible for performing the work processes at various stages of analysis and decision within DOD change, the leadership and roles of PPBES process participants frequently changes causing the system to become somewhat incomprehensible and confusing to even those who are highly experienced, having worked within it for long periods of time. Given all this real or potential change and uncertainty it is no wonder that new entrants to the system often are confused and almost overwhelmed by its complexity.

NOTES

1. In many aspects of public work, means are ends in themselves. Equal employment laws and merit system protections in the civil service, competition in contracting procedures, and the assurance of due process in justice are examples of means as ends.
2. OODA Loop refers to the combat operations feedback loop of Observe—Orient—Decide—Act developed by Colonel John Boyd. For more on how this process applies to public and private organizations and DOD, see Jones and Thompson, 1999 and 2007. Applying Boyd's concept they characterize the OODA loop as a reinvention cycle (1999) or a realignment process (2007).

CHAPTER 5

CONGRESS AND THE DEFENSE BUDGET

From the Cold War to the War on Terrorism

INTRODUCTION

In earlier chapters we provided templates and detailed analyses of the federal and defense budget processes. The purpose of this chapter is to follow-up this material with analysis of how Congress influences national defense policy and resource management through deliberations and decisions made in the budget process. The chapter focuses on policy setting and budgeting in the period 1989 through 2012, an era in which defense policy and the budget were reshaped beyond all expectation.

National defense policy and budget strategy suffered two great shocks as a result of historic events that occurred on May 1, 1989 and September 11, 2001. The first was the dissolution of the USSR and the end of the bipolar Cold war world with its heavy emphasis on strategic nuclear deterrence, strategic air and sea power, forward deployment to surround the enemy, and procurement, troop, and training programs geared to meet a single, powerful adversary. As this threat dissolved so did the need for some amount of defense spending. By the middle of the 1990s, defense budgets had come

Financing National Defense: Policy and Process, pages 153–200

down by about one-third from peak Cold war levels and strategists were struggling to think beyond the doctrine and force structure that had guided and serviced American defense posture for more than 40 years—from the rise of the Iron Curtain at the end of WWII to the fall of the Berlin Wall.

At the beginning of the 21st century, defense planners continued to grapple with what it meant to be the primary military power in the world. A moderate political consensus emerged in the late 1990s to recognize that some parts of the force structure had been underfunded and cut back too far at the end of the Cold war. With the election of President George W. Bush, plans for a slow but sizable increase in defense spending began to emerge. However, in the aftermath of the shock of the terrorist attacks on the U.S. on September 11, 2001, this buildup was accelerated as a new enemy shifted the threat scenario to a fundamentally different perspective.

In this chapter we chronicle the defense policy and budget dialogue between the Pentagon and Congress during the period of decline and retrenchment in the 1990s followed by the war on terrorism expansion to meet an even more menacing worldwide threat in the decade of the new millennium. Our purpose is not merely to describe the ebb and flow of support for defense spending. Rather, we examine this period in the content of a theoretical assumption grounded in the work of noted budget authority Allen Schick. His thesis is that budgeting by the U.S. Congress fits a model of improvisation under conditions of policy uncertainty. Relative to our previous definition of the budget process as incremental, following the tenets of Wildavsky's theory of the budget process, Schick's hypothesis provides another perspective on the ways in which congressional policy definition and budgeting, and the budget process itself, may be examined and explained. Does the theory of improvisation contravene that of incrementalism? Does one hypothesis provide a better model to guide inquiry than the other? Alternatively, is it possible that a combination of two hypotheses provides a more complete contingency theory of budgeting? To answer these questions first we must examine the evidence in detail. Secondly, we may draw conclusions based upon this assessment. We begin by reviewing the legislative process for analysis and funding defense budget.

THE CONGRESSIONAL BUDGET PROCESS AND DOD

The defense budget begins as the product of executive branch deliberations and is presented in the Budget of the United States that the President transmits annually to Congress. Congress then reviews this proposal and decides what to fund. The Constitution provides sole authority to Congress to tax and spend. No such authority is available to the executive. Congress approves DOD programs and the funding to carry them out annually.

The President's Budget is merely the starting point. A number of decision points draw attention during the congressional review and approval process; these are sometimes confused by the news media and the public. Essentially, the annual congressional defense budget process may be divided into three phases: passage of the Budget Resolution (not a law), passage of the Defense Authorization Act (law), and passage of the Defense Appropriation Act (law).

The Budget Resolution

The Budget Resolution (BR) represents the budget plan of Congress. Recently the media have taken to calling this the budget, but only the appropriation bill actually provides money that may be spent. The Budget Resolution is a commitment by Congress to itself as to how much it will tax and spend in the coming appropriation process; it is a resolution, not a law, and, as such, it is meant to be used as a tool to guide congressional budget negotiation and enactment. The BR provides prospective limits for what will be spent on defense in the current year and what might be spent on defense in the next several years. Deficit avoidance strategies from 1995 onward contained congressional Budget Resolutions with five to seven year guidance. While the years after the fiscal year under consideration for approval are seldom realized as such, observers can look at the trend line and draw their own conclusions.

For example, the 1997 BR included a decrease in defense spending for 1998 (FY99). While this was the judgment of Congress in 1997, the enlightened observer would have viewed this decrease with some skepticism, given DOD problems in meeting its readiness goals after the drawdown budget years of the early 1990s and the flat years of the mid-1990s. In the 2000s the BR reflected the need for increasing defense spending when it was approved by Congress. In some years Congress has not found sufficient consensus to approve and pass a BR. Thus, while it provides useful information, the BR is not the defense budget. It does indicate what is likely to be the top limit for defense spending in the current year, and the BR outyear profile indicates how Congress presents the prospects for future year defense spending.

According to the guidelines provided in the Congressional Impoundment and Control Act of 1974 as amended, the Budget Resolution is supposed to be passed by April 15th, but Congress rarely meets this deadline. Typically, it is June or July before Congress passes the BR that provides taxing and spending targets for all fiscal committees and subcommittees. When the BR is passed, the media tend to refer to the amount approved as the "defense budget." In reality, after the passage of the Budget Resolution, attention turns to the defense authorization bill.

Defense Authorization and Appropriation Acts: Conflict and Compromise

The authorizing committees for defense are the House and Senate Armed Services Committees (HASC and SASC). The authorization bill is important because it approves defense programs, focusing especially on proposed new programs. While the intent of the BR is to provide a guide to overall federal and defense spending, the focus of the authorizers is on specific programs. However, authorization legislation may include a variety of monetary and non-monetary policy directions for DOD. For example, the authorization bill may include the annual military pay raise, authority to begin development of a new weapon system, or authority for new military construction projects, such as a barracks or family housing. It may also suggest to the President positions on treaties, arms limitations, nuclear weapons research and the use of American troops. Without the authorization, DOD is not permitted to begin spending money on new or existing programs, even if it has provided money for them in defense appropriations legislation. To begin new programs, DOD needs both authorization and appropriation. To maintain their oversight and leadership role the authorizing committees have gradually expanded their scope. This is evident in table 5.1 indicating the spread of annual authorization requirements.

The third step in the congressional budget process is appropriation. This is done by the House and Senate Appropriations Committees (HAC

TABLE 5.1 Addition of Annual Authorization Requirements

Year	Public Law	Programs Added
1959	86-149	Procurement of aircraft, missiles, and naval vessels
1962	87-436	RDT&E for aircraft, missiles and naval vessels
1963	88-174	Procurement of tracked combat vehicles
1967	90-168	Personnel strengths of each of the Selected Reserves
1969	91-121	Procurement of Other Weapons
1970	91-441	Procurement of torpedoes and related support equipment; Active duty personnel strengths of each component of the Armed Forces
1973	92-436	Average military training student loads of each component of the Armed Forces
1975	94-106	Military construction of ammunition facilities
1977	95-91	National defense programs of the Department of Energy
1980	96-342	Operation and Maintenance of DOD and all its components
1982	97-86	Procurement of ammunition and "other" procurement
1983	98-94	Working capital funds

Source: Tyszkiewicz and Daggett, 1998: 36, from Senate Committee on Armed Services, Defense Organization. S.Prt.99-86, 1985: 575.

and SAC) and their sub-committees for national defense. Appropriation committees assess and approve spending for programs—money by appropriation title as noted above. Generally, defense appropriations act budget authority (BA) is provided to DOD in lump sums for a specific appropriations account e.g., $6,535,444,000 for Navy aircraft procurement and very specific language about how this money can (or cannot) be spent. Other language of this nature is found in the committee reports accompanying the appropriation acts. These reports direct DOD provide detail on the intent of Congress regarding how appropriation amounts are to be spent. Where appropriation and report language does not differ from the DOD budget request, then whatever the justification material presented by DOD for various programs, accounts, and items is taken to be binding and Congress assumes DOD will carry out the program as it was presented (Tyszkiewicz and Daggett, 1998: 40). Most of the actual funding for the Department of Defense is provided in the defense appropriation act and the military construction act.

In some years the defense appropriation bill is passed on time before the beginning of the fiscal year. However, often this is not the case. In 2000, the defense appropriation bill was passed on time. As can be seen in figure 5.1, it was approved by voice vote (vv) in the subcommittees in each chamber. The full Appropriations Committee passed it by voice vote in the House and the Senate committee vote was unanimous. Substantial majorities passed the bills on the floor of each chamber in early June. The Conference Committee report was issued about a month later, on July 17 and passed each chamber within 10 days. The President signed the bill into law on August 9, 2000. The bill was early because there was substantial agreement on what needed to be spent for defense, the budget was in a surplus position, and a presidential election would take place that fall. Because of its size, the defense appropriation bill is usually the last to be considered and is usually late. From 1970 to 2003, the defense appropriation bill had been passed and signed by the President or before the start of the fiscal year six times in 1976, 1977, 1988, 1994, 1996 and 2000 (Reynolds, 2000:11). This was the case several times in the decade that followed but the norm was and continues to be late approval by Congress, leaving DOD to operate on what are termed Continuing Appropriations (CRAs), to be described and analyzed subsequently in this chapter. This confounds DOD budget execution.

Without an appropriation of some type, little or nothing can be done. Thus, the annual appropriation act is the true defense budget. It contains the dollars to fund people, weapons, and supporting expenses. Both the authorization and appropriation bills may contain a dollar figure for a defense program, but the appropriation bill contains the real money, while the authorization bill number may be viewed as a ceiling for the program, unless it is exceeded by the amount enacted in appropriations. Still, the

Bill No.	Subcommittee Approval		Committee Approval		House Passage	Senate Passage	Conference Report	Conference Report Approval		Public Law
	House	Senate	House	Senate				House	Senate	
Defense	5/11/00	5/17/00	5/25/00	5/18/00	6/7/00	6/13/00	7/17/00	7/19/00	7/27/00	P.L. 106-259
H.R. 4576	(vv)	(vv)	(vv)	(28–0)	(367–58)	(95–3)	House Report 106-754	(367–58)	(91–9)	8/9/00
S. 2593			House Report 106-644	Senate Report 106-298						

Figure 5.1 Defense Appropriation Bill in 2000—Passed on Time. *Source:* Library of Congress. 2000. Defense Appropriation Bill in 2000—Passed on Time: 93.

appropriations act is not enough to initiate spending. Matching authorizations have to be approved by Congress eventually. Further, appropriations must be matched (cross-walked) with what DOD has asked for in the President's budget. Additionally, critical information on congressional intent must be gleaned from careful reading of the Conference Committee Report that accompanies the defense appropriation act. Sometimes language (statutory) is inserted in an act about how money is to be spent, or in some cases even reduced in one area and increased in another. As noted, this type of language also may appear in the committee reports. Typically, these cues from Congress are not subtle but often require legal review before DOD can commence spending.

For example, in 2000 the Conference Committee took funds from the C-17 program and put them in a revolving fund so that the Air Force could not spend them on fighter aircraft. The Conference Committee also cut $1 billion from the LPD-17 ship construction program but in the report allowed the Secretary of the Navy to move up to $300 million from other budget accounts to cover unexpected increases in this program. (Congressional Quarterly Almanac 2000: 2–51). More subtle cues also may be provided directly in hearings, where legislators indicate what they think DOD ought to do.

Statutory language is law and must be obeyed. Report language does not have the force of law, but when DOD fails to heed non-statutory report directives, inevitably some official will be called before committee to explain why. Questions and suggestions made in committee hearings by members of Congress, assisted by their staffs, are not binding and DOD need not fully implement them when they run counter to the needs of the military departments and services (MILDEPS). However, completely disregarding congressional "suggestions" may result in lengthy interrogations at hearings, requests for reports, and a tightening of thresholds for reprogramming dollars in budget execution if Congress loses faith in DOD stewardship. When DOD deviates from what appropriation committees have explicitly approved, these actions must be justified to Congress. Even if members forget, their staffs do not. This is a key part of committee and members' staffs jobs, i.e., to "assist" or control the management of DOD to satisfy constituent interests of members of Congress.

Consequently, multiple sources must be considered to understand the defense budget including the conference committee reports that accompany the defense authorization and appropriation acts. Defense insiders routinely deal with "fences, floors, and ceilings" with respect to congressional actions and directives, meaning that Congress has required specific amounts of money within an overall appropriation to go exclusively to certain programs (a fence), that not more than x dollars be spent on it (a ceiling), and not less than y dollars be spend (a floor). DOD managers treat what are called "congressional interest items" with great care because they know that if they don't, their entire organization will be called to task for their errors.

All Defense Spending Does Not Come From the Defense Appropriation

In any specific year, the Defense Appropriations Act provides most of the funding executed by DOD, e.g., approximately 90% (the exact percentage varies by year). Other funding for defense is appropriated in the Military Construction Appropriations Act, the Energy and Water Development and Appropriations Act (mainly for nuclear weapons activities and base closure), the Housing and Urban Development and Independent Agencies Appropriation Acts, e.g., the Federal Emergency Management Agency, support of National Science Foundation logistics in the Antarctic, the Commerce, Justice and Department of State Appropriations Acts (mainly for defense related activities of the FBI and for Maritime Security programs). Further, these appropriations may be enhanced by supplemental budget appropriations, as we analyze in chapter 6. In addition, creation of the Department of Homeland Security shifted funding for some national defense activities into this appropriation.

Once the defense appropriation bill is approved, it would appear that Congress has done its job on the budget unless it has to act again in response to a Presidential veto. So, here we might conclude, now we have the budget. However, as explained below, this is not quite the case. Authorization and appropriation acts differences must be reconciled before DOD can begin to spend. Adding confusion to the congressional process, sometimes the authorization bill is passed after the appropriation bill, but this is not the intended process. Programs that are appropriated, but not authorized, may not be executed.

It was not always this case. Until the mid-1980s, if a program gained an appropriation, but had not been authorized, it was assumed that the appropriation included an implicit authorization, for surely the logic ran, those who voted to appropriate the program meant to authorize it too, otherwise they would not have appropriated it. Thus, appropriation included implicit authorization. However, in a strict legal sense this concept was found lacking and perhaps deservedly so, given the size of the both the defense authorization and appropriation bills and the number of items covered in them.

It is an understatement to observe that not all members of Congress know all the provisions of the defense authorization and appropriation bills. Moreover, given the lack of membership overlap between the authorizing committees and appropriations sub-committees for defense, it is hard to argue that the appropriations bill carries the implicit will of the authorization committees. The appropriation bill can still pass into law even if authorizers vote against part or all of it but it cannot be spent without the authorization. The law is clear on this point. Title 10 of the U.S. Code, the body of law that governs the Military Departments and the Department of Defense, indicates in Section 114 that funds must be authorized before they may be spent.

Moreover, the rules of the House and Senate each prohibit appropriating funds for programs that have not been authorized. (Tyszkiewski and Daggett, 1998:44) However, some of this clarity is dissipated in actual practice when appropriations bills, "... often provide funds over and above amounts provided in authorization bills" and for activities which were not mentioned either in the authorization bill or the report that accompanied it and when legal opinions have consistently held that appropriations bills may provide more or less money than has been authorized for a particular program (Tyszkiewski and Daggett, 1998: 44) The Government Accountability Office (GAO) long ago indicated that the legislation passed most recently prevails over earlier legislation (this is customarily, but not always, the appropriation bill), and that more specific provisions prevail over less specific provisions (GAO, 1991).

What makes this whole congressional budgeting system workable is that the laws themselves are rarely in conflict over the same item or issue because the authorization and appropriation bills aggregate totals in categories that are more general rather than by discrete items. The language of the reports accompanying the bills carries the details clarifying what and how many programs or units are funded by appropriation category, and this language is subject to the brokering and clarification that goes on between members and the two houses of Congress and their staffs and DOD representatives during the legislative budget process.

As far back as 1998, a particularly acrimonious debate over defense authorization led to a letter from Senator Robert Smith of New Hampshire explaining the consequences of not passing the defense authorization bill (Congressional Record, 1997: S11817-8). Smith commented on a variety of items that would impede the good management of the Department of Defense and defense programs, ranging from delaying the construction of family housing and other military construction projects to the absence of authority to expand counter-narcotics programs in South and Central America, to the absence of authority to accelerate advance procurement programs, the lack of authority to increase military personnel retention bonuses. What is threatened by the mismatch between authorization and appropriation bills is found primarily in the acquisition area in lack of funding for new program starts and expansions or decrements to old projects. While the dollar amounts may seem small in many instances, they are the marginal dollars that may prevent new ideas from implementation, good programs from expanding, and programs that have outlived their usefulness from being reduced or eliminated.

The authorizing committees provide DOD a forum for advocacy. These committees look both at macro and micro defense needs, examining defense force structure, overseeing DOD operations, and looking for ways to fund projects that will benefit their constituents.

The following Exhibit is based on a conversation with a staff member of the Senate Armed Services Committee. It traces the defense authorization bill cycle through bill passage. The comments might not hold for all committees or all years or all staffers, but it does provide an insight into a process that is not normally seen.

The Authorization Cycle from the Staff Perspective

The material in the following section was provided by a Senate staff member. It shows how staffers view the budget process from the perspective of an insider to this process.

The cycle starts with State of Union address. The first round of hearings is pretty cut and dried; the same witnesses appear before the committee each year: Secretary of Defense, the Chairman of the Joint Chiefs, the Secretaries of Army, Navy Air Force, etc.

The subcommittee hearings are different every year, dependent on issues. Staffs work as facilitators, call 5 or 6 people, ask them what they would say if asked certain questions, then decide who to use, form panels of witnesses and script the hearings so that the issues are brought out and both sides aired. Staffs suggest to members the questions to ask and what the answers should be so that expected issues are highlighted.

The hearings are valuable. They educate the Senator on a complicated issue; he learns when he gets truth, when he is being hoodwinked. The hearing builds a record to take new action. It is also used for oversight: has the agency done what it was asked to do last year? The hearing involves briefing books, statements by members and witnesses, questions, answers. Most of the time it is hard for staff to get ten minutes with the Senator so the briefing around the hearing is a two hour block of time to go over in detail what the issue is about, educate the Senator so that in mark-up he will remember this issue and be prepared to make a decision on it, or so that he can be put back into the picture by referring to the discussions at the hearing when the issue was first considered.

Markup follows hearings. This is usually in June, by subcommittee and by full committee. The subcommittees are given a number by the committee chairman that they have to attain. Staffs sit where witnesses sat, answer questions, give advice to their members, but they cannot vote. Member of Congress themselves must decide. It can take from 20 minutes to 4–5 hours. Staffs have to agree on markups and what has been decided. Go late on Thursday night, put a report together, put a bill together, proof with members and staff, send to the Government Printing Office and file with Clerk of the Senate. Staff position is that bill is perfect when it comes out of committee and no amendments are needed, but amendments are always offered because Senators have to have a record for re-election.

Staff tries to fend off amendments, instead give them language in the report, or a sense of the Senate resolution, or promise to take it to conference. The first couple of days serious amendments are offered, after that it is cats and dogs and the committee/staff tries to kill them. Staffs talk around, try to find out who is going to offer an amendment, but without giving them any ideas.

Floor debate on our bill is usually scheduled right before August recess; otherwise they would talk about defense issues all fall. Armed Services bill is a bill that will pass, so people try to add amendments. A Senator who gets an amendment accepted has a three-for: he gets to talk about it when it is accepted, again when the bill passes, and when it goes to conference with the House. He can talk about it three different times.

After floor passage, the bill goes to conference with the House. Members tell staff to negotiate a potential solution on 90% of issues, take them back to the member to be blessed. House staffers are specialists and it is a really tough task to prepare Senate member to go head to head with a staffer who is a real expert. When you get agreement, the staff member has to speak up and say "we understand the agreement is . . ." in order to make sure all agree on what has been agreed. Have to build in enough time in the conference process so people can talk, ventilate issues, but not too much time. At this point also the appropriations bill is right on the heels of the authorization bill, like passing notes through a hole in the wall between committees. If authorizers propose too much for a program according to appropriators, authorizers may change the shape of program a little to spend less.

Big ticket issues drive the conference. Most of the other issues (80–90%) can be settled by staff and blessed by members.

Precision in conference committee work is very important. When bill is in the hearing stage in committee, if errors are made they can be fixed on the floor (or if "bad" provisions inserted), floor mistakes can fixed in conference committee work with the House, but the product of the conference becomes law; has to be perfect, language has to be clear, precise, correct. To fix it requires passage of another law.

Next the bill and the conference report go back to Senate. The Conference Report is not amendable and must be voted up or down.

My job is to make the Senator look good: I might say "here are the pros and cons; I'll give you a recommendation if you ask me." As the Soviet threat vanished, the tendency to pork and earmarking seemed to increase. Then 9/11 came along and the pressure to respond was intense for both chambers of Congress. But respond they did.

Source: Jones and McCaffery, 2008. U.S. Senate, Armed Services Committee staffer.

The product of the Senate Armed Services Committee is the Defense Authorization bill. The sense here is that a getting a bill passed is a very significant achievement, like a team making it to the Superbowl.

When the Senate Armed Services Committee was chaired by Senator John Warner (R-VA), this Senator was particularly effective in initiating pay raises and pension benefit changes for the military services, helping to mitigate the consequences of defense budget shortfalls of the Clinton Administration. For servicemen and women, the authorization committees provide the first indication of how quality of life issues such as pay, benefits, and family housing will be treated. Contractors and manufacturers also are very interested in authorization committee work, for they too get early indication of when and how much a program or weapons contract might be changed.

While most programs in defense are authorized annually, the number of uniformed military personnel to be employed is handled differently. A permanent personnel end-strength level is authorized, then the appropriators fund the dollars to support the level of employment authorized. From time to time the end strength numbers are changed up or down by the authorizers as needs warrant, based on recommendations from DOD and the executive branch. The appropriators must fund this authorized strength in the annual appropriation bill. The authorizers may also approve annual pay increases, but sometimes the appropriation process results in Congress providing DOD with new money for only half or some portion of the pay raise, forcing DOD to fund the other half itself by finding funding from other appropriation accounts, excluding MILPERS but including O&M.

Other Opportunities to Influence Defense Spending Allocation

Members of Congress have many opportunities to cast votes for or against defense spending and defense legislative liaison personnel have many locales in Congress upon which to focus their energies. Tyszkiewicz and Daggett (1998: 31) found 22 stages of congressional action where votes are cast on parts of the defense budget, including sub-committee and committee votes to report out each bill (5 in each chamber), the final floor vote on the bill, and the vote on the final conference committee report for the bill (6 in each chamber). Other appropriations bills that have defense dollars in them would increase the number of voting opportunities, e.g., the

Military Construction or the Energy bills. In table 5.2 we present this information somewhat differently. The legislative liaison groups in DOD have to monitor votes in as many as 38 places throughout the legislative year, beginning with budget committee action, proceeding through the authorizing committees (Armed Services) and the Appropriations Committees. DOD usually has a supplemental bill to worry over and since Congress usually does not finish appropriations on time, each session usually sees one or more Continuing Resolution Appropriations. Any of these events could be critical for defense programs, thus must be monitored, and in many cases responded to with appeals and lobbying efforts. On the legislative side, only committee members vote in committees, (and sub-committee members on sub-committees), but all members vote on final floor action and on accepting conference committee reports. Thus every Congressman has at least ten opportunities to vote on ordinary defense bills; this expands to 13 or more when supplementals and one or more CRAs are added.

During one congressional session Senator Diane Feinstein (D-CA) made 15 major roll call votes on Defense issues in the Senate from April through December. The votes included two budget resolution votes, one reconciliation vote that was tied to a defense reduction, four supplemental appropriation votes, two votes on the military construction appropriation, five authorization votes and one appropriation bill roll call vote. The Senator does not list all the actual roll call votes, just the final actions. This list was easily accessible on her web page so that her constituents could see how she voted on important defense issues.

If an observer were to step inside each stage and make a count based on individual votes taken, the voting opportunity of legislators is enlarged significantly when amendments to bills are considered or as items are voted up or down in the sub-committee and committee mark-up process. Moreover, the critical vote on a bill could come on one of dozens of amendments that are offered, if the coalition that can successfully pass an amendment

TABLE 5.2   Milestone Votes on the Defense Budget

	Opportunities for Votes on Defense						
	Budget Res. (BR)	Armed Service Auth	DOD APPN	Milcon APPN	DOE APPN	DOD Suppl.	CRA
Subcommittee		H, S	H, S	H, S	H, S		
Full Committee	H, S	H, S	H, S	H, S	H, S		
Floor	H, S	H, S	H, S	H, S	H, S	H, S	H, S
Conference Report Approval	H, S	H, S	H, S	H, S	H, S	H, S	

Source: Authors, 2011. Data derived in part from McCaffery and Jones, 2004: 150.

to a bill can hold together for the vote on the main bill. Although we only show three in the exhibit, it is also useful to note that tracking the total defense budget in Congress involves tracking five appropriations bills, as well as the Budget Resolution. Moreover, a supplemental appropriation is also a possible venue for change, as the current year supplemental could affect in some way current-year programs in the following year.

CONGRESSIONAL APPROPRIATION PATTERNS

Aaron Wildavsky's incremental model included the precept that agencies are expected to be program advocates and as a result, the executive branch would ask for more that it had last year and, taken together, Congress would reduce that request. (Wildavsky, 1964: 13–35, 63–84) Moreover, the argument ran, Congress would be a marginal modifier, reducing the budget request some, but not a lot, in the aggregate. LeLoup and Moreland's work on the Department of Agriculture budget found this to be true. From 1946 to 1971, they found that Congress reduced the annual request by 2% on average. From 1980–1989, Congress changed the Agriculture request an average of 5.9%, more, but still what might be called marginal. (McCaffery and Jones, 2001:132) The exhibit below covers the defense appropriation bill from 1980 to 2000. Congress reduced the request in 16 of the 20 years. Note that the House and Senate never arrived at the same number for defense, thus a conference committee was always a necessity and even where numbers were close, the two chambers could have funded dramatically different defense profiles.

TABLE 5.3 Defense Appropriations 1980–2001: President's Request, House, Senate and Final Enactments and Change from Request

Fiscal Year	Request	House	Senate	Enacted	Change from Request
1980	132,321	129,524	131,661	130,981	−1,339
1981	154,496	157,211	160,848	159,739	+5,242
1982	200,878	197,443	208,676	199,691	−1,187
1983	249,550	230,216	233,389	231,496	−18,054
1984	260,840	246,505	252,101	248,852	−11,988
1985	292,101	268,172	277,989	274,278	−17,823
1986	303,830	268,727	282,584	281,038	−22,792
1987	298,883	264,957	276,883	273,801	−25,082
1988	291,216	268,131	277,886	278,825	−12,391
1989	283,159	282,603	282,572	282,412	−747
1990	288,237	286,476	288,217	286,025	−2,211
1991	287,283	267,824	268,378	268,188	−19,095
1992	270,936	270,566	270,258	269,911	−1,025

Fiscal Year	Request	House	Senate	Enacted	Change from Request
1993	261,134	251,867	250,686	253,789	−7,345
1994	241,082	239,602	239,178	240,570	−512
1995	244,450	243,573	243,628	243,628	−822
1996	236,344	243,998	242,684	243,251	+6,907
1997	234,678	245,217	244,897	243,947	+9,268
1998	243,924	248,335	247,185	247,709	+3,785
1999	250,999	250,727	250,518	250,511	−488
2000	263,266	267,900	263,932	267,795	+4,529
2001	284,501	288,513	287,631	287,806	+3,305

Note: Table by Stephen Daggett Foreign Affairs Defense and Trade Division CRS.
Sources: For FY 1950–74 Department of Defense FAD Table 809, issued Oct. 21, 1974;
FY 1975–82 and FY 1989–99 annual Appropriations Committee conference reports;
FY 1983–88, Department of Defense Comptroller, annual reports on congressional action
on appropriations requests (FAD-28 tables); FY2000–01, House Appropriations Committee.

TABLE 5.4   Percent Increase from Prior Year Appropriation: 1981–2000

Fiscal Year	Pres Request	Final Approp	Pres Req/ App cy-1	Cong change	Total change
1980	132,321	130,981			
1981	154,496	159,739	17.95%	3.28%	21.24%
1982	200,878	199,691	25.75%	−0.59%	25.16%
1983	249,550	231,496	24.97%	−7.80%	17.17%
1984	260,840	248,852	12.68%	−4.82%	7.86%
1985	292,101	274,278	17.38%	−6.50%	10.88%
1986	303,830	281,038	10.77%	−8.11%	2.66%
1987	298,883	273,801	6.35%	−9.16%	−2.81%
1988	291,216	278,825	6.36%	−4.44%	1.92%
1989	283,159	282,412	1.55%	−0.26%	1.29%
1990	288,237	286,025	2.06%	−0.77%	1.29%
1991	287,283	268,188	0.44%	−7.12%	−6.68%
1992	270,936	269,911	1.02%	−0.38%	0.64%
1993	261,134	253,789	−3.25%	−2.89%	−6.15%
1994	241,082	240,570	−5.01%	−0.21%	−5.22%
1995	244,450	243,628	1.61%	−0.34%	1.28%
1996	236,344	243,251	−2.99%	2.84%	−0.15%
1997	234,678	243,927	−3.52%	3.79%	0.27%
1998	243,924	247,709	0.00%	1.53%	1.53%
1999	250,999	250,511	1.33%	−0.19%	1.13%
2000	263,266	267,795	5.09%	1.69%	6.78%
		Average change	**6.03%**	**−2.02%**	**4.00%**

Source: Authors, 2011 drawn from Table 5.3.

On average, the President during this period asked for an average of 6.03% more for defense each year over these years and Congress cut this request by an average of 2.02%. The outcome is a modest 4% year over year change. However, these averages mask the Reagan build up of the early 1980's (81–85) where the President asked for increases up to 25.75% over the previous year's appropriation. These increases were supported in 1981 and 82 by Congress and then cut substantially, but the result was still a substantial increase. The defense drawdown began in 1987 when the outcome was 2% below the previous year's appropriation. The 1990s clearly show cuts to defense appropriations, with the President asking for decreases in 1993, 1994, 1996, and 1997. Congress supported the President and cut more in 1993 and 1994 (made the decreases larger), but added money back to the defense budget in 1996 and 1997 (diminished the President's cuts). To simply indicate that the President asked for increases on average of 6% over the period while Congress cut on average of 2% does not do justice to the complexity of the story, nor does it adequately describe the turbulence of the period. In Figure 5.2 this turbulence can clearly be seen. The defense drawdown is apparent in the 1990's. In 1993 and 1994, what Congress appropriated was below the previous year's base, and so was the President's budget

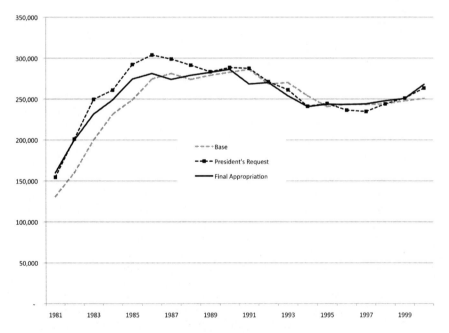

Figure 5.2 Turbulence in Defense Funding. *Source*: Authors. 2011 derived from Tables 5.1 and 5.2.

request. Appropriations would stay at or slightly above the base from 1995 through 1998, although in these years we can see that Congress added some money to defense while the President sought further increases.

HISTORICAL BACKGROUND
ON CONGRESSIONAL BUDGETING

Allen Schick has asserted that the federal budget process in the 1980s could best be characterized as a case of improvisational budgeting, where the nature of the annual budget process could not be predicted (Schick, 1990: 159–196).[1] According to Schick, leaders in Congress and the Executive branches adopted whatever strategies and process steps they thought necessary during the 1980s to approve the annual appropriation bills. In most years, appropriations bills were passed late; continuing resolution appropriations were a common occurrence and some years were funded by omnibus continuing resolutions. In five of ten years from 1980–1989, more than half of the appropriation bills were funded for the whole year by a continuing resolution appropriation; this included all 13 in 1987 and 1988. Over the course of the decade, 63 of a potential 130 bills were funded for a whole year in a continuing resolution appropriation (computed from Schick, 1990: 181). Summit meetings were held between key Presidential advisors and congressional leaders with varying degrees of success; some summits succeeded and some did not. On the whole, argues Schick, outcomes from these summits were minimal and the problems budget makers faced after the summits and in succeeding years were much the same as those faced before budget summits were held.

In this environment, budgetary gimmickry was pervasive. Such deceptions ranged from a simple shifting of paydays in the Department of Defense to meet spending caps to overly optimistic GNP and revenue estimates. The budget process usually began slowly with Congressional Budget Resolutions passed late, an average of 53 days late from 1980–1989 (calculated by authors, 2003, based on Schick, 1990: 174). Much of the appropriation end game was deliberately moved into the new fiscal year where a sense of fiscal urgency helped decision makers reach compromises. In this turbulent environment, Schick observed that CRAs that were usually funded at the cost to continue current levels of operations actually gave agencies more money than they had had the previous year and more money than was asked for by the President's budget (Schick, 1990: 181). They became real appropriation bills rather than placeholders.

Another change of consequence concerned the President's budget which, in Schick's view, had lost its status as an authoritative statement of what the President indicated was needed to fund the executive branch. Instead, according to Schick, the President's budget had become an opening bid in some sort of extended budget auction where true needs and motives were difficult to perceive, all the better to facilitate bargaining over key programs during the remainder of the session. Consequently, it is not surprising that timeliness was a victim of improvisational budgeting. The process began late, with late budget resolutions, and failure to adopt both authorization and appropriation bills on time. In some years, the appropriation bill preceded the authorization bill, making the guidance-giving role of the authorization bill problematic.

The view that the President's budget is merely a starting point for Congress, or that it is unnecessary (e.g., dead on arrival), is disputed by those who view the budget as the primary document and point in the budget process where the President has the opportunity to state his policy and program priorities—that will inevitably differ from those of many members of Congress. (McCaffery and Jones, 2001: 97–108)

Incremental and Improvisational Theory Differences

A simple incremental model of the budget process would include aggressive agencies asking for more for their programs, OMB cutting them back somewhat to include all priorities within available resources and Congress acting as a marginal modifier. In the classic period of American budgeting, this might have meant the House as masters of detail and guardian of the purse cutting deeply while the Senate acted as a court of appeal and restored some, but not all, of the cut made by the House. Outcomes would show the Presidents budget request being the best predictor of the outcome of the struggle. This picture changes somewhat for the 1990s.

The budget processes of the 1990s and 2000s and in 2011, have been improvisational, with constant changing of normal budgeting routines in various ways including an abundant supply of continuing resolution appropriations, appropriations passed before authorizations, summits and government shutdowns. Some of this improvisation may be seen from the funding outcomes. In six of the years of the decade 1990 to 2000 the DOD appropriations signed into law were less than the President's Budget request (PB), with cuts ranging from $19B in 1990 to $488 million in 1998. In four of the ten years, DOD appropriations enacted exceeded the President's budget request, with increases ranging from $6.9B in 1995, to $9.2B in 1996, $3.7B in 1997 and $4.5B in 1999. Since 2001 most presidential defense budget requests have been passed at or above what was requested, especially when supplemental appropriations funding military warfighting are included in total defense appropriations.

CONGRESSIONAL BUDGETING: RECURRENCE OF THEMES AND TRENDS

While Schick's analysis pertains to the 1980s, his conclusions still seem accurate when considered in 2012. Budget Resolutions continue to be late and, in some years, no budget resolution has been passed, rather each chamber adopted a resolution to guide itself irrespective of what the other chamber did. Appropriations have been late and the defense authorization bill sometimes preceded the defense appropriation bill. Partisanship remains high, sometimes bringing the budget process to a halt. In many years the budget process has not operated on time, but often this has been the result of a greater policy divergence between the President and Congress or between the two houses of Congress.

The turbulence of the 1990s culminated in the bitterly contested presidential election of 2000 with the nation divided into red and blue states and many democrats believing the election had been stolen. When President George W. Bush was elected in 2000 and the House and Senate elections gave his Republican Party majorities in both houses of Congress, conditions were ripe for increased consensus on defense policy and budgets. A federal budget surplus and recognition that defense had been cut too deeply in the 1990s provided rationale for both large tax cuts and increases for defense. However, the 9-11 attacks put the nation into a new war on terrorism that resulted in significant increases in defense appropriations and spending.

The narrowness of a party majority in the Senate following the elections of 2000 and again in 2010 made the House, where the Republicans had gained the majority, the driver of defense spending increases. However, in 2001 the Republican defense hawks were defeated by the Republican tax cut hawks and President Bush gave policy priority to a tax cut. If effect the debate over the size of the tax cut and its timing, fought out in the first half of the year, took both policy decision space and dollars away from a potential defense re-capitalization. Despite the fact that the President had suggested adding as much as $48 billion to defense during the 2000 campaign (Congressional Quarterly Weekly, Feb. 10, 2001: 337), he submitted a placeholder defense budget in April contingent upon a review of defense needs by then Secretary Rumsfeld. Bush asked for more for defense in June, but mostly for operating expenses, not recapitalization. As debate over the tax cut proceeded, Congress put off any wholesale defense changes (CQA, 2001: 7.3), and Rumsfeld said that the administration's plan was to stick with the Clinton budget of $310 billion for FY2002. Secretary Rumsfeld's attempt to lead the Pentagon into a new era angered the service chiefs and some in Congress as he proceeded with transformation plans that focused on technology and less on conventional weapons (CQA, 2001: 7.4). The outcome for FY2002 was a Presidential request of $319.5 billion and a defense appropriation of $317.6 and increase of 10.7% over FY2001's $298.5b., with most of the increases going to the operating accounts rather than recapitalization.

Before final action on the FY 2002 defense act could be taken, the terrorists struck New York and Washington, D.C. with devastating effect. The response from the President and Congress was swift and purposeful, including an extraordinary emergency supplemental of $40 billion introduced on September 14th and passed on September 18th. President Bush vowed to respond to the terrorist threat with whatever was necessary for as long as was needed to find and punish those responsible for the attacks. Congress responded on both sides of the aisles with unanimous support for further increases in defense spending to counter the terrorist threat. The divisiveness of the 1990s was replaced quickly, and for a limited time, with unanimity on policy and budget direction. Consensus prevailed as the President mobilized forces and initiated a war on the Taliban regime in Afghanistan. The President requested and Congress appropriated both defense budgets and budget supplementals to fund what has turned out to be a long war in Afghanistan.

The terrorist attack that struck the World Trade Center also struck the Pentagon, causing more than 100 casualties and extensive damage to one side of the building. That day produced many heroes, in New York City, in Pennsylvania where passengers took over a terrorist-hijacked flight at the cost of their own lives, and at the Pentagon. In late August, rumors had begun to circulate that Rumsfeld was bogged down in his transformation efforts, had lost the confidence of Congress and the service chiefs and would be the first cabinet member to resign. After 9-11, Secretary Rumsfeld and then Secretary Gates and Secretary Panetta went from the role of average Secretary of Defense in peacetime to, in effect, Secretaries of War, during the recovery period after 9-11 and during the campaigns in Iraq and Afghanistan.

As reflected in congressional elections in 2006, 2008 and 2010, the wholesale consensus of support of Presidential policy, including that of President Obama, on defense began to weaken but it did not break. Despite the political unpopularity of the war in Iraq internationally, and even given eventual increasing dissatisfaction in the U.S. over the war in Afghanistan that was reflected in some serious head-scratching and debate in Congress, in the end congressional support for warfighting was unwavering. From the passage of a defense supplemental appropriation of about $76 billion in April 2003 to fund DOD for costs incurred in the successful campaign to oust Saddam Hussein from power to passage of a cost of war supplemental in 2010 Congress found ways to support U.S. foreign policy objectives. From the fall of 2001 on as the nation moved to a war fighting mode, funding for defense came from a stream of sources including regular appropriation bills and supplemental appropriations. However, support for the war on terrorism was not universal during the decade after 9/11. Some Democrats ran on an anti-war platform in 2006 and in part this allowed them to gain a majority in both houses of Congress, somewhat to the surprise of the President. The elections of 2008 brought both the presidency and the Congress to Democrats, but this condition lasted only until 2010 when general dissatisfaction with the economy elected a Republican majority to the House of Representatives.

When viewed from the perspective beginning in the mid-1990s, the budget process moved further into improvisational routines. In our view, these improvisational routines occurred because of a number critical policy decisions. The first was the Gingrich House Republicans and their Contract with America from 1994 to 1996 that eventually provided a roadmap to a balanced budget and attempted to force President Clinton to follow it. The second event was the victory of the tax cut Republicans in the summer of 2001 that limited defense recapitalization and domestic discretionary spending. The effect of the tax cut was exacerbated by the puncturing of the dot.com stock market bubble and the recession that followed the events of 9/11, reducing government tax revenues. The third event was the terrorist attacks of 9/11 and in response the proactive war policy of the Bush administration in Iraq (not Afghanistan). By 2006, according to opinion polls a majority of the American public no longer supported U.S. efforts there. The next critical set of decisions were those made by Presidents Bush and Obama along with Congress to pass large spending authorizations in 2008 (the Troubled Assets Relief Program or TARP) and 2009 (the Recovery and Reinvestment Act) in response to the collapse of the secondary housing market, housing prices generally and the recession that followed beginning in late 2008. The latest set of events influencing departure from routine budgeting as of this writing was the acrimonious debate over passage of the federal debt limit increase that led to collapse of the US and foreign stock markets, reduction of the nation's perceived credit-worthiness and consequent increase in US Treasury borrowing costs, thereby exacerbating public concern over the size of the U.S. federal government by increasing it in terms of the costs of future borrowing. The displeasure of the American public towards Congress and the federal government generally increased measurably as a result of congressional delays in acting to keep the government from defaulting on debt and other payments.

Thus, since the mid-1990s to the present the congressional budget and appropriation processes have had to work around these events and decisions. This was accomplished by using improvisational means.

Budget Resolution Turbulence

An important indicator of budgetary turbulence and improvisation may be found in the treatment of the Budget Resolution (BR). This resolution is developed by the House and Senate budget committees and begins the budget process in Congress. The Budget Resolution was established to precede the authorization and appropriation bills and set limits on spending, both as to totals and to functional areas. The limits were derived annually by the committees using the Presidents Budget as a starting point together with estimates from the Congressional Budget Office (CBO) as to GDP, inflation, unemployment, and revenue and spending projections and from testimony presented to the budget committees by various administrative officials such as the Director of OMB, the Secretary of Treasury and various agency officials such as the Secretary of Defense. The Budget Resolution provides advisory guidance for Congress and does not have the force of law; however a point of order can be raised on the floor of either house (procedures vary between the two houses of Congress) when a bill exceeds its BR limit. In such cases, a 60 percent super-majority is required to defeat the point of order. In effect, if passed, BR targets can become substantial guides for the budget process. This is intended to provide discipline to the budgetary process but often this goal is not met. Appropriators are supposed to meet the budget resolution targets for each spending bill. During discussion and amendment on the floor, a point of order may be raised against any amendment that would send a particular appropriation bill over the BR targets. While the point of order can be overturned, the 60% super-majority of votes necessary to do so is not easy to gain. All things considered, the budget resolution functions can serve as a useful starting pointing and continuing reference for scorekeeping once the appropriations bills are negotiated.

Notwithstanding its importance, the budget resolution has been subjected to the same turbulence as the rest of the budgetary process. It has usually been late, and by a significant margin. It was on time once in 1999 and once it was early as the Democrats rushed to get out ahead of their President in 1993. In 1998 and in several years since no budget resolution was ever passed. As we have noted, the budget resolution is not a law; rather it is a compact by which Congress agrees to limit its spending tendencies by providing freedom within the amounts provided, but sanctions when they are exceeded. However, despite that it isn't law the degree of debate over passage of the BR, and whether it is passed at all, provides an early indication about the consensus, or lack thereof, of Congress heading into the authorization and appropriation processes.

With all this turbulence, the temptation is to denigrate the importance of the budget resolution. This would perhaps be premature. The budget resolutions of 1995 and 1997 had great symbolic value as they set out pathways to a balanced budget for four years ending in 2002. In 1997 particularly, congressional Republicans intransigence about showing a pathway to a balanced budget forced President Clinton to make several changes in his budget estimates to reach the same goal. In 1993, congressional Democrats thought that they were making an important statement about their readiness to do budget business in a new era. In 1998 and a number of times in the 2000s the budget resolution was never completed because some participants thought the guidance provided by the 1997 resolution was good enough, and because the House and Senate Republicans could not agree on policy directions. Debates over the BR often have been bitter and prolonged battles and indicate that the Budget Resolution has meant something to the participants, even if they could not agree on the numbers. Thus, it can be argued that the Budget Resolution emerged from the 1990s as an important and useful budget mechanism. However, the same conclusion could not be stated as definitively after the period 2000 to 2010. In 2011, at a time when the to houses of Congress were divided along paty lines the BR was passed as an indication of congressional intent to reduce spending or at lest the rate of spending growth. This signaled a serious disagreement over budget cutting in the 2011 budget process for FY2012 and for subsequent year appropriations.

Two observations about the BR and congressional budgeting under some type of spending ceiling seem relevant. First, progress in changing the budget process to consider both cuts as well as increases is not linear. While incrementalism may indicate that budgets change by a relatively small percent in comparison to the previous year's base, the budget process itself is not predictable; what will lead to a good or bad budget year is not easy to discern. The fireworks around the budget resolution do provide a hint, but only a hint of what is to come. Consensus and cleavage count. When consensus exists, the budget year is smooth; cleavage results in a disrupted budget process. Sometimes the cleavage is between the President and Congress, sometimes between the parties in Congress, sometimes between the two chambers and sometimes within one chamber. Secondly, when cleavages exist, surpluses matter, but deficits appear to matter more.

Passing the budget resolution on or before the 15 April deadline is difficult. Delays in establishing spending levels in the budget resolution would seem to inevitably lead to late appropriation bills. However, late defense appropriation bills often have not been the result of late budget resolutions. It is important to remember that the budget resolution is constructed by budget function, and the number it carries is a guide or limit to total spending for defense, the 050 function in the President's budget. Thus, the budget resolution does not directly correspond to the defense appropriation bill. For FY 2002, the defense function was actually funded by 8 different appropriation bills, notwithstanding that almost all of it (99.9%) coming from the defense appropriation bill (92.8%), the military construction bill (2.9%) and the Department of Energy bill (4.2% for defense related atomic energy purposes). Nonetheless, the budget resolution allows us to see how much

spending is planned for the total defense function. This may be compared
to the display by function in the federal budget to provide a rough judgment about how good a guide the BR is for the budget process.

Next we delve into at how well the budget resolution disciplines Congress
by comparing the budget resolution to the defense authorization bill. In a
few years the outcome of the budget resolution and the defense authorization bill show a high congruence. What this means is that the party that controls Congress have maintained their discipline. To be useful in the budget
process, the budget resolution must come near its intended deadline.

For the 1990s, the budget resolution seems to have served as an effective congressional tool. In six of ten years, it was close to the final total
authorized for defense. In four of these ten years, more congruence may
be found between the budget resolution and the authorization bill than between the President's budget and the authorization bill. One of the stories
of this decade seems to be the rise in importance of the budget resolution.
Conversely, during the period 2000 to 2012 for most years we would draw
the opposite conclusion. We explained this period as a time of sometimes
extreme turbulence, change, and improvisational budgeting routines. The
budget resolution needs to be further interpreted before it is a good guide
for the appropriations process. This must be done by the Appropriations
Committees in the House and the Senate. These committees are given a
sum of discretionary budget authority by the budget resolution and it is
their job to divide it among the 13 appropriation bills. The individual bill
targets are called 302b targets. The 302a amount is the total amount provided in the budget resolution to the appropriations committees.

Taken altogether, changes in the annual Congressional budget process
from 1990 through 2011 often have been striking. First, the budget resolution began to operate on a hit or miss basis. The Congressional Budget
process, as reformed in 1974 and afterwards,[2] provides agents and a process
to set overall goals and targets. (Henliff and Keith, 2005: 14, 24–25) In addition, the Budget Resolution sets sectoral allotments (defense, education
et al.) and often carries sense of Congress provisions, e.g., that Congress
ought to pursue a balanced budget, and stipulations about process including paygo rules, spending ceilings and firewalls.(Heniff, 2005: 23–24) However, passage of the BR has become a highly uncertain event. In 2002, 2004,
and 2006 the House and Senate could not agree on a budget resolution.
In 2003 and 2005 they did. In 2002, the Senate failed to complete a budget
resolution for the first time since 1974.[3] In 2004, the Senate again could not
bring itself to agree with the more conservative House budget policy outlines, although it did produce and pass its own budget resolution. (Heniff,
2005: 5) In 2006 the Senate attached a "deeming" resolution to the spring
defense supplemental for use as its budget resolution.

Some might argue that since budget resolutions do not get passed every year, they must not be very important. This does not seem to be case. Had it not failed in two of four recent years, the case could be made that the Budget Resolution has become increasingly important. For example, from 1997 on, the number of declaratory statements has increased (it is the sense of . . . that . . . a balanced budget should be sought). It seems clear that the budget resolution has come to provide useful guidance not only for Congress in its aggregate spending, taxing, and deficit totals and its future year fiscal policy expectations, but for the executive branch, capital markets, and the public. Testifying before Congress, Bill Frenzel, Guest Scholar at the Brookings Institution and former member of Congress explained: "Setting that agenda and work plan is one thing the Budget Act has accomplished. There have been some years in which the Congress works on little else but the Budget and the Appropriations Bills that flow from it. When no Budget is passed, Congress just gulps and then revs up the spending machine."

Much of the criticism of the Budget process is that it has overwhelmed the legislative process. To me that is a positive development. The budget provides coherence and order to the process. The legislative process needed some order and discipline. Prior to 1974, each committee worked on whatever it felt like working on, unless the majority leadership could persuade it to handle pressing issues. The result was not exactly whimsical, but neither was it in any sense orderly." (Frenzel, 2005) Also, Allen Schick credits the budget resolution with driving major tax cuts; of the five major tax cuts in the last 25 years, two have taken place during this period of budget system turbulence:

> In general, the policy changes voted in the budget resolution have been more dramatic on the revenue side of the ledger than on expenditures. Major changes in revenue were triggered by the budget resolution in 1981, 1990, 1993, 2001, and 2003. Smaller changes were driven through Congress in half a dozen other years." (Schick, 2005: 6)

From our perspective, losing the budget resolution targets for bringing along the individual appropriations bills would be a setback, but only that. The far more important loss would be to the overall guidance in the current and future years of overall fiscal policy. History would suggest that the budget process will be rescued, that it is only a manifestation of current political divisions, but this is not a certainty. Both chambers have processes to 'go it alone,' and have done so, with the House using a deeming resolution to deem its budget resolution as the Joint Resolution when the Senate has failed to act (e.g., in 2002), and the Senate using its own budget resolution or budget committee marks and seeming to depend on 'winning' in the end game with scrutiny of the individual appropriations bills or in support or opposition to individual tax policy changes at the conference committee stage.

There also have been some other important process improvisations in-dependent of the BR. First, in the 2000s most of the budget process and spending controls established in the 1990s elapsed. These include paygo provisions to control entitlement spending growth, and spending caps on discretionary appropriations, mechanisms that came into the budget pro-cess with the Budget Enforcement Act of 1990 and a budget agreement of 1997 thaat gave discipline to the budget process during the late 1990s. Thus not only did the congressional budget process process become more uncertain after 2000, it became far less disciplined. Recently efforts have been made to reinstitute some of these provisions, but the jury is still out on their effectiveness. Also, the House made efforts to reform the committee structure so that by 2005 the two chambers were operating with a slightly different committee structure. The patterns of the recent past indicate that Congress is adept at improvising, with consolidated appropriation bills or by attaching what used to be carried in separate bills to conference commit-tee reports for another bill at the conference committee vote stage. What this asymmetrical structure does is make it harder to compare budget leg-islation in the House and Senate, thus further decreasing the transparency of the budget process.

Another change that had to be absorbed was creation of the Homeland Security department and the corresponding appropriations bill to fund it has affected the structure of Congress. At first, everyone wanted a part of homeland security and the White House identified 88 committees and subcommittees of Congress that exercised some authority over homeland security policy in 2002; in the House this included ten of 13 appropriations subcommittees.(Scardaville, 2002) This was rationalized somewhat in the appropriations process with a homeland security appropriations bill and homeland security subcommittees for the House and Senate Appropria-tions Committees. All of these changes have impacted the appropriations process and have had to be digested.

Turbulence between Authorization and Appropriation Committees

Another way to assess how Congress provides coordination and control for its disparate elements is to examine the timing between passage of au-thorization and appropriation bills. Traditionally, the authorization bill, at least in the eyes of the authorizers, provides critical guidance for the ap-propriators, particularly on the macro-items in the budget, e.g., how big should the military be, how many carriers should the Navy have, what is the right balance between strategic and tactical air, does the pay and benefits system sufficiently reward uniformed members of the services, and is the retirement system adequate?

When the authorization bill is passed later than the appropriation bill, the timing of the guidance is impaired. Simultaneous timing allows the authorization bill to provide timely guidance for the appropriators on defense policy questions. Nonetheless, it appears that the authorizers cannot be depended on to provide timely guidance for the appropriators. To some extent this failure occurs because the authorizers also see themselves as giving the President guidance over programs as well as treaties, strategic deterrence, and peacekeeping activities, notification of Congress when American troops are about to be committed and so on. In some years the authorization bill stalls due to issues with the President and thus sometimes the authorizers are caught between an internal guidance role to the appropriators and an external guidance role to the President. Their average outcome is hard to interpret, but it should be remembered that guidance on a thorny issue successfully given one year might last for many years.

However, even when the authorization bill is late for a good reason, this is still a source of turbulence for the appropriators who might look to the authorizers for guidance as well as for the authorizers who want the appropriators to pay attention to their guidance.

Inability to Budget: Continuing Resolution Appropriations Turbulence

One of the most telling indicators of legislative turbulence is the passage of Continuing Resolution Appropriations (CRA). When Congress fails to pass an appropriation bill before the fiscal year begins, it must enact a continuing resolution appropriation to fund agencies until the full appropriation is passed. The funding level may be set at last year's rate or any rate that Congress desires. These are usually passed quickly and are not controversial. They usually preclude new program starts and new initiatives. In defense this sometimes penalizes procurement programs that get off to a late start and thus may run into cost overruns due to delays in production or contract execution.

Budget Process Inversions

The events described above characterize an appropriations process under stress, one that may be said to be inverted. Unusual events are occurring and usual events are happening out of the usual order. This has been manifested in several ways. First, when DOD appropriations bills are one of the last bills to be taken up in Congress or are passed very late after the beginning of the new fiscal year, DOD suffers. After 2001, the DOD appropriations bill went first, and sometimes it was almost the only bill passed on time. However, this did not mean the authorization bill was accorded the same level of precedence.

The task of the authorizing committees is to build programs and authorize changes, particularly in personnel levels and benefits and in procurement. Authorizers like to think of themselves as giving guidance to the appropriations committees, therefore a normal or 'good' process should exhibit a pattern where the authorization bill precedes or parallels the appropriations bill.[4] This happened in calendar 2001, after the 9/11 terrorist attack. In subsequent years the authorization bill has been dramatically behind the appropriations bill. For comparison, in the 1990s the authorization bill preceded the appropriations bill twice, was signed on the same day twice, and was within ten days twice and 20 days twice in ten years; thus 80% of the time in the 1990s, the authorization bill could be said to be close enough to the passage of the appropriations bill to provide guidance. (McCaffery and Jones, 2004: 164) Since 2000, only in calendar 2001 was this same profile evident; this was a year when Congress wanted to make a statement of business-as-usual.

A second sign of inversion of the budget process is when the DOD appropriations bill is been passed on time, e.g., in 2003 and 2004; it was passed on time only twice from 1990 through 1999. This is an inversion in Congress' normal way of doing business.(McCaffery and Jones, 2004: 164) A third sign of inversion is the failure to pass the rest of the appropriations bills other than defense; eventually these are considered must pass legislation because the government must be funded, but in this era after 2001 the majority of appropriations bills have been passed as consolidated appropriations bills, used because Congress could not resolve policy conflicts, despite the consensus on defense.

Later in this book we argue that the use of consolidated bills and turmoil around the budget resolution have resulted in a diminished budget process and less efficient spending outcomes as the prevalence of earmarks has increased and aggregated to a level where important sums of money are allocated without a useful priority setting process. (Doyle, 2011; Savage, 1999). We argue that this new budget era has led to significant opportunity costs and has contributed to the accumulation of annual budget deficits.

A fourth sign of inversion concerns the dramatic changes in defense supplementals. These used to be small; now they are large. Moreover they are now provided inside and outside of the appropriations bill. When inside the DOD appropriations bill, they are designated emergency funds and not necessarily called supplementals in scoring, but DOD spends them the same way. Also supplementals have been supplied for defense and homeland security issues (i.e., these portions are spent outside of DOD) and some defense appropriations bills have also carried appropriations for other agencies. In 2004, Congress included in the regular appropriations bill $25 billion in emergency funds for early FY2005 costs of operations in Iraq and Afghanistan and another $1.3 billion for such activities as wildfire firefighting in the Western states (mostly California), humanitarian assistance for the Sudan, Iraq embassy security and conventional security funds for Boston and New York.[5] In the 1990s these would probably have been done in a separate supplemental bill. Also the $40 billion 9/11 supplemental was passed swiftly, but spent out slowly and only a small portion of this went to DOD ($14 billion in 2001 and $3.86 billion in 2002 (Carter and Coipuram, 2005: 24–25), although it was all intended for national defense and homeland security.

Another complication involves the contingent emergency designation. In 2002, Congress passed the summer supplemental for $28.9 billion, but designated $5.1 billion as contingent emergency spending, leaving it to the President to accept some, all, or none of the $5.1 billion. Within two weeks of signing the original bill, the President turned down all the $5.1 billion in contingent emergency funds. (CQA, 2002: 2–40) However, since this time funding for warfighting has come exclusively from supplemental appropriations, although in reality spending in support of some warfighting has often been drawn from the base budget, e.g., money to finance U.S. operations against Libya in 2011.

A fifth inversion occurred in 2003, when a supplemental was tapped to provide money for the main defense appropriations bill. The April 2003 supplemental created an Iraqi Freedom Fund of about $16 billion. Meanwhile Republican appropriators shifted about $3 billion from the Bush defense budget request to domestic discretionary spending, but this left the defense bill short and later in the summer appropriators shifted $3.5 billion out of the IFF to fill out the defense appropriations bill. (CQA, 2003: 2-83, 2-27) The result here was that the funds meant for reconstruction of Iraq in the wartime supplemental ended up increasing domestic discretionary accounts and ongoing operations within DOD normally funded by the main appropriations bill. Also in 2003, there were two supplementals, one in the spring and the other in the fall. The first paid for ongoing operations; the second appears to have contained funds for future operations. The result of all of this turbulence makes it hard to track supplemental and emergency funds and the boundary between normal DOD operations and emergency and wartime operations (normal appropriations and emergency and supplemental appropriations) has become blurred as has the boundary between defense appropriations and total spending for defense, homeland security, and foreign aid and assistance efforts in support of the war on terrorism. And finally, some things happened out of sequence. In passing the consolidated appropriation bill in February 2003 to clean up the 2002 appropriation process, Congress provided DoD with $10 billion in emergency funds mostly for intelligence matters; this normally would have been included in a supplemental bill, for example the one which would be passed in April of that year. In addition, the consolidated bill in 2003 also carried mandatory funds (mainly for the HHS accounts) of $397.3 billion. Since this included a small increase in the mandatory programs, the consolidated bill also had to pay for it with a paygo provision cut of .0065 across the board for most discretionary accounts, including those funded in the consolidated bill and in the previously passed appropriation bills. (CQA, 2003: 1–5).[6] All of this illustrates a period of great stress for the budget process.

Pork in the Budget Wars

Each year the defense appropriation bill funds projects of dubious merit, inserted at the request of lobbyists, state and local governments, and various contractors who profit from federal spending. These projects tend to be location-specific and highly visible. In the non-defense area, they take the form of courthouses, highways, airports and the like and are traditionally referred to as "pork." Procedurally, a project is inserted in an appropriation bill as an adjustment to a specific line-item; hence, it may also be called an "earmark." Pork, according to the Webster's dictionary, is a government appropriation that provides funds for local improvements "... designed to ingratiate legislators with their constituents." Some observers argue that pork is a characteristic of democracy in action, that pork projects are part of the price of passing legislation; votes for a bill are bought by inserting items of benefit to particular members or groups of members so that they will support the bill. In this way, the interests of the many are protected by making a side-payment to the few in order to provide the necessary votes to pass the more general measure.

Defense also has it share of pork barrel projects. Sometimes these items are related directly to defense, sometimes they are not. Further, it is important to recognize that much pork spending comes at the request of DOD and contractors. And significant amounts of pork funding takes place outside of defense. Schlecht (2002: 1) found these bits of non-defense pork spend in the FY 2002 defense appropriation:

$1.3 billion for environmental restoration.
$7 million for HIV prevention in Africa.
$150 million for the Army breast-cancer research program.
$85 million for the Army prostate-cancer research program.
$19 million for international sporting competitions.
$3 million for aid for children with disabilities.
$1 million for math-teacher leadership.

$4.5 million for a cancer research center.

$2.6 million for the Pacific Rim Corrosion Project.

$2 million for the Center for Geo-Sciences.

Schlecht concedes that some of these programs may be worthy of funding, but that none of them can be said to be directly related to the defense of, "... our country or the safety of our soldiers on the modern battlefield," and finds that, "Such pork-barrel spending wastes billions of taxpayer dollars and undermines the combat effectiveness of our military." (Schlecht, 2002:1). As a percent of a $2 trillion budget, the amounts cited above may seem trifling, but these dollars accumulate into the billions of dollars, as we will explore later, dollars misused or applied to priorities that do not serve well all or even large numbers of the people of country.

One of the watchdogs of wasteful spending is the Washington DC based institute, Citizens Against Government Waste (CAGW). This group was founded in 1984 by the late Peter J. Grace, Chairman of President Reagan's Commission on Cost Control (Grace Commission), and by newspaper columnist the late Jack Anderson. CAGW has grown to a membership of more than one million. CAGW refers to itself a public interest lobby against government waste. Every year it issues a summary of pork projects called the Pig Book. CAGW has defined pork as "... all of the items in the *Congressional Pig Book Summary* that meet at least one of CAGW's seven criteria, but most satisfy at least two:

- Requested by only one chamber of Congress;
- Not specifically authorized;
- Not competitively awarded;
- Not requested by the President;
- Greatly exceeds the President's budget request or the previous year's funding;
- Not the subject of congressional hearings; or
- Serves only a local or special interest." (Citizens Against Government Waste Pig Book, 2005: 3)

For CAGW, process is important. A project that is requested by the President, subjected to Congressional hearings, is specifically authorized and appropriated, serves a general interest and is competitively awarded might avoid the label of pork.

Pork Funding

Where does the money come from for these pork projects? Ultimately, of course, it comes from the taxpayer, but in the immediate case Congress finds ways to provide the money, sometimes out of thin air. A study by Wheeler (2002) of the FY2002 military construction bill found that the military construction sub-committees of the House and Senate Appropriations Committees used two gimmicks to provide dollars for such projects as museums, gyms, warehouses, water towers, day care centers and so on. First, they assumed that the value of the dollar would rise against foreign currencies, thus making projects in foreign nations cheaper, by about $60 million. Second, they wrote into the MILCON bill a 1.127% across the board cost reduction in all projects in the bill. This appeared to save $140 million and provide space for additional projects while keeping the bill under the $10.5 billion cap allocated the House and Senate Appropriations Committee chairmen for the bill. Wheeler found that California was the number one beneficiary of these tactics, gaining $25 million for the 8 projects added that were not requested by the President, followed by Texas and West Virginia. Wheeler notes: "...the top eleven recipients of added military construction projects consisted entirely of senior Democrats and Republicans in the House or Senate who just happened to be the sitting Chairmen or Ranking Minority Members of the Appropriations and Armed Services committees and subcommittees who handled the Department of Defense and its military construction budgets." (Wheeler, 2002) Writing in the fall and winter of 2001 and in the aftermath of the 9-11 terrorist attack, Wheeler was concerned that spending on pork would divert funds from reaching necessary defense items. He also studied the defense appropriation bill for FY2002 and found Congress using gimmicks to reduce O&M funding in order to make space for projects they favored.

Pork barrel spending also manifests itself as over-investment in certain weapons systems produced in the home state of a powerful committee chair, floor leader, or long-term defense authorization or appropriation committee member. Here, the pork involves over-production of missiles, or aircraft or whatever. They have a clear defense benefit, but too many are bought; that is what makes them pork.

Sometimes pork battles are not about the weapon system *per se*, but about in whose district they will be built, i.e., who will be the beneficiary. Other types of pork emerge in the budgetary process as typified in omnibus CRAs that are filled with additional funds for many programs and activities that have nothing to do with defense.

How may pork be identified? In this chapter we have provided a number of ways to think about pork outcomes, in terms of its cost, where it is spent, to whom the benefit accrues and to what extent the good provided is a public good and can not be provided by the private sector. The essential problem with pork is that it is funded by tax dollars drawn nationally and dedicated to projects of somewhat dubious merit or low priority where the benefit could be quite local. In some cases, the project is a substitute for something the private sector could do. Many defense functions are close to pure public goods and can only be supplied by government. Others are not. Annual defense funding exhibits a mix of true public goods of high merit and some not so meritorious projects that support local interests.

Earmarks and Pork

Study of changes to DoD appropriations bills indicates that the pork question is more complex than is normally thought for defense. Some projects clearly benefit local areas and are at the request of the local members of Congress. Some clearly are of national interest, even if at the request of local representative. What is commonly called pork gets into DOD bills because Congress keeps an eye out for things that will benefit them and their re-election records; however DOD also campaigns for 'pork.' The battle of the budget does not stop inside the Pentagon when the President's budget is printed. When the President's budget is presented to Congress, the battle for congressionally inserted pork has just begun, but the struggle within the pentagon has been going on for some time, at least since the spring of the previous year with the beginning of the military department budget process and goes on through the congressional session as the services campaign to get things included in the appropriations bill right up to the vote on the conference committee report on the appropriations bill.

A MILDEPS may appeal on provisions in the appropriation bill through conference report deliberations, if the treatment of an item in one chamber differs from that in the other. Then too, DoD testifies, responds,[7] informs, and educates members of Congress all during the Congressional budget process. Inside DOD, various unfunded priority lists are created and maintained and when it appears Congress is interested in funding something close to a Unmet Priority List (UPL) item, DOD representatives will attempt to get the item torqued toward the UPL list item and fully costed with language inserted that gives DOD some flexibility in how it administers the program. These ideas may be surfaced at various committee venues, conversations, and in responses to questions asked by Congress or, as politics in Washington go, at a variety of social settings. Thus, while Congress appears to take the blame for all these earmarks, it is clear that that they are not necessarily totally congressional; DOD may find a sponsor to carry a needed project.

Pork projects are not necessarily taken from SECDEF's UPL either. Pork may also result when a member of Congress or a staffer visits anything from an aircraft carrier to an obscure military base and is given the routine command brief, a brief which just happens to include a need which may have been denied in the DOD budget process. This need will help the base and the local area, so it is no wonder it strikes the member of Congress as a good idea, one which he may take up and pursue at the committee stage in hearings and markups. Thus for a base commander who is knowledgeable, there are wrestle back provisions in the budget process. He or she may lose within DOD, but what is to prevent him from salting the commander's brief for a visiting representative from Congress with a 'nice to do' project that did not make it past the next step upward in the administrative budget process?

If the Senator or congressman wants to take on an issue as a personal item and add it to the appropriations bill later, how is the base commander to protest? In the Pentagon, this item may not be anywhere close to an important priority and did not make it in the budget request, but if it appears that the Congressman has a good chance of getting it included in the appropriations bill, DOD will work to see that it is fully costed (if a new building, that maintenance and personnel costs are included) and that to the extent possible the item is changed to help include some item that is a DOD priority. From the DOD perspective, these earmark plus ups are of a lower priority than the items Congress is pushing out of the appropriations bill to make room, thus from DOD's perspective Congress is substituting inferior for superior goods. It is probable DOD will have to ask for some or most of these excluded items in the following appropriations cycle, or perhaps in the supplemental bill.

Also, we have said nothing up to this point about the defense industry and individual corporations, yet it is clear from our earlier discussion of 'iron triangles' the defense corporate world is an interested and eager participant in the policy process, especially in the procurement and the RDT&E and procurement accounts and in some cases the development, production, fielding and continued sales of new weapons systems. This is, after all, where the money is. Here pork manifests itself when Congress holds production lines open when DoD recommends it already has enough of a certain weapon system or when Congress inserts additional numbers of an aircraft in the appropriation bill even though DoD has not asked for any in its request. Naturally the affected corporations and their subcontractors and suppliers are only too happy to let individual Senators and Congressmen know how important these systems are to the national defense effort.

Recently these efforts have started when the budget is still in the military departments in the late summer, before the POM and budget have been finalized. For weapons systems, it is particularly important to win in the POM, since the POM controls the future year defense plan and may well be the difference between five fat years and a series of years so lean as to put the company out of business (or into someone else's hands). With stakes this high, it is a small wonder that any decisions are taken on merit. In our judgment, the majority of the DOD appropriation is presented and passed on merit, but there is a large and seemingly growing part of it that seems to respond to the rules of pork, inserted in appropriations bills through earmarks.

MACRO-CONCERNS OVER THE DEFENSE BUDGET

Another area to study for budget process inversions is how Congress treats the President's defense budget request. Conventional wisdom suggests that Congress is a marginal modifier of Presidential budget requests and will support wartime budget efforts of the President. The closest analogous historical period to the current era is the Viet Nam era. Below is a record for that time of what the President requested and how he was treated in Congress.

The bills for fiscal years 1965 and 1974 were constructed and passed in a peacetime environment. Calendar 1965 marks the first war year; note the high levels of support in Congress for the war effort in 1966 and 1967. Thereafter, Congress began to make substantial cuts to the DOD bills. Conventional wisdom says that consensus around defense disappeared in the late 1960's; what this profile illustrates is that Congress modified the President's budget before Viet Nam (FY1965) and afterward (FY1974); supported the war in the beginning and became disenchanted later (FY1969 et seq.). The first Nixon bill would have been in 1969 and Johnson would have signed the 1968 bill, as a lame-duck President who chose not to run again because of the war issue. On average, the President received about 97% of what he requested in this period.

In the modern era, the President seems even more successful. On average, the President received over 90% of what he requested from 2001

TABLE 5.5 Wartime Budget Request: Vietnam

Signed[a]	FY	President requested	Congress gave	President received	Congress impact
8/19/1964	1965	49.014	47.22	96.34%	−3.66%
9/29/1965	1966	47.471	46.752	98.49%	−1.51%
10/15/1966	1967	45.248	46.887	103.62%	3.62%
9/29/1967	1968	57.664	58.067	100.70%	0.70%
10/17/1968	1969	71.584	69.936	97.70%	−2.30%
12/29/1969	1970	77.074	71.869	93.25%	−6.75%
12/28/1970	1971	75.278	69.64	92.51%	−7.49%
12/15/1971	1972	73.543	70.518	95.89%	−4.11%
10/13/1972	1973	79.594	74.372	93.44%	−6.56%
12/20/1973	1974	77.2	73.7	95.47%	−4.53%
			Averages	**96.74%**	**−3.26%**

[a] Data derived from Congressional Quarterly Almanacs from 1964 to 1973.

through 2010, excluding one year (2004). Often during this period Congress added funds to the Presidential request. Conventional wisdom suggests that Congress is a marginal modifier of executive budgets; that three months debate over a budget resolution and countless hours of hearings, testimony, and debate over the appropriations bills beginning in February and ending in November or December, or in exceptional cases in January or later, all result in very small increments of change to what the President has requested. However, a closer examination of the budget process reveals that conventional wisdom may be incomplete. Speaking about budget reforms Schick comments:

> The present role of the President is informal and political, and arises out of the fact that he can veto appropriations and reconciliations bills, as well as other budget-related measures passed by Congress. The President already exerts considerable influence on congressional budgeting, and in some years he is the dominant player. The exuberant hopes of 1974 that the budget resolution would be a declaration of congressional independence from the White House have been dashed by the realities of American politics. Yet, even as a political partner, the President does not get all that he wants. (Schick, 2005: 12)

In fact, one year in the 2000s (2004), the President got about half of what he wanted for defense. Long ago, Aaron Wildavsky defined the concept of budget base: "Base is the general expectation among the participants that programs will be carried on at close to the going level of expenditures..." (1964: 17) Wildavsky also noted:

> Budgeting is incremental, not comprehensive. The beginning of wisdom about an agency budget is that it is never actively reviewed as a whole every year.... Instead, it is based on last year's budget with special attention given to a narrow range of increases or decreases. Thus the men who make the budget are concerned with relatively small increments to an existing base. (Wildavsky, 1964: 15)

Research since Wildavsky focused on incremental change and generally found the executive getting most of what he wanted in the budget with the legislative branch playing the role of marginal modifier.[8] This is not an accurate picture of the recent budget process. In 2004, after the legislative process had finished, of $24.3 billion in year to year change in defense appropriations, the President got half and Congress got half; if the cut is scored with the money Congress controlled, then Congress directed $14.1 billion of the $26.2 billion requested, or 53.8%. Moreover, when cost of living adjustments and fuel and medical care cost increases are counted, they may well soak up the rest of the year to year change in the President's share. Thus it is possible that Congress, so-to-speak is dictating almost all of the programmatic year to year change, but doing it in a piecemeal fashion, earmark by earmark.

For students of congressional government, what we report is both good and bad; good in the sense that all that effort by Congress actually leads to important outcomes, but bad because of the nature of earmarking process. Earmarks set funds for a specific purpose, use, or recipient. These can be useful additions or specifications for budget spending, but the process lacks transparency and this means that it is hard to know what the earmark does and what is its benefit?

Earmarks as Inversions

Earmarks are inversions in the budget process because they replace the logic of the many with the preferences of the few. Austin Clemens (2005) of the budget watchdog organization *Taxpayers for Common Sense* analyzed the FY2005 defense appropriation bill and found that most earmarks were added while the bill was in subcommittee, committee, or conference committee. In the Senate version of the appropriations bill, Clemens attributes 65% of the earmarked funds to members of the appropriations subcommittee on defense, 11% to other appropriations committee members and 24% to Senators not on the appropriations committee. Committee leaders did better than members. States with representation on the appropriations committees did better than states without; this was particularly true in the House, where states with representation on the subcommittee averaged $316 million and states with no representation on the committee (either full or subcommittee) averaged $22 million. (Ashdown, 2004) Clemens worries about this process because of the lack of review to determine the merit of these earmarks and because "very few of them are discussed in detail in the bill, contributing to poor transparency in the appropriations process." (Clemens, 2005: 2) A further irony is that the process shows Congress as very active in changing Presidential budget requests on a piecemeal basis while backing off from a harmonized plan to guide its changes, the budget resolution.

Some may believe that the fact that earmarks escape control at the budget resolution level is all right because earmarks do not amount to much money. This was perhaps true in the past; it is not so now. The FY2005 $12.2 billion earmarked in the defense appropriations bill adds up to a significant amount of program space. If the earmark total belonged to a country, it would have ranked 14th in the world in defense spending in 2000. In fact, the U.S. Congress earmarks more money in defense than Israel, Canada, Australia and Turkey each spend on defense. (McCaffery and Jones, 2001:72)

On a total budget basis, *Citizens Against Government Waste* (2009)[9] has tracked earmarks and finds a considerable growth in their number since 2001 when there were 8341 earmarks worth $20.1 bill to 2004 where there were 13,997 earmarks worth $27.3 billion; this a growth of 67.8% in number and 35.8% in dollar value in four years. This amount is more than the budgets for such agencies as Commerce, Energy, Interior, State, EPA, NASA, and NSF measured against their 2004 outlays. (Office of Management and Budget. 2005: 76) In terms of the power of the budget resolution to allow Congress to set the increase at the macro level for domestic discretionary spending, Allen Schick says, "Through the budget process, Congress has effectively decided the annual amount of increase in discretionary appropriations." (Schick, 2005: 9)

Our perspective is from the micro-analytic side. Our analysis shows Congress not only dictating the amount of change permissible under the budget caps, but also going inside the different appropriations bills and, using the earmarking process, pushing out items requested by the President and replacing them with items desired by some or a few members of Congress. (Ashdown, 2005)[10] When Congress cannot find enough items to throw out of an appropriations bill, it uses across the board reductions in one or another accounts, based on supposed savings or improved methods, or reduced inflation expectations or some such pretext.

Earmarks are not new. Clemens traces earmarks in the 1969 and 1979 defense appropriations bills where there are many fewer earmarks, but substantial sums of money–$5.6 and $8.9 billion, respectively. (Clemens, 2005: 5). Senator John McCain (R-AZ) has been on an anti-pork crusade for some years, (McCain, 1997)[11] but neither he nor Citizens Against Government Waste or Taxpayers for Common Sense seem to be making much of an impact.

Specific events during wartime over the past decade have underscored the complexity of the earmark issue. From 2003 on the Pentagon has moved more slowly than Congress has wanted in delivery of such force protection items as armored humvees and other armored vehicles and Congress has intervened, for example, by increasing funding for Humvees by $5.2 billion more than the Pentagon requested from 2003 through 2007. (Morrison, Vanden Brook and Eisler, 2007: 1–2) Compared to total DOD procurement spending for this period, about $470 billion, the $5 billion was not much, but it was helpful, it did speed armored humvees on their way to the battlefield (Jones et. al., 2010), and it seemingly helped change the mindset of the acquisition bureaucracy early in the process. Of the episode, the staffer's boss Representative Duncan Hunter (R-CA) said, "The acquisition bureaucracy was not in the war…We have an acquisition bureaucracy (at the Defense Department) that just doesn't respond quickly enough to the needs of the warfighter." Senator Mary Landrieu (D-La.) originally sought an earmark to help a company in her state that produced armored police vehicles (Textron's ASV) which proved highly resistant to explosive devices. Landrieu pleaded with the Army to use the vehicles, but the FY2004 budget included no money for the vehicles: "They send me an executive budget with this zeroed out," Landrieu recalls, "I hit the roof." She kept the program alive by inserting earmarks in appropriations bills. Says Landrieu, "They were getting ready to shut the line of these vehicles down completely.... It was on its last breath and I literally was so determined (that) I said I was not going to allow it." For the next two years Landrieu helped insert $700 million for the vehicles in the DOD appropriations and kept the program alive even though the Army budgeted nothing for it until FY2007. Looking back, Landrieu says, "I can say in my life there's one really great earmark."

Defense experts warn that congressional intervention imposes a problem in terms of intervention into both the DOD and congressional budget process. Rather than letting the DOD acquisition process work to reveal the best solution and lawmakers to add money for real needs to the DOD budget, it has to come from somewhere, and in doing so it is shortchanging the development of future weapons systems to pay for immediate needs. The same tension exists within the DOD programming and budget process; future weapons may be sacrificed when the emphasis shifts completely to what the warfighter needs now. Outside observers are often frustrated with the pace of change as well as the decisions that the Pentagon makes; thus, in some respects the earmarking process may function as a pressure relief valve, doing some good in the short run. However history has taught us that defense is too complex an arena to be managed by committee and the decision process needs to honor the long run as well.

It is no coincidence that since 9/11 during an era of large appropriations for defense has led to an increasing use of earmarks. Passing large bills often is a case of taking the bad with the good; in this case the bad is extensive earmarking. Defense is often considered veto-proof because of its size and mission, but its size and status are also an invitation for earmarking. The large consolidated appropriations bills were also an invitation to earmarking due to their size and to the fact that they were the best compromise going between a bad bill and no bill at all. The result seems to be a degradation of budget process outcomes, because earmarking is adding up to sizeable amounts of money through aggregation of smaller sums, thus escaping both the discipline of Presidential budget review at the department and OMB level and Congressional budget review at the macro-level in the budget committees. Anyone sensitive to the concept of opportunity cost can not help but be critical of this outcome. With the earmarking process, there is no choice of spending $27 billion more on education or environmental protection or defense, rather bits and pieces add up to an unanticipated, but large, total, as if some manic grocery shopper were filling his cart with all sorts of things that might be useful individually, but not add up to a well-balanced diet.

Given the size and frequency of pork dollars and their persistence despite well-intentioned efforts to rectify the situation by some leaders and some followers, pork persists, as if to say "we won and now we will enjoy the spoils." In any case, pork is a fact of life and so it seems is the earmarking process despite valiant efforts to the contrary. In conclusion, pork barrel spending is a two-sided coin. On one side, it is wasteful spending, on the other local economic development. On the definition of pork, clearly where you sit determines where you stand. (Jones and Bixler, 1992: 11–12; see also Wildavsky, 1988: 42)

THE EFFECTS OF PARTISAN POLITICS ON CONGRESSIONAL BUDGETING

Part of the current crisis around budgeting and the debate over deficit control in 2011 and 2012 is due to policy conflicts and part is due to process problems. Almost all of this has been seen before, in the Viet Nam era for example, or the budget debacle of 1995–96. What has not been seen before is the consistent use of large consolidated appropriations bills accompanied by a growing tendency to earmark funds. Earmarks have been around before, inside and outside of the defense bill, but the size and critical nature of the defense bill makes them difficult to stop. The same things happen in the large consolidated omnibus appropriations bills.

These outcomes are a setback for rational budgeting and efficient allocation of resources for public purposes. It is clear that this earmarking process needs to be brought within the budget process discipline. Prospects for a good outcome in this respect are not good. The earmarking process does not appear to be a triumph of the rank and file over the leaders, or a repeal of committee decisions by floor events. Rather it appears to be a slackening of the discipline of the 'cutters,' those committees and individuals interposed in the appropriations process to control the appetites of the spenders. This is a serious weakening of the appropriating process. Moreover, if it is allowed to continue in obscurity, then it may not necessarily be self-correcting. That is to say, if the guardians of the purse are the ones who benefit most greatly from earmarking (appropriations sub-committee and committee members and other leaders) then there is little reason to suppose that they will fight to restore the budget process to a more rational and open pattern.

What also may be appearing in the improvisational patterns of the budget process is a reflection of an increasingly divided country where what used to be just a matter of dollars has become a matter of principle and, not only that, but these matters of principal are transmitted to the national scene from increasingly safe districts whose representatives need not compromise to stay elected. All of this has been exacerbated by a prolonged recession, rising U.S. debt and political party conflict in the period after 2010.

It may be argued that the majority party has little need to compromise over budget issues so long as they reflect the values of their districts, districts which are becoming increasingly divided into two camps. Moreover, when party control of the levers of government is split, this also means that compromise will be harder to attain, since what is compromised may be seen as a matter of principal and not dollars. This could be a partial explanation for the wreckage of the budget process over the last decade.

When people in Red and Blue states have significantly different beliefs and these beliefs are rooted in values related to strongly held ideas focused on partisanship, ideology and/or religion, then the art of compromise is more difficult to pursue. This clearly affects the budget process which rested on a slowly moving consensus about the ends of society from the end of WWII until the mid-1960s and then again from the Ford Presidency 1974–6) until perhaps the mid-1990s, ending perhaps the Republican's Contract with America in 1995. During this time, the main outlines of American society were generally agreed upon by both parties, and in the budget process, legislators could decrease conflict by claiming issues were just a matter of dollars. Since this period, with the exception of support for warfighting in response to the terrorist attacks of 9/11, dissensus has become the political norm in the nation and in Congress.

TURF BATTLES IN DOD

Military departments compete with each other for missions and this can result in a public squabble both in DOD and Congress. The source of competition is mission responsibility. The military service that has the mission will get the money, and this may mean control of that mission and all the weapons it takes to perform that mission for the next decade. For example, the Air Force core mission involves tactical fighters and strategic bombers, the Navy core mission involves carrier groups, marine deployment groups, and ballistic missile submarines. The Army core missions revolve around infantry, artillery and helicopters and armored vehicles. In the 1990s, turf battles were seen in appropriation struggles over the Apache and Comanche helicopters, M-1 Abrams tanks and THAAD missiles. The Air Force concerns were focused on the B-1 and B-2 bombers, the F-15 Eagle, F-16 Falcon and F-22 Raptor fighter jets as well as the C-17 cargo plane. The U.S. Marine Corps is defined by their amphibious warfare abilities included within the LHD-7 and LPD-17 ships in addition to the AAAV and Bradley assault vehicles.

Turf battles arise where one service challenges another either for a peripheral mission or for a core mission. For example, thematic debates during the 1990s concerned such big ticket items as the V-22 Osprey, SDI, B-2 bomber, Seawolf submarine, aircraft carriers, tank production, F-18 E/F, and other costly weapons programs. Such weapon systems essentially define a branch of service and the cancellation of any one of these money-generating programs would put into question the purpose or mission of the respective branch of service using or defending that program.

The debates between military services over funding for major weapons programs are a source of heated discussion with service reputation, readiness and constituency employment implications hinging on the outcomes of the contested items. The emerging trend toward joint weapon systems is the natural result of spreading the support for a weapon system across services, resulting in a more stable platform made less vulnerable to cuts due to its widely distributed support across services and districts alike.

In 1999 a turf struggle occurred over funding for the US Navy F/A-18E/F Super Hornet and the Air Force F-22 Raptor. Each represented billions in potential new procurement dollars, in addition to representing its respective service's future strategic requirements. Since both were required for their respective services, they were both funded with the Super Hornet getting an increase from $1.35B to $2.1B and the Raptor received an increase of $331M. It may be that for the Navy and Air Force a similar conflict may arise over replacing Legacy and Super Hornets by 2013 with what has become an increasingly expensive Joint Strike Fighter (F-35).

In 2007 a turf battle erupted over control of the highly successful Predator and Global Hawk aerial drone programs with the Air Force pushing to become the executive agent for drones that fly above 3,500 feet. This move was opposed by the Army, Navy, and Marine Corps who feared that it would make the Air Force responsible for the acquisition and development of unmanned aerial vehicles for these latter services, including the Army's Sky Warrior program. (Financial Times, 2007) What the services fear is loss of control and a product which will not quite fit their missions as well as if they had sole control over its development. In any case, turf battles erupt almost every year.

CONCLUSIONS

It would seem that to a great extent in defense budgeting the phrase, "What goes around, comes around" applies all too often. We might ask, however, whether any of this is new? The answer is emphatically, "No!" In the 1790s, members of Congress complained about making lump sum appropriations with no accountability mechanisms in place, and "that is to say" clauses in appropriation bills that directed money to be spent on specific items or locations (McCaffery and Jones, 2001: 54–55). Members of Congress are elected by their constituents in part to "bring home the bacon." In this context, the appropriations for national defense provide great opportunity. Pork barrel spending and protection of turf are standard elements of resource competition in democratic political systems. As long as specific constituencies want their piece of the federal budget and defense portion of the pie, such dynamics will be a prominent part of the defense budgeting game. As we have noted elsewhere, some degree of strategic misrepresentation is taken for granted in budgeting in all contexts (Jones and Euske, 1991). To think it otherwise is naive relative to the norms, values and history of our budget culture.

In period 1990 through 2011 the normal complexity of the budgetary process was embellished with improvisational behavior by the participants. This made it hard to find similarities from one year to another. It is true the same products were produced, ultimately, including the budget resolution, the defense authorization bill and the defense appropriation bill, but the process which produced these was marked by great diversity from one year to the next. The President is always in the lead, but sometimes Congress hears a different melody, and like a headstrong dance partner begins to veer off in a different way and the result is improvisational budgeting.

Sometimes it appears as if the only worthwhile decision-forcing mechanism is the end of the fiscal year and in some cases eve this has not been enough to force discipline into congressional budgeting. Negotiators seem to take things more seriously the farther into and then beyond the fiscal year they get. Moreover, spending and taxing decisions are high-visibility decisions that can be the basis for a good re-election campaign. This means that legislators must make sure they get their position in front of their relevant public. It also means that while some debates are just a question of dollars and thus easy to compromise. Other budgetary issues including how to reduce or control spending, deficits and debt raise large philosophic questions and are not easily resolved. Inevitably disagreements on the federal budget eventually raise serious questions about the U.S. defense profile and funding. In facing complex and expensive budgetary issues the means for reaching traditional budgetary compromise are not easy to attain.

Any current analysis has to consider both the good and bad news about the effectiveness of the congressional budget process. The good news is that eventually defense budgets are passed and funding has increased through 2011. The bad news is that cuts in future defense budgets appear inevitable. At the same time it is imperative that the mission of DOD be reduced commensurate with budget cuts. Whether this will be the case over the next five years is a serious question.

In this regard, it appears that when the path to definition of defense policy is not clear, improvisational budgeting is an inexorable result. However, improvisation does not result only from dissensus; it also typifies consensual decision making. When consensus was present as it was after the attack of September 11, 2001, improvisation also occurred and Congress showed remarkable ability to move quickly in support of President Bush in prosecuting the war on terrorism through approval of funding for enhanced defense spending and wars in Iraq and Afghanistan.

The good news in the years following 2001 was the consensus on the need to fight terrorism on individual battlefronts wherever they arose in a war where U.S military action to counter terrorism increased around the globe. The bad news is that the consensus developed since 9/11 appeared to begin to unravel and deteriorate as the Presidential and congressional elections approached in 2012. Defense issues that had been treated in bipartisan manner prior to this point became increasingly politicized, dividing Republican and Democrats. The good news was that even with this partisan split, overall support for defense and the war on terrorism persisted. The bad news was that the gap in funding for replacement of aging military weapons platforms remained large after the spending increases from 2002 through 2011. The gap in overall weaponry to replace obsolete assets including ships, aircraft, tanks and other infantry weaponry, bomber aircraft or alternative delivery systems including non-piloted and space based weaponry remained significant.

The process employed by Congress to determine defense policy and budgets is highly disaggregated, disjointed, conflictual, competitive, and inefficient. National interests compete with local interests and in many instances the compromises struck result in less than optimal allocation of scarce resources. However, as has been demonstrated in this chapter and subsequent to the events of 9/11, Congress has been up to now capable of responding to contingency in a relatively rapid fashion despite a considerable degree of visible and not so apparent decision dysfunctionality. We would observe that such confusion and inability to cooperate among the members of Congress is part of the price paid for reaching agreement in policy and budget negotiation in democratic political systems.

Finally, what may we conclude with respect to alternative theories of budgeting? Is Schick's improvisational budget theory a better model than Wildavsky's incrementalism to explain congressional budgeting in the period 1990 to the early 2000s? Schick's argument has considerable merit. In fairness to Schick, he did not explicitly propose his theory as a better model than Wildavsky's. The competition between these alternative models is one we choose to assess.

Support for the Wildavsky model is evident. The overall budget of the federal government has continued to increase by relatively small increments. In addition, while defense spending did not increase consistently over this period, the cuts and increases to the defense budget during this period were never huge relative to the size of the base in any particular year. Cumulative decreases from 1990 to 1998 and increases from 2002 to 2011 were substantial, but defense (and federal) budgeting for those periods appears to fit Wildavsky's model. Both policy decisions and spending decisions in this period were made on the margin in adjustments to the base. Therefore, it would appear that quite a bit of evidence exists in support of Wildavsky's thesis. However, the congressional budget process and budgetary behavior during the period examined was very obviously improvisational. It would appear that Schick too is right. The process innovations in this period were not incremental—they deviated from the base as defined by how the budget process operated in periods of greater defense policy and budgetary consensus, e.g., the 1950s, or even in the more contentious and inflation plagued period of the 1970s.

How may we reconcile the perspective of these two scholars? It is our view that both scholars are right. The key conception supporting this judgment is the difference between a focus on substance versus process. Wildavsky based his theory on the evidence that budget numbers and the programs in budgets are modified incrementally from the base by Congress in annual budget deliberations. On the other hand, Schick observed the entire congressional budget process over more than a decade and found significant deviation from standard practice and process. Our examination of congressional budgeting supports Schick's assertion that to accommodate dissensus (a term Wildavsky apparently coined to describe budget disagreement) a pattern of significant improvisation and innovation emerged. However, our analysis adds to Schick's theory. We conclude that congressional budgeting since the 1980s has been improvisational in responding to contingency, whether the budget share for defense was increasing or decreasing, and despite whether the annual budget was in deficit or surplus. Consequently, we conclude that the theories of incrementalism and improvisationalism combined lead to a more sophisticated understanding of federal and defense policy setting and budgeting. Finally, we believe that despite what is written in law, budgetary life as it is practiced indicates that there is no model budget process. Rather, there is a responsibility to produce appropriations and Congress modifies the process it uses to fund the federal government to some degree almost every year to accomplish this end. However, in many years including 2011 citizen patience with this process grows thin before Congress eventually acts.

A September 2011 public opinion poll (New York Times–MSNBC, 2011) indicated that only 12% of U.S. citizens thought Congress was capable of solving the nation's debt and economic problems. This statistic was alarming. To some extent it could be explained by public frustration resulting from three years of recession with no real end in sight, and a variety of other problems including high unemployment and serious underemployment, a creeping loss of living standard in a vanishing American economic middle class, the high amount of poverty in the nation (more than 20% of the population was estimated to be living in poverty), the absence of recovery of the housing market accompanied by the increased pace of house foreclosures, worries about the financial soundness of some very large U.S. banks, the rising costs of health care, the deteriorating condition of K–12 education, the spiraling costs of higher education, the financial weakness of state and local governments, worry about the size and the consequences of dealing with what the public was told was an unacceptably large amount of federal government debt, concern about the fiduciary soundness of Social Security and other pension programs and plans, deep worry about the potential effect of debt default of financially weakened Eurozone nations, distress about the inability of the U.S. leadership to end two highly costly war operations, fear of the continuing threat from terrorists attacks in the U.S. and abroad, stress caused by a spate of school and other public shootings and public attacks by insane "lone wolf" heavily armed mass murderers—and added to all of this the horror from having to witness or for many to cope with a series of natural disasters and terrible weather in many areas of the country, and so on.

What was shocking about these poll results was the revelation of a profound absence of confidence in the ability of elected officials at the national level in Congress, and the President (whose confidence ratings also were down as measured by this poll), to solve the many serious problems faced by the nation. This degree of lack of confidence had not been seen over a prolonged period of time in the U.S. since the early years of the Great Depression (e.g., 1929–1933). From one perspective the degree of partisan political party bickering in Congress might have been enough alone to seriously discourage the general public about the reliability and integrity of members of Congress. From another perspective the poll results may be viewed as a more profound expression of frustration that always accompanies economic recession. Evidence from recessions in the 1970s, 1980s, 1990s and the early 2000s suggests this is part of the explanation for the unpopularity of Congress. And perhaps the most disconcerting aspect of the poll results was the evident feeling of hopelessness among a large portion of the U.S. population. While some observers referred to the words of Franklin D. Roosevelt spoken in his first inaugural address at a low point of the depression in 1933, "... the only thing we have to fear is fear itself" (Roosevelt, 1933) in attempt to steady public fears, it appeared that this message was not persuasive to a highly disturbed U.S. citizenry. Further, the prospect of deep federal government spending reductions encouraged some but worried others, depending upon their degree of dependence on the government and their political ideology about the appropriate role of government in the economy and society.

NOTES

1. See Allen Schick, The Capacity to Budget (Washington, DC: Urban Institute Press, 1990): 159–196.
2. Major reforms included the Balanced Budget and Emergency Deficit Control Act of 1985 (commonly referred to as "Gramm-Rudman-Hollings" and the Budget Enforcement Act of 1990 (BEA). These reforms were modified or extended through less comprehensive measures in 1987, 1993, and 1997. For FY1991 though FY2002, there were statutory limits on discretionary spending (e.g., defense, education et. al.) and paygo provisions for direct (mandatory) spending (e.g., social security and revenue legislation).
3. The Senate passed built and passed budget resolutions in 1998 and 2004, but it did not go to conference with the House and pass a joint resolution. This means the Senate itself had its own resolution to guide its appropriations work, thus 2002 remains unique.
4. The Congressional Quarterly Almanac prefaced its summary of the appropriation process in 2002 (FY2003) by saying, "As is customary the (DOD) appropriations bill followed the outlines of the authorization bill." (CQA, 2002: 2–9). This was notwithstanding that the authorization conference committee report was approved 28 days after the appropriation conference committee report that fall. The same statement appeared in the 2004 volume, when there was a 48 day difference.
5. Also the delay of the 2002 appropriation bill was basically caused by disagreement about how to control the second $20 billion of the $40 billion passed on 9/14; opinions differed as to how to retain appropriate control over the funds rather than writing a blank check. About a third of this bill went to DoD. (CQA, 2001: 1–11)
6. This consolidated bill was about $785 billion, about 1/3rd of total federal spending for FY2003.
7. In 2000, the House Armed Services Committee leadership asked the Joint Chiefs to speak up and support more money for defense from the $12 to $15 billion in unmet needs identified on the Joint Chiefs unmet priority needs list. (CQA, 2000: 2–40).
8. For example, this is the picture LeLoup and Moreland supplied. Others, including Roy Meyers penetrated within the base to show active decision making taking place there, but incrementalism with a dominant executive has proved a durable organizing concept. (See LeLoup and Moreland, 1978: 232–239; Meyers, 1994: 1–60).
9. CAGW publishes an annual Pig Book. Editions from 2000 to 2005; these were consulted in this research.

10. Ashdown argues that at least part of the earmarks included in the 2005 DoD appropriation bill were paid for in the spring 2005 Emergency Supplemental bill; he cites the operations and maintenance cuts made in the fall of 2004 and then replenished in the emergency bill in 2005. (See Ashdown, 2005) www.taxpayer.net/TCS/PressReleases/2005/3-03defensedatabase.htm

11. Senator McCain (R-AZ) has taken on this issue since July, 1997; see the list of 95 press releases and talking points about pork // mccain.senate.gov/index. cfm?fuseaction=Issues.ViewIssue&Issue_id=27.

CHAPTER 6

SUPPLEMENTAL APPROPRIATIONS FOR NATIONAL DEFENSE AND THE FEDERAL GOVERNMENT

INTRODUCTION

Supplemental appropriations provide rapid additional funding to augment the current year base budget, usually for national defense contingencies including warfighting and for natural disaster emergency response and humanitarian relief. Supplemental appropriations have been critical for the Department of Defense in the past decade to pay for the post 9/11/01 costs of wars in Iraq, Afghanistan and military operations elsewhere in the war against terrorism. Recently and somewhat regrettably, the supplemental appropriation process has evolved to include more of the complexities of the regular appropriation process. For example, both the President and Congress may suggest when a supplemental is a dire emergency and thus beyond spending discipline in terms of whether it adds to annual budget deficits and total federal government debt where their costs have not been offset by either new revenues or reductions in government spending. Some supplementals have paid for non-emergency activities, others have been financed by offsetting funding decreases in the current year while others have resulted in spending in future years. Compared to the normal ap-

Financing National Defense: Policy and Process, pages 201–234
Copyright © 2012 by Information Age Publishing
All rights of reproduction in any form reserved.

propriations process, supplementals are usually but not always passed expeditiously. Defense supplemental appropriation amounts tend to be more precisely priced while disaster supplementals tend to be more open-ended lump sum estimates. All supplemental appropriations carry their share of pork projects that too often are completely unrelated to the purpose of the appropriation.

In the 1990s, the major focus of supplementals was to meet emergent needs resulting from disasters and defense concerns. As the budget year unrolls, unpredictable events occur. Some of these are met by transfers or reprogramming of dollars already appropriated, but others are of such a magnitude that only an additional appropriation will suffice. Supplemental appropriations fill these needs. This chapter is about the complexities of supplemental appropriations and the process that enacts them into law.

Supplementals are an old tool, but since 9/11 their number and size have changed the funding magnitude and importance of supplementals to support warfighting (see table 6.2 at the end of this chapter for a complete summary over the period 2000 to 2010). In this chapter we focus on defense supplementals to explore the relationship between DOD and congressional uses of supplemental appropriations. Some of our data has been drawn from interviews with budget analysts and comptrollers in the Department of Defense and the Department of Navy and from congressional appropriation and other documents. In some cases these sources are not cited by name due to preferences for anonymity.

THE PURPOSES OF SUPPLEMENTAL APPROPRIATIONS

Supplemental appropriations may provide for natural disaster aid to states, regions and directly to individuals for blizzards, floods, drought, fires, and hurricanes. Supplementals met emergent military missions in the 1990s beginning with the Gulf War in 1991 and Bosnia in 1996 and continuing in the financing of warfighting in Afghanistan through 2011. In addition to military missions, the Department of Defense also has been tasked with providing aid in time of disaster to foreign nations. Although not much is made of it, since the 1990s the U.S. military has became an of emergency response force for humanitarian relief, fulfilling a variety of police, fire and emergency rescue functions when tasked to do so by the President with the concurrence of Congress.

Federal departments and agencies other than DOD also receive and execute supplementals in international emergency assistance efforts. Often a supplemental aimed at disaster relief also carries with it capacity-building technology, and technical assistance to get the work done on schedule and without losing funds to corruption, fraud, and misuse. Thus, funds from supplementals have been spent both in and outside the United States as policy makers have reacted to events suddenly thrust upon them, to provide direct aid and sometimes to help the other nations to build or rebuild their technical capacity that is expected to benefit these nations and the U.S. in the future.

THE SUPPLEMENTAL APPROPRIATION PROCESS
AND RELATED POLICY ISSUES

Before the passage of the Congressional Budget and Impoundment Reform Act of 1974, the federal government regularly used supplementals for provision of funding for day-to-day agency operations, even including pay raises for federal employees. However, in recent years supplementals have been mainly used "to provide funding for unanticipated expenses–although there is sometimes an argument about the whether the requirements should have been anticipated or not." (Tyszkiewicz and Daggett, 1998: 42–43). The Congressional Budget and Impoundment Control Act of 1974 attempted to control the use of supplementals by providing that anticipated supplementals must be included in the Presidents budget and that an allowance be set aside in the budget resolution for anticipated supplementals. The Budget Enforcement Act of 1990 imposed spending caps on federal spending which meant that a supplemental that exceeded the cap only could be passed if it could be matched with offsetting spending reductions or revenue increases, or both, or if it were deemed to be a *dire emergency* supplemental in which case it would simply be funded out of deficit spending.

During and since the 1990s, Congress and the President have at times tried to offset supplemental appropriations with rescissions or cancellations of budget authority provided in earlier appropriations bills that have not been obligated, or have been deemed unnecessary. Because Presidential rescission budget cuts have not paid for all supplemental costs, dire emergency provision also have been used to finance rapid response actions as required. Since the 1990s, many if not most supplementals have been passed under the dire emergency clause and thus have avoided spending cap and deficit neutral discipline. No such escape clause existed prior to 1990 but in every year from 1991 through 2010, part or all of the supplementals passed by Congress have been designated as emergencies. Most were passed in response to real emergencies; some were not or significant portions of these dire emergency supplementals have not been used to finance actual emergency responses. This has been the case independent of which political party held the majority in Congress and the Presidency, i.e., elected officials do not differ in their support of dire emergency supplementals strictly along predictable party lines. In this respect the supplemental review and enactment process differs to a certain degree from the regular appropriations process.

The executive branch generally but not always controls the timing of when the supplemental appropriation request will be submitted by the President to Congress and OMB and departments/agencies estimate its initial purpose, amount and how quickly it is needed. This is not to say that Congress has no role in determining the purpose and need for the supplemental; it does. However, in terms of process the Executive branch initiates the supplemental funding request to Congress. The President (OMB) also may suggest what we term "bill-payer" types of supplemental appropriations, i.e., the funding requested is to pay the costs of some type of emergency assistance for which money already has been spent or is in process of spending as services are delivered. For national defense most supplementals in fact repay money already spent form the DOD and military department and service base budget. However, strictly speaking some "billpayers" may never be acknowledged for such use. Also, the President cannot decide which supplementals are to be deemed of the dire emergency type. The President may ask Congress for dire emergency designation for a supplemental request but Congress must concur with the provisions of the request in enacting the supplemental. Emergency designation is a shared power. Congress can add projects and funds not requested by the Executive branch to both supplementals and regular appropriations and can suggest that some supplementals be considered for emergency designation even when the President has not requested it. During certain periods of time Congress has routinely done this; about 33% of supplemental dollars were designated as emergencies in the 1990s. (CBO, 1999: 4) In turn, the President (assisted by OMB and the Treasury) has to officially accept Congress' designation of a supplemental as an emergency or dire emergency before the dollars appropriated may be released for obligation, and has to sign the supplemental, regardless of type, to make it a law. Thus, the end of process control rests with the President but it is rarely used for rejection in that Presidents typically sign off on supplementals as appropriated by Congress in part to gain political credit and public recognition for having approved the measure into law to meet a need that in many cases is highly newsworthy and covered extensively in the news media.

In the twenty-six years from 1974 through 1999, sixty-one supplemental appropriations were enacted, totaling more than $430 billion dollars.[1] This is an average of $7.05 billion per supplemental and $16.5 billion per year. The largest emergency supplemental was for the Persian Gulf War in 1991 and totaled $42.6 billion. Two other supplementals were passed that year totaling just over $1 billion (Godek, 2000: 36). Conversely, in 1974 a supplemental was passed for $4.75 million. Two more small supplementals were passed in 1974, one for $8.77 million and another for $8.66 million.

(Godek, 2000: 36) Twenty supplementals were passed from 1974 through 1979, including five in 1978 alone. Due to attempts to hold down spending to curb the deficit and, perhaps because of fewer natural disasters, fewer bills passed each year during the 1990s; only one supplemental was passed in 1995, 1996, 1997, and 1999. Starting with the Gulf War supplementals, emergent military missions often made the defense portion of the supplemental a key policy issue where the debate focused not only on funding but also on the Presidents responsibility to inform Congress before engaging in peacekeeping missions that involved use of military force. In 1991, 1992, 1999, and since 2001 defense needs have constituted more than half of the funding provided by all supplemental appropriations.

Supplementals Usually Pass Quickly

Supplementals allow the federal government to show that it is willing and able to respond to urgent needs, both at home and abroad, and that it is capable of doing so quickly. This is a useful corrective measure to counter the slowness of the regular appropriation process, which is often delayed by political party bickering and gridlock and where regular appropriation bills often are passed late after the new fiscal year has begun. Supplementals have a clear seasonal distribution. In most years where supplementals have been approved, Congress has introduced and passed supplementals within a four-month period or less. In general, disaster relief and most defense supplementals are passed quickly. This may not seem particularly quick unless it is considered relatively; passage of normal annual appropriation bills can take up to ten months or longer from the time of introduction by the President to passage by Congress.

SUPPLEMENTALS FOR NATIONAL DEFENSE

When the President decides to commit forces to war, to evacuate embassy staff in a foreign nation, to rescue American citizens, to deploy forces in an emergent peacekeeping action, or to deploy defense forces in a humanitarian enterprise, he creates an unbudgeted expense for the defense budget. The President authorizes action by the Department of Defense, which executes the mission using whatever funding, equipment, weapons platforms and personnel are available, oft-times including military reservists and the National Guard. If the mission is small and very limited, the expense may be absorbed within the DOD budget. For larger and more extensive missions, DOD executes the mission and pays for it typically by 'borrowing' against its fourth quarter appropriations and execution expense budget, especially from operations and maintenance (O&M) accounts. DOD then depends on the President to request and Congress to pass a supplemental appropriation soon enough so that whatever has been borrowed can be replaced so that the original fourth quarter spending program can be executed as planned.

Conventionally, if a mission requirement appears or continues from the previous fiscal year into the fall or winter of the subsequent fiscal year, the DOD Comptroller, with OMB's permission, waits to submit the supplemental request to OMB until after the budget process has begun in the following spring so that Congress does not confuse the supplemental funding needs with those asked for in the regular appropriation request imbedded in the President's budget. This means in some cases that since the regular congressional budget process begins in February, the defense supplemental may be submitted in late March or April to request funding for operations that were performed during the previous fall or winter. In such cases budget execution is seriously confounded and DOD comptrollers at all levels must be sure that legal accountability of accounts is maintained. This often requires a significant degree of creativity and flexibility within the military and civilian comptroller community. This type of circumstance and responses to it are topics unto themselves. (See Philips, 2001, Appendix B in McCaffery and Jones, 2001: 423–440)

DOD and Incremental Costs

In passage of the Budget Enforcement Act of 1990 by Congress the Department of Defense was directed to report only the incremental costs of carrying out contingency or emergency operations of all types including warfighting. This was clarified in 1995 and extended in February, 2001 to incremental costs "above and beyond baseline training operations, and personnel costs." (DOD Financial Management Regulations, 2001: 23–26) Thus, the military departments and services (MILDEP) may identify incremental costs related to military operations, military and civilian pay, clothing and equipment, reserve activation, operational training, supplies and equipment, facilities and base support, airlift, sealift and inland transportation. Before this detailed guidance, each MILDEP had some degree of latitude in developing incremental costs. However, as supplementals began to be passed for warfighting post 9/11 within a year or two the Office of the Secretary of Defense (OSD) and Congress began to develop additional justification and reporting requirements that have complicated the preparation of supplemental budget requests. This complexity and the cost of compliance for DOD have grown extensively in almost every year since the end of 2001.

Historically, DOD has tended to absorb the costs of very small missions. The incremental cost doctrine has increased this tendency to some extent which varies by department and service. Thus in some missions an incremental cost bill could have been specified and requested but DOD has decided to absorb it, typically because the cost of the mission is small enough that absorbing it does no serious harm to executing the defense program planned for in the original defense budget. In some instances, in the Navy for example, diverting personnel and ships from training exercises for a small and short-term rescue and humanitarian relief missions does not cost much and the actual activity might be close enough to the training mission to provide a training benefit, albeit different than planned. Moreover, it is a fact of life that the President is not going to ask for, nor will Congress provide, funding for many such small missions unless it will provide pork for a sufficient number of constituencies to warrant the attention of Congress. For larger and more elaborate peacekeeping and humanitarian relief missions, incremental costs are priced and the OMB comptroller, in essence, submits the bill to OMB and then via the President to Congress as part of a supplemental appropriation request.

It is clear that the incremental pricing requirement has resulted in more precise pricing of defense supplementals. Also, it may also have resulted in slight decreases in spending for programs in the base budget as DOD absorbs small incremental costs rather than asking for their reimbursement, especially as noted for small operations that can be considered similar to baseline funded training exercises.

DIRE EMERGENCY SUPPLEMENTALS AND SPENDING RESTRAINT

DOD supplementals were impacted by the pay-as-you-go provisions of the Budget Enforcement Act of 1990 (BEA) and subsequent actions by Congress and the President. The BEA (passed primarily to balance the annual federal budget) required that spending above allowable spending caps provided in the Congressional Budget Resolution be deficit neutral, meaning that either a revenue source had to be found or funding taken from another program through a rescission to finance new operations and programs. Program financed in accordance with the spending cap requirements then would become bill-payers for the programs authorized and other programs or activities in some instance, at the discretion of the DOD comptroller community and with the permission of the Under Secretary of Defense or SECDEF, or in some instances the Assistant Secretary/DOD Comptroller. Alternatively, the President could ask that the supplemental be classified as a dire emergency and if Congress concurred, the supplemental amount would be relieved of the bill-payer necessity and the allowable spending cap would, in effect, be increased. This method was used depending on the willingness of Congress to support it. In addition, beginning in 1991, the emergency designation began to be used to for increasingly more supplemental appropriation justification in Congress, and even for some funds in regular appropriations bills.

Consequently, by the end of the decade of the 1990s and into the 2000s the result of this evolution was the emergence into the budget process of many large supplemental appropriations bills with mixes of emergency and non-emergency funding, and regular appropriations with emergency funding provisions embedded into them. Designation of part of a supplemental as emergency or dire emergency was a further complication, as was the mixing of the emergency designation into regular appropriations bills. Moreover, often the emergency funding lasted for more than one year (e.g., warfighting "bridge" supplementals); thus a supplemental with warfighting emergency funding affected not only the current budget execution year but the next fiscal year as well. In many cases emergency supplementals have carried spending for programs and activities that would seem to fit far more appropriately into regular appropriation bills. Moreover, in some cases they also have created ongoing expenses which become part of the budget base as opposed to one-time emergency funding appropriations. As emphasized previously, using emergency supplementals in this manner tends to confuse an already complex budget process.

Critics have argued that the obvious solution is to put the second year of emergency funding in the regular appropriations bill. Others observe that if and when spending caps are set unrealistically low, spending is deemed emergency spending simply to get around the caps, and that caps that are more realistic would lead to fewer or smaller supplementals and less emergency spending. The budgetary politics of the 1990s was a history of spending caps first imposed from 1990 through 1995 with the Budget Enforcement Act of 1990 and then extended in 1993 and 1997, for a five-year period through 2002. The net effect of these caps was to put the discretionary sector of the budget under very tight discipline, so that budget growth was capped at less than inflation for most agencies. After 2002 these spending caps were no longer in force, leaving Congress to spend without restraint and without concern for the total debt generated by annual budget deficits, until the middle of 2011 when Congress, particularly House Republicans, suddenly became highly interested in the size of the debt as part of an early electoral campaign strategy directed toward the general election of 2012. However, unless imposed by a joint select congressional committee scheduled to meet in November and December, appropriation caps in the discretionary part of the federal budget will not have been imposed to constrain spending for almost a decade regardless of which party was in power in Congress or the Presidency. In addition, the absence of spending caps has provided some advantages; it assisted DOD in making it far easier to gain congressional passage and presidential approval of large supplemental appropriations to cover the costs of war and the ongoing battle against terrorism.

With an ordinary supplemental, one not deemed a 'dire' emergency, the President and Congress have to find funding sources to pay for the bill. This could include sources inside or outside the defense budget. A bill-payer outside the defense budget, for example surplus money in the food stamp account, means that a supplemental would be funded outside defense.[1] Sometimes DOD is required to pay a portion of a supplemental from DOD base budget funds. For example, in some years Congress has authorized a pay raise for the military and DOD civilians but appropriated funding sufficient to pay only half of or some portion of the increase, forcing DOD to find money for the other portion, which is a common element of trickery in budgeting. (Jones and Euske, 1991)

In the Navy budget, offsets to pay for money appropriated in supplementals might come from delaying property maintenance or ship repair and overhauls, categories where spending can be delayed into the next fiscal year without too much damage to fleet readiness or program execution. This is in the case where the U.S. is not engaged in active warfighting, which has not been the case since late 2001. We may note that in such cases DOD and Congress are, in effect, cost-sharing the supplemental. Congress uses this technique to drive down the cost of the supplemental. Thus, even though the supplemental has been accurately priced, Congress may refuse to pay the whole bill and require a contribution from DOD, which, on its part, may borrow against the future by delaying construction or maintenance programs already funded in the budget into the future. These programs and activities have won approval through the executive budget process and in appropriations acts passed by Congress, but they can be forced to compete again for funding. In effect, supplementals financed in this manner have forced budget cuts in currently approved programs and put them in jeopardy. For DOD such cost-sharing when Congress under-funds a war funding supplemental solves one problem in the present but creates other present and future budgetary needs. When this occurs another effect is to put different factions in Congress in competition against each other. Those members whose constituents already benefit from what is in but then cut from the base will oppose supplemental funding composition that cuts the DOD base budget if cut hit "their" programs. Alternatively, members whose constituents will benefit from new supplementals funded indirectly and in part by DOD will support the supplemental.

Congress always has a series of choices to make when considering passage of a supplemental appropriation, e.g., to fully-fund or cost-share; to find bill-payers to provide rescissions; to declare all or part of the bill a "dire emergency." A mixed approach is possible. For example, a 1999 supplemental appropriation for $15 billion was a hybrid; it was offset by $1.7 billion cut mainly from the food stamp account while the remainder of the $13.3 billion was declared and passed as a dire emergency, thus excluding it from having to be deficit neutral.

Both the enabling statutes that created supplementals and conventional wisdom suggest that dire emergency status only should be used for supplemental bills for genuine crises, but these conventions are not always observed or met when Congress acts. To escape spending cap restrictions prior to 2002, and presumably to do so starting again in 2012, all of the supplementals passed in some years have been designated as dire emergency spending, but this also has been the case for an additional billions of dollars for regular appropriations. Using the dire emergency designation to avoid spending cap or any other type of discipline such as that related to balancing the annual federal budget in the regular appropriations bills has for more than twenty years set a precedent that has been fully embraced by both political parties and by Congress and various Presidents. This may be viewed as both an evasion of budget discipline and as a safety valve, particularly since the amounts approved in supplementals comprise a very small percentage of total federal spending. Whatever the justification, since the mechanism was available, Congress decided to use it, and use it increasingly during the period 2001 through 2011.

Logically, it would seem improper to include emergency designated funds in regular appropriation bills that are, after all, for the following fiscal year. Since most appropriation bills are passed after the start of the fiscal year, Congress does have an opportunity to address emergency needs within current appropriations bills. This may help explain the rise of this practice as well as why supplementals are generally a spring and summer phenomenon.

While the emergency designated sums are very small compared to total spending, the emergency designation device is rather widely used among appropriation subcommittees and in agency appropriation bills. For example, in 1999, while Defense was the largest beneficiary of the emergency designation, twelve appropriation subcommittees had some emergency money within their purview (only the District of Columbia failed to benefit) and eighteen cabinet level agencies gained dollars through the emergency designation process, with NASA, Veterans Affairs, and the Department of Education being overlooked in this year. NASA was the only agency never to receive emergency designated money in the 1990s. (CBO, 1999: 3)

The temptation is to say that Congress and the President learned over two decades starting in 1990 how to get around all restraints and spending discipline by designating some funding in regular appropriation bills as emergency funding. Additionally it may be noted that when supplementals pass early in the session, and with the tendency to pass only one supplemental per year, and considering that Congress typically does not get serious about its appropriation work until August, September or later, the ability to designate dollars in regular appropriations bills for emergency purposes presents a convenient safety valve. However, it complicates our understanding of the supplemental appropriation process when emergencies are found in regular appropriation bills and non-emergencies in supplemental bills, e.g., where non-war related maintenance, spare parts, and family housing funding in the defense supplementals have been funded in some supplementals. It also impairs the transparency of the budgetary process and goes against efforts to balance annual budgets or to reduce the national debt.

SUPPLEMENTAL HISTORY

Supplemental spending bills are not new. The use of supplemental appropriations began with the second session of the first Congress in 1790 and continued through the 1800s and 1900s and on into the 21st century. (CBO, 2001:1) They frequently included additional appropriations for agencies that had overspent their appropriations and as such became known as deficiency appropriations. This practice became so routine that the House Appropriations committee divided them into general deficiency bills and urgent deficiency bills. In the 1870s when anti-deficiency legislation was passed, critics of this process accused Congress of under-funding the regular appropriation bills to appear frugal stewards of public funds, and then, "...after elections were over, make up the necessary amounts by deficiency bills." (CBO, 2001:1)

The Antideficiency Act of 1905 attempted to control deficiency spending by giving the Treasury Department the authority to apportion funds to agencies to reduce the need for supplementals. This power was further refined by the Budget and Accounting Act of 1921 and the Antideficiency Act of 1950 that encouraged agencies to set aside reserve funds for emergencies, and limited supplemental appropriations to legislation enacted after the Presidents budget had been submitted and for emergencies relating to the preservation of human life and property. (CBO, 2001:2) However, the issues surrounding supplementals have remained contentious. In 1966, the Joint Committee on the Organization of Congress again raised the issue of lawmakers projecting an image of economy by under-funding the regular appropriation bills with the tacit understanding that they would later pass supplemental appropriation bills (CBO, 2001:2).

In the 1970s, supplementals primarily affected mandatory entitlement accounts including unemployment benefits and increased food stamp funding. In 1977, discretionary programs were supplemented with a $9.5 billion program for job training to counteract the recession of 1973–1975. Most of the other supplementals for the 1970s were for federal pay raises, programs that were authorized after their appropriation bill had been passed, and expenses related to natural disasters such as blizzards, floods, droughts, and forest fires. Supplementals declined in the 1980s, to a low of .1% of budget authority in 1988, mostly as a result of the struggle with the deficit and the requirement for offsets to pay for supplementals. Mandatory entitlement programs were again high: two-thirds of the supplementals in this decade were for mandatory payments, primarily for support of farm commodity programs due to worse-than-expected conditions in farm commodity markets. Supplementals also were applied to the food stamp program, unemployment insurance, and various higher education programs.

In the 1990s, only 6.3% of all supplemental appropriations went to mandatory accounts. (CBO, 2001:10) Most of the discretionary supplemental appropriations in 1991 were for military operations Desert Storm and Desert Shield. Domestic spending dominated discretionary supplemental appropriations from 1993 to 1998, but defense spending re-emerged as the largest category in 1999 and 2000 because of peacekeeping missions in Bosnia and Kosovo. Since 2001 supplementals have been used for two primary purposes: to fund humanitarian relief associated with natural disasters and to fund the costs of war.

Academic research on supplementals has been sparse. In his classic *The Politics of the Budgetary Process* (1964) and in subsequent revisions, budgetary guru Aaron Wildavsky mentioned supplementals, suggesting that, "Congressmen get headlines for suggesting large cuts, but they often do not follow through for they know that the amounts will have to be restored by supplemental appropriations." (Wildavsky, 1984: 23; 1997: 50). Wildavsky observed that appropriations committee members may even talk to agency officials about areas they can cut and then restore later in supplementals, but he did not pursue this issue further. However, Christopher Wlezien did, examining the relationship between appropriations and the supplemental process based on data from 1950 through 1985. Wlezien found that regular appropriations and supplemental appropriations were linked in a two-stage process with certain accounts under-funded in the regular appropriations bills and replenished through the supplemental process. Wlezien associated these results with party identification and found that Republican Presidents were more likely to engage in strategic under-funding and make it up later with supplementals during the period under study.

Wlezien explained that the two-stage process he discovered in the appropriations process from 1950 to 1985 occurred in a single stage in the early 1990s, with the politics of bargaining between the President and Congress, "... largely confined to regular appropriations," (Wlezien, 1996: 62) primarily as a result of changes in the budget process. Observation of the supplementals process since 2000 up to the present day indicates that supplementals and regular appropriation bills each have their own arenas for institutional bargaining. With supplementals the bargaining issue often is focused on size (dollar amounts), and about dire emergency designation, finding billpayers and, of course, pork. Supplementals also have been the locus for bargaining about the limits of executive power, with Congress suggesting that the President must get advance approval from Congress before engaging in peacekeeping and warfighting operations. Regular appropriations also have been delayed by issues of size but mostly, since 2000, the arena of negotiation for these bills has been filled by disagreements related to policy, political ideology and party politics. Noteworthy is the fact that even as the size of the federal debt grew rapidly during the first ten years of the new millennium, deficits and debt were hardly ever issues of partisan debate. Both parties appeared to spend and ignore the consequences except in terms of benefits provided—but not costs incurred.

We do not agree with Wlezien that the old two-step process has disappeared. This was true fro the period he studied but not more recently. Largely, but not completely, the defense spending supplementals of the latter years of the 1990s and from this time on provide the evidence to support our view. It appears that some of the initiatives funded for DOD in supplementals should have been funded in the base budget as continuing expenses, such as reorganization of the Army command structure in the mid-2000s, and that the pattern since 2001 has been for Congress and the President to pass increases in funding for the DOD base budget through regular appropriations and then add money through the emergency supplemental process for warfighting, just as in the old two-stage process.

NON-DEFENSE SUPPLEMENTALS

Excluding the Department of Defense, the Federal Emergency Management Agency (FEMA) has been the second largest recipient of discretionary supplementals, receiving substantial amounts of supplemental budget authority. Supplementals for FEMA primarily represents appropriations to the FEMA disaster relief account to pay for relief efforts in the wake of hurricanes, floods, earthquakes, and droughts. From 1992 on, the Small Business Administration also received funding for disaster loans, as did the Commodity Credit Corporation for aid to farmers for crop losses. In 1999, funds were also provided to compensate farmers through price supports for certain commodities. (CBO, 2001:17)

For disaster assistance that has involved longer-term aid such as loans, loan guarantees, or other forms of financial assistance other than direct federal emergency aid, passage of the supplemental may take longer and reflect different party viewpoints on what is the best way to handle longer-term disaster relief. Thus, generalizations about supplementals are difficult to make other than that they are small compared to total budget spending and result from unforeseen events. Their symbolic value is high. It was clear after the terrorist attacks in September 2001 that the American people expected their government to protect them and help with disaster relief. This was obvious in the quick passage of the $40 billion supplemental of financial aid related to the World Trade Center and Pentagon attacks in four days in September 2001 and the relatively quick passage of the $65 billion Operation Iraqi Freedom supplemental in 2003.

NON-EMERGENCY SUPPLEMENTALS AND OFFSETS

When DOD has had to offer up offsets in a non-emergency supplemental, the usual practice was for the military departments to share equally, each coming up with one-third of the offset. If the costs of the mission are equal, this is fair for all. However, the Military Departments have different structures and in joint missions, one service might be able to accomplish its mission more easily and more inexpensively than another service. If they all are asked to offset the bill on an equal shares basis, a tradition within defense because all share the pain equally, one military department may help pay for expenses in another.

For example, since the Navy is forward deployed at times its costs may be lower than the Army and Air Force but, if it has to offer up a full third of the offset, the result could be that Navy money will flow to pay Air Force or Army bills. Let us suppose the total cost of a supplemental is $6 billion, with the Navy share $400 million in incremental costs and the Army and Air Force costs each being $2.8 billion. If Congress decides that DOD has to offset this cost and the DOD comptroller asks each Military Department to offer up $2 billion in offsets, the Navy would gain $400 from the supplemental but contribute $2 billion to the offset for the supplemental. The Navy might have to cut its fourth quarter operational tempo–flying and steaming hours- so that the Army and Air Force can more fully fund their fourth quarters. Much of this depends upon how DOD leaders, including the Comptroller, decide to allocate shares of the cost burden. For example, the decision might be made to share the costs equally or a differential cost pattern might be imposed, taxing one military department more heavily than the others because budget execution reports indicate that some slack is evident in its budget due to program under-execution.

Defining and measuring true incremental costs is almost always difficult when the incremental cost is small (it isn't easy when it is large either). For example, if Navy flying hours increase 3% over the normal yearly allotment because of incremental costing related to emergencies, the DOD Comptroller tends to urge the department to absorb it, because it is so small. This tendency is even more apparent when it comes to incremental maintenance costs due to emergency operations. As one source said, "If flying hours go up 30,000 in a year due to a contingency operation, that is a small number on a 1,000,000 hour flying hour program. Because we are out there flying and steaming anyway, it is hard to argue for these small percentage increases." He added that absorbing costs was all right for the ships already deployed, but the non-deployed units would not be getting enough training dollars to get to the right readiness level unless the incremental costs of the emergency operation were put back in the budget.

While an emergency tasking cost might seem quite small as a percentage of the overall defense budget, it is a much larger percentage of the bill-paying budget account. This is usually the fourth quarter Operations and Maintenance (O&M) account where the opportunity exists to change the pace of spending. What this means is that if the money is not restored, a sizeable amount of planned and budgeted program will go unfunded and unexecuted, most of it in force operations and direct support.

Each quarter in the O&M account mandatory expenses of a contractual or semi-contractual nature must be paid and these cannot be arbitrarily foregone. Amounts required to support military exercises, training missions or force protection have to go forward and be executed. Money cannot simply be 'robbed' from these accounts without serious consequences. However, if this money in the O&M account it may be borrowed but it has to be repaid within the same fiscal year. Such is common practice.

SUPPLEMENTALS AND BUDGET CONTROL

We have explained in earlier sections of this chapter that spending control was largely lost and not perceived as politically necessary from the late 1990s until mid-2011. The Budget Enforcement Act of 1990 changed the supplemental process by requiring incremental costs to be specified clearly and by imposing deficit control measures that involve spending caps, billpayers and/or dire emergency designation. These changes appeared to have changed the nature of supplementals as well as the supplemental process. Insofar as the old two-step supplemental process is concerned, the budget ceilings set aside all a functional area could receive for a year, so it would be a waste of time to under fund an area and expect to come back later and fund it in a supplemental; another program might have used up the space under the ceiling in the meantime and a sequestration (a congressionally mandated reduction of budget accounts implemented by the President, typically as across-the-board cuts) could be ordered for an appropriation that breaks the ceiling subsequently. In 2005, the defense appropriation bill broke the ceiling on discretionary spending and resulted in an across the board cut, although the bridge supplemental was held harmless from this by declaring it a dire emergency.

Thus, the Budget Enforcement Act of 1990 created and enforced a pay-as-you-go discipline for entitlement program spending after 1990 for supplementals not designated as "dire emergencies."[2] However, irrespective of the funding technique used, since the late 1990s the supplemental has become for the most part a focus of attention as a rapid response mechanism and coercive deficiency strategizing (i.e., budget cutting) became less evident a focus of Congress. In response to the financial collapse of late 2008 and continuation of expensive war operations, Congress and two Presidents have supported large spending increases for economic recovery and the costs of warfighting. And, as has been the case throughout history, supplementals and debt have financed both of these demands for spending

However, we believe evidence demonstrates that supplementals have not been nor are they now like regular appropriation bills or Continuing Resolution appropriations. An appropriation bill is a forecast of what is to come, a promise that X units of work will be accomplished with Y dollars for personnel and supporting expenses. However, no matter how carefully budget reviewers labor over the numbers, the budget remains a forecast and the corresponding appropriation is similarly imprecise. Supplementals, be they for defense or disaster relief, are different. Disaster-focused supplementals are not like regular appropriation bills because they are basically lump-sum appropriations focused on a goal, remediation of the disaster. The funding mix may resemble what was done for similar disaster situations in the past or it may be set up as what is recognized as an initial endowment with more funding to follow, once the magnitude of the disaster is measured. Moreover, the symbolism of governmental response to the crisis is important. The funding provided may or may not be enough for the task; what it is designed to do is get aid to a problem quickly, in the current fiscal year, in order to start the healing process. Rough estimates of what will be necessary are made, but a disaster supplemental cannot have the careful budget quality numbers that go into an appropriation request.

Defense supplementals are different from both appropriation bills and disaster-related supplementals. A defense-focused supplemental is unlike an appropriation request because the defense supplemental is a statement of costs for services actually and already performed. Cost estimates are somewhat more accurate and predictable given that Presidential national security policy has been established for a set of given objectives and activities. Cost estimates, although not easy to derive for war supplementals, are generally more accurate than an appropriation bill can be, given that budget execution takes place so long after appropriations are negotiated in Congress. The size of these two types of different appropriations also is an important distinguishing variable. Even the best regular appropriation estimate in most cases cannot have the accuracy of a one year (or less) defense supplemental; after all, one is a forecast, the other is a bill for the incremental costs of services rendered.

Some defense supplementals have funded ongoing military activities that last the duration of a fiscal year, or more. These are more like a regular appropriation bill in their duration, but better than an appropriation bill in their precision. Moreover, the defense supplemental with its itemized and constrained estimation methodology for pricing and given the focus on incremental cost is at the opposite end of the scale from the lump-sum disaster supplemental or the regular appropriation. The war funding supplemental is retrospective and pays for actions that for the most part already have been done while the latter two types of appropriations are prospective in providing aid in a disaster situation or the programs, plans and activities financed through the budget base appropriation.

In summary, supplementals respond to unbudgeted and largely unpredictable events. They are relatively small in most but not all cases compared to total budget base spending, but they pay for what is needed in defense and other areas and have great symbolic value in defense and in humanitarian aid and disaster relief. Events in recent years have only added complexity to the analysis of supplementals. From 2001 to 2003 DOD gained its war related funding in a stream of regular appropriation bills and in supplementals, including those of July 24, 2001 ($6.74 of $9b.), September 18, 2001 ($3.4b of $20), May 21, 2002 ($14 b. of $21), and April 2003 ($62.6b of $75). None of these provided funding only for defense, but all had significant defense money in them. July 2001 saw passage of a "normal" supplemental; the rest have been reactions to wartime emergencies, including the cost-or-war supplemental in April 2003 and thereafter through 2010. (See Appendix A) Wartime supplementals that have been passed almost routinely since 2002 need separate study because they depart from the model presented thus far in this chapter.

WARTIME SUPPLEMENTALS

So far we have argued that supplementals fund the current year for emergency operations, that disaster supplementals are estimates and defense supplementals are more accurate because they are paying back for actions that already have occurred. Thus, our theory is for defense supplementals is as follows: a contingency or emergency occurs and the President sends the military to help or fight. DOD and the MILDEPS borrow against 4th quarter funds and the supplemental pays back the accounts from which the funding was drawn in time for DOD to execute its 4th quarter budget as originally planned (given that there always is some difference in execution from plans no matter how good they are; see chapter 7 on budget execution). While this was true of the humanitarian, disaster-relief, and peacekeeping activities of the 1990s, it was not true for either the Vietnam era nor for the Iraq/Afghanistan supplementals. Thus, our simple scenario for supplementals in peacetime (and peacekeeping), does not apply in time of war.

In essence, the DOD appropriation (as well as DOE for nuclear purposes) builds a defense capability to deter war and, as ready as possible, to go to war. Recent history suggests that when this action process is put to use, the costs of war are partially or almost completely funded through a supplemental. This also was the case for WWII, Korea, and Vietnam. For example, two supplementals were passed in 1941, the first in March and the second in October before the U.S. entered the war. Wartime supplementals were also passed in 1942 and 1943.[3] The October 1941 supplemental was the Lend-Lease Act which allowed the U.S. to provide ships and other support and supplies to Great Britain while still remaining out of the war on an official basis and thus in a position that technically was still neutral. (Evans, 2006) This was of great symbolic significance to our friends and enemies alike and resulted in much needed help for Great Britain. World War II was pursued subsequent to the attack on the Pearl Harbor, under the aegis of a Declaration of War in which the U.S. pledged its full treasury and assets to seek a successful outcome. (Miller, 2006) The magnitude of the effort for WWII dwarfs anything since that time, as can be seen in the exhibit below. No conflicts since then have been fought under a declaration of war, but they all have involved supplemental funding. In general, supplementals helped finance the early stages of these conflicts and then ongoing costs were included in regular appropriations bills as soon as cost projections could be made. Korea is a perfect example of this; the wartime supplementals passed for FY1951 beginning in September of 1950 tripled the size of the DOD budget in 1952. However, as noted, Iraq and Afghanistan operations have been financed (technically and politically but not completely in terms of the DOD budget) by supplementals.

For the war in Vietnam the Johnson administration used a mixed model of a supplemental for FY1965, a budget amendment for FY1966 and regular appropriations and supplementals for FY1967 and FY1968. In the early 1990s funding for operations in Somalia, Southwest Asia, Haiti and Bosnia were provided in supplementals; from 1996 through 2001 war funding was provided through the regular defense budget and appropriations bill passed either before or after one or more supplementals. Since 2001 supplementals have been used to finance war operations.

It should be noted that these latest conflicts, and operation Desert Storm in Kuwait and Iraq in 1990 and 1991, changed the focus of U.S. defense

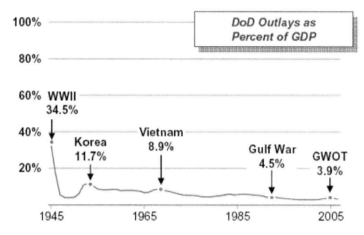

Figure 6.1 War funding as a percent of GDP. *Source*: Department of Defense, 2007. FY 2007 Supplemental Request for the Global War on Terror. Washington, DC, February, 2007: 2.

posture in that it included creating and maintaining bases around the Persian Gulf as well as increased deployments for the U.S. military including Navy carrier groups to the eastern Mediterranean, Persian Gulf and Indian Ocean. Thus there are continuing costs from supplemental initiation that then need at some point to be built into the regular defense base budget.

Current Era War Supplementals: Large and Frequent

Current era war supplementals have been large in dollar terms and what they financed, significantly larger than those of the Vietnam era or the early 1990s, and the defense base budget has expanded significantly as well. From 2004 until 2007 it became almost normal practice to pass two supplementals, a bridge supplemental in the fall with the DOD appropriation bill and another larger supplemental in the spring, both for current fiscal years. In many calendar years or fiscal years in the 2000s there were two wartime supplementals. The first bridge supplementals were passed as separate sections in the DOD appropriation bill (e.g., Title IX in the FY2006 and FY2007 appropriation laws) and were intended to ensure that DOD did not run out of money before the end of the fiscal year. In later years it was intended to bridge DOD over until the spring wartime supplemental became available. In the summer of 2004, the first bridge supplemental was clearly aimed at ensuring DOD did not run out of money in the current fiscal year.[4] The supplemental was to be available immediately to pay for shortfalls in the final months of fiscal 2004. It passed with the FY 2005 regular appropriation bill, which the President signed on 5 August 2004. The Pentagon and Congress disagreed on the amount of money needed. The conference report was filed just hours after GAO said pentagon would face a $12.3 billion shortfall before the end of the fiscal year. DOD officials had estimated they would be short by $2.8 billion and proposed to make that up by reprogramming. Congressional appropriators had insisted all along that the shortfall would be greater. While Congress was more generous with funds than was the DOD leadership, Congress added controls over where the money could be spent. Congress restricted all but $2 billion of the $25 billion to specific accounts, thus DOD was not given carte blanche about how to spend the money. Moreover, the administration did not get all that it wanted; transformational programs in the Army and Air Force did not fare well and lawmakers worried about a procurement shortfall and warned that there were more big ticket items in the pipeline than DOD would be able to afford. Congress also supplied specific remedies to battlefield weaknesses that the pentagon had not requested, adding $1.5 billion not requested by DOD to deploy more armored vehicles and replace other equipment. (Congressional Quarterly Almanac, 2004: 2–14)

In 2005, a $50 billion bridge supplemental was added to the FY2006 appropriation bill to tide DOD over until the new supplemental was submitted in February 2006. During the summer, appropriators warned that Army would be out of money by August. Nonetheless the appropriation bill and its bridge supplemental was the "last and arguably most contentious" appropriation bill passed that year, with the conference report cleared on December 22nd. (Congressional Quarterly Almanac, 2005: 2–14) The supplemental carried additional equipment for Army, Air National Guard, and Army Reserves. Also eight billion was earmarked for equipment lost in combat including humvees, trucks and radios and $1.46 billion for testing and fielding new equipment to thwart roadside bombs (IEDs). (Congressional Quarterly Almanac, 2005: 2–18) The FY2006 appropriation bill carried an across the board cut of 1% applied to all discretionary spending other than Veteran's Affairs and emergency spending including the DOD bridge supplemental and disaster related funds. This cut was deemed necessary because the discretionary appropriations total exceeded the agreed on temporary discretionary budgetary caps set in the Budget Resolution. Although this resulted in a $4 billion cut from the DOD appropriation bill, the cut did not affect the bridge supplemental because it was "dire emergency" spending. Part of the discussion in 2005 was about the Army's request to fund the first two years of its command structure realignment efforts using supplemental funding (about $10 billion) imbedded into emergency supplementals for the war in Iraq. Proponents argued that this was "... too urgent a task to risk the delays and pitfalls of the regular budget process." On this point defense experts noted observed that there were immediate requirements that were in process of validation in the field in Iraq and Afghanistan. Essentially what the Army was doing was playing catch-up with the fact that the Army's force structure had not changed as it should have after the end of the Cold War in 1990. (Schatz, 2005: 510.)[5]

The 2006 supplementals were interesting for three reasons. First, in the main defense appropriation bill process, Congress cut the defense request by about $4 billion in order to provide more than the President had requested for some domestic programs. This move did not affect defense because "appropriators made up the difference in the emergency section of the bill," (Congressional Quarterly Weekly, 2006: 3331) in the bridge supplemental. The appropriation bill (including the bridge supplemental)[6] was signed on 29 September 2006.

Secondly, in February the main supplemental for DOD was sent to Congress as a separate bill, but merged by Congress with hurricane relief spending. The majority of the bill was for defense ($70.4 out of 94.6 billion) and all of it was designated emergency spending, meaning it would escape any spending cap discipline and no billpayers would need to be found, other than future taxpayers who have to pay down the national debt. Thirdly, a deeming resolution was included setting discretionary spending limits for the Senate because Congress could not complete the normal budget resolution process. (Congressional Quarterly Weekly, 2006: 3336) The Senate cleared the supplemental on 13 June 2006 and thus could use the deeming resolution in the supplemental for its deliberations on all of the upcoming appropriation bills (including the defense bill).

Lastly, the deliberations over this supplemental took longer than expected. There were some efforts to give the president policy guidance over the war and some fiscal conservatives in the house objected to the inclusion of items not related to war and hurricane relief, but enactment of the measure was never in doubt because it included money for troops in wartime. The House voted for the bill 351–67 and the Senate 98–1. The President signed it on June 15th.

By this time the wartime supplementals had become mini-appropriation bills, with the President submitting a bill and Congress changing it substantially through hearings, markups and floor debate processes, including debate over policy guidance, using the emergency provision to escape spending caps, inserting add-ons and earmarks into the bill and in the bridge supplementals using some smoke and mirrors to shift money from defense to domestic programs while making defense whole again either in the same bridge supplemental or in the regular DOD appropriation. The main difference is that the independent supplementals were passed rather quickly in the spring and the bridge supplementals lent an air of immediacy to the main appropriations bill as it supported forces in the field in time of war. They were all 'must pass' legislation. In form these supplementals were somewhat similar to DOD appropriation bills, with justifications, familiar

TABLE 6.1   War on Terror Supplementals FY 2001–2007

Year	DOD $ Appropriation	Year to Year % Change	Main $ Supplementals	Bridge Supplementals	Ratio (%) (to DOD base)	Ratio (%) with Bridge Supplementals
FY01[a]	297	3.13%	14		4.71%	
FY02	328	10.44%	17		5.18%	
FY03	375	14.33%	69		18.40%	
FY04	377	0.53%	66		17.51%	
FY05	400	6.10%	101	25	25.25%	31.50%
FY06	411	2.75%	115	50	27.98%	40.15%
FY07	435	5.84%	93.4	70	21.47%	37.56%

[a] The FY01 supplemental was the DOD share of the supplemental passed in
response to the attacks on the U.S. on 9/11/2001. The U.S. then began a
bombing offensive on the Taliban in Afghanistan on October 7, 2001.
Source: Computed from DOD FY2007 Emergency Supplemental Request: 1, taken
from defense appropriation acts, FY2001–FY2007.

line items, and comparison to the past and the future. Also it could be
argued that by 2005 more of the supplemental costs were for future costs,
as opposed to incurred costs, and thus a departure from our model. We
acknowledge that some future costs would certainly be payback for past actions, e.g., in warfare platform including vehicle and aircraft replacement.

The exhibits below (Figure 6.2) show these wartime supplementals from
FY 2001 through 2007. Data on all supplementals including war supplementals are included in Attachment A to this chapter.

In the case where wartime supplementals are passed to pay for future costs
of war, they became mini-appropriation bills for a specific purpose, not for
the whole function of defense, but rather targeted at a specific goal: winning
the war on terrorism which soon came to mean winning the war in Iraq.
These supplementals also had other characteristics of appropriation bills.

Wartime Supplementals are Complex

As with the normal defense appropriation bill, the wartime supplementals have become more complex. For example, in the 1990s a supplemental might simply have asked for reimbursement of operations and maintenance costs, but the supplemental presented to Congress in the spring of
2007 was very complex, with some money for standard operations of DOD,
some money for allies, some procurement money to rebuild force struc-

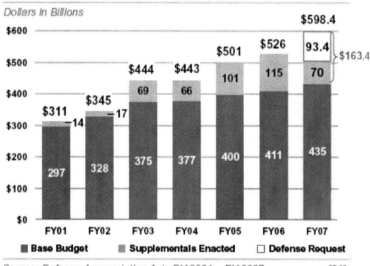

Figure 1. DoD Base Budget and GWOT
FY01 to FY07

Dollars in Billions

Source: Defense Appropriation Acts FY 2001 – FY 2007   55-06

Figure 6.2 (Figure 1): GWOT War Spending. Source: Department of Defense, 2007. FY 2007 Supplemental Request for the Global War on Terror. Washington, DC, February, 2007: 1.

ture (reconstitution), some money to defeat IEDs, and even a little pot of money that commanders in the field could tap to buy supplies and or use to provide for humanitarian relief (CERP).[7] The lesson here is that wartime supplementals are not simple.

The exhibit below (Figure 6.3) is testimony to the complexity of these wartime supplementals.

Wartime Supplementals: Normal Budgetary Logic and When it Applies

Normal appropriation bills often use historical comparison as a basis for analysis. This practice has carried over into the wartime supplementals. For example, the 2007 supplemental justification was presented with comparisons to the bridge supplementals for both 2006 and 2007 and the main 2007 supplemental.

This part of the supplemental request reflects the changing conditions of the battlefield, with the total for body armor increased 775% over the 2006 levels. Nonetheless there was some money for force protection in the

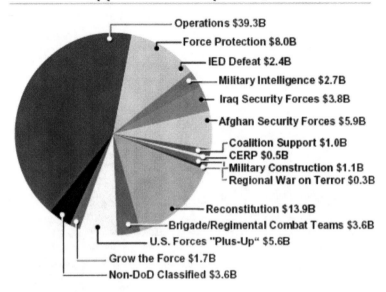

FY 2007 Supplemental Request: $93.4B

- Operations $39.3B
- Force Protection $8.0B
- IED Defeat $2.4B
- Military Intelligence $2.7B
- Iraq Security Forces $3.8B
- Afghan Security Forces $5.9B
- Coalition Support $1.0B
- CERP $0.5B
- Military Construction $1.1B
- Regional War on Terror $0.3B
- Reconstitution $13.9B
- Brigade/Regimental Combat Teams $3.6B
- U.S. Forces "Plus-Up" $5.6B
- Grow the Force $1.7B
- Non-DoD Classified $3.6B

Figure 6.3 Wartime Supplementals Complexity. *Source*: Department of Defense, 2007. FY 2007 Supplemental Request for the Global War on Terror. Washington, DC, February: 11.

bridge supplemental in 2006, in the regular supplemental for 2006, in the bridge for 2007 and in the regular supplemental for 2007. Some would argue that this is a good case for including this kind of funding in the base budget because it was a known and continuing need, had a track record, and would lend itself to analysis which might lead to an increase or decrease in the category.[8] Others would point to the percentage growth in the category and argue that this is precisely what a supplemental needs to fund, a need whose dimensions were either unknown or changing so fast that it could not wait for the regular appropriation bill.

Budget control staff such as those working in OMB and CBO and some DOD comptrollers would argue that such a huge percentage increase would be deserving of scrutiny and perhaps reduction, a normal part of the budget process. Warfighters would be expected to argue that forces were currently at risk and the task at hand is to meet the need and not worry about 'beancounter' concepts like 'percentage rate of increase.' At the end of the day, the 2006 and 2007 supplementals did look a lot like the regular defense appropriation bills and not like the stereotype of the 1990s defense rescue and peacekeeping supplemental.

As opposed to our model for regular supplementals, both years had funds for military construction, research and development, and procurement;

these are investment accounts as opposed to readiness accounts aimed at current operations. Spending on investment accounts is another indication of the complexity of wartime supplementals. Our model suggested that supplementals pay back for actions that were already over. As can be see from the exhibit above, current wartime supplementals have amounts in them devoted to long-term tasks to be undertaken in the future. Also these are of significant size and of high growth. For example, in both years the procurement account is third largest in dollar size. For FY2007, the procurement accounts primarily went to reconstitute the force, for such things as armored vehicles, trucks, and helicopters that were either destroyed or worn out, but some of those funds went for such things as the next generation strike fighter (the Air Force asked for two in one request), with the argument made that as the current generation planes were "used up" in Iraq there was no production line open to buy current models of these, and that reconstituting therefore meant stepping up to the next generation of equipment.

Costs may be differentially borne

Currently the majority of the costs of war are borne by the Army. Stanford's Larry Diamond has observed, "America is not at war. The U.S. Army is at war." (Friedman, 2007)[9] Of course, the Marine Corps and Special Operations Forces (e.g., SOFs from NAVSOC and USSOCOM) would dispute this view. Thus while the Navy and Air Force have been arguing for next generation ships and airplanes, Army recapitalization and transformational initiatives may come to grief under the sheer weight of past wartime activi-

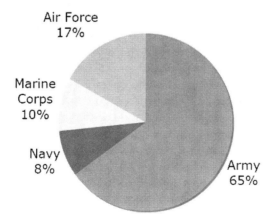

Figure 6.4 Supplemental Shares by Service. FY2006 Supplemental Breakdown by Service (From: Office of Management and Budget, FY2006 Estimate #3 Emergency Supplemental Appropriations) in Evans, 2006: 37.

ties in Iraq and past and current operations in Afghanistan. However, in fact, spending in Iraq also squeezed Navy and Air Force accounts because there was not enough money left over to pursue modernization activities, be they transformational or simply recapitalizing by buying more modern aircraft and ships. This illustrates another departure from our model of supplementals: the burden of the activity may be unequally borne within DOD, but with the passage of time it eventually impacts all DOD programs and financing.[10] Our model simply suggests that the President tasks DOD to do something and the military department that carries out the task is (or should be) reimbursed for that task on an incremental cost basis, i.e., is reimbursed for services rendered. However, wartime supplementals are not so simple.

Wartime Supplementals May Have Long Term Consequences

Our model of supplementals assumed an event or events that were quickly concluded and reimbursement was made for incremental costs of activities already concluded, over a known time period where cost factors could be derived, accumulated, and submitted for reimbursement. This is not the case with the current wartime supplementals (nor was it the case in Vietnam). In some respects, these supplementals are like a rolling declaration of war, with a promise to pay incrementally over time. As the graphic below indicates, the U.S. involvement in Iraq and Afghanistan may last years, perhaps beyond the deadlines for force withdrawal set by the Obama administration. After all, the first Gulf War did not end when conflict ceased in Kuwait in February, 1991; the U.S. has kept troops in the region at sea and on land bases, and has patrolled no-fly zones over Iraq and then Afghanistan since late 2001, and subsequently in cooperation with the UK and France (NATO) over Libya until 20 October when the former leader of this nation, Col. Moammar Gadhafi, was killed. Presently, it is not possible to predict with much confidence when operations in Iraq, Afghanistan and elsewhere in the region will end with a high degree of certainty although in 2011 President Obama reassessed and then set new target dates for withdrawal. However, it is not clear what the preferences of those national governments involved will be in the future as withdrawal deadlines approach.

From 2006 and up to the present it has become clear that the war efforts in Afghanistan and Iraq were and will continue to be sizeable and ongoing and that significant funding from Congress both in supplementals (if Congress and the President agree, which was in question as of late 2001) and in DOD regular appropriations (even more contentious an issue for FY2012 and beyond) would continue to be necessary, and that despite the passage of past large supplementals and the fact that the DOD base budget also has

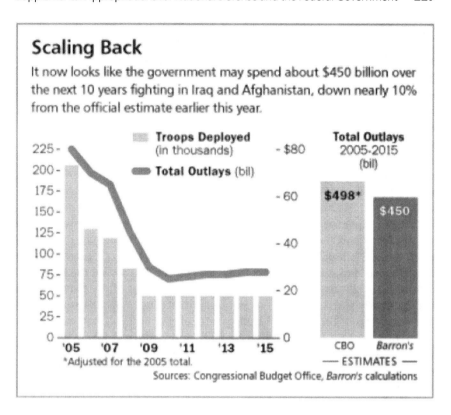

Figure 6.5 Supplementals Have Long Term Impacts. *Source*: Authors, 2011 including data from Evans, 2006.

grown rapidly, albeit not as fast as it did during the Vietnam build-up from 1964–69, DOD needs badly to invest in new and replacement weapons platforms and weaponry. President Obama announced that almost all U.S. forces would be withdrawn from Iraq by Christmas, 2011 on October 21, 2011. Still, some DOD spending in support of Iraq security and recovery will continue. Congress has become restive to this situation and called on several SECDEFs to task for not including known costs of war in the base budget. In a sense, Congress has been encouraging DOD to do what SECDEF Mc-Namara and his successors ultimately came to do during the Vietnam era: include the known and predictable costs in the base budget and ask for the unknown and emergency financing in war supplementals.

To some extent the budget process is always subject to gamesmanship and political maneuvering, and experts disagree about the levels of present and future funding and the need for supplementals versus regular appropriations. However, to Pentagon PPBES participants anything that has made it into the budget has survived fierce competition against items that were not funded, and reprogramming means not doing things that were of high enough priority on someone's list to make it through the appropriating process. Reprogramming means that someone's valued program is cut. Everything in the MILDEP and SECDEF's budget is tied back to an intended expense. To displace money from one item to another means a shortage in the first account. Some things do not get done and the cool discussion of reprogramming masks the real impact on programs and people of reducing or even canceling programs on short notice and within the budget year. Consequently, from this view war costs always should be paid for using supplemental appropriations. Also it is clear that any war financed through war supplementals is a partially funded war.

Starting in 2007 some partial-funding cleavages began to spring up concerning funding for operations in Iraq. Elsewhere we have discussed strains on the Army and the other services, but the partial funding syndrome applied to the National Guard as well. For example, a congressional commission chartered to study the effect of the war on the Guard and the Reserves, reported, "Right now in the United States, 88 percent of the Guard is not combat-ready when it comes to equipment... (and) ... that's worse than the worst days of the so-called hollow force in the late 70's and early 80's". The story went on to report that by 2007 approximately 200,000 Guardsmen had served in Iraq and Afghanistan, despite these shortages and despite the fact that they received less training than professional soldiers. (Dreher, 2007: A-8)

It also is clear that since 2007 a majority of citizens in the country had lost faith in the mission in Iraq and the supplemental process illustrated that fact as a Democratic Congress provided more money than the President requested, but also provided guidance in the supplemental about when and how to end the war in Iraq. In some ways the supplemental process had began again to resemble the main appropriation process. The vital question in 2012 is whether any more war supplemental appropriations will be proposed by the President and, if so, approved by Congress? It seems likely that efforts on the part of the President and OMB, and some members of Congress to cease introduction and passage of war supplementals will succeed. The impact of this change will have serious implications for the DOD and MILDEP base budgets.

What lessons may be drawn from analysis of war supplementals in the period 2001 to 2011? First, DOD officials have been forced by Congress to be more forthcoming about the costs of war; basically the supplementals have funded it and supplemental amounts are a matter of public record. Second, as the war operations in Afghanistan and against the forces of terrorism worldwide have continued some blurring has occurred between what has been funded in war supplementals and in DOD budget base appropriation. Ultimately, despite DOD base budget increases and large supplementals, the result appears to be either under-funding for the war effort or under-funding the rest of the defense requirement, especially the Operations and Maintenance (O&M) accounts and some acquisition programs.

This last point is more complex than it seems at first glance. If we consider that funding for the war in Afghanistan for example, the warfighting equipment that our forces have used and over-used and expend in supplies (e.g., gasoline and ammunition) should continue be financed in a war supplemental. Clearly transportation of forces to and from the operating theater should be in the supplemental, or at worst, funded specifically in the base if supplementals are politically unacceptable. Now to add additional complexity: how much of the training of forces should be funded as warfighting (time and equipment costs), some, e.g., 50%, or all? How much of the maintenance of bases where forces train or how much of the time of the instructors and materials and supplies support tail should be included in a supplemental or clearly funded in the base? From the view of DOD a substantial amount of such costs should be funded by supplementals but have not been and may not be in the future.

When helicopters used in operations or in operational training are damaged and repaired, clearly the immediate repair should be charged to warfighting costs, but what if the repair is accompanied by an upgrade that extends the service life of the asset by a few years? Should that be charged to warfighting? The same question may be asked of other types of equipment, including aircraft and parts that wear out and cannot be replaced by the same type because the production line has long been closed. Is the supplemental the appropriate place for this kind of expense, or will it be funded and specifically identified in the base, and if not, why not?

It seems that defense is not a separable public good in economic terms when the mission and operations extend over a long period of time. This is part of the tooth to tail argument where for every person in combat there are many military and civilians working in support roles in the training, equipping and supporting infrastructure, including medical care, family housing and more. It is no wonder then that during the period of war from 2001 to the present the base DOD budget has increased even while war supplementals have been passed on a regular basis. What should be funded in wartime supplementals or the base budget is a judgment call made by a number of different participants in the defense budget process. There is some logic to support the view that the supplemental and the main appropriation bill should be treated as a package as long as the casts of war are funded by some means in the budget process.

Finally, as shown in Appendix A, supplemental appropriations for war financing and other purposes were used repeatedly during the decade from 2000 to 2010, and for FY 2011 (not shown in the appendix). However, as we have noted, as of 2012 war supplementals may likely be a thing of the past, although the budget process in always subject to the political demands that fuel it. Further, if the base DOD appropriation is cut and war operations continue, this will reduce to a significant extent (some say "gut") many other DOD appropriation accounts. Further, if (or when) this occurs, figuring out what has been spent from the supplementals that actually has augmented the base and visa versa will be a difficult and time consuming process.

CONCLUSIONS

In this chapter we explored how supplemental appropriations have been used in both the non-defense and the defense areas of the discretionary budget. These are our findings. First, compared to the normal appropriation process, supplementals are usually passed expeditiously. Secondly, contrary to popular perceptions, supplementals do not always result in supplements. In fact, a supplemental package may result in a net decrease due to offsets. Thirdly, some supplemental bills also may supplement future year budgets, an action commonly thought to be in the province of future year appropriation bills. Fourthly, supplementals are commonly thought of as a tool the executive branch uses to supplement budget authority to meet a current year need but, in practice, Congress may substantially increase the size and scope of the supplemental bill and may take the lead in proposing the supplemental. Fifthly, although supplementals are commonly thought to be for emergency supplements for the current year, some supplementals are aimed at non-emergency and future year uses. This brings up another relevant budget process fact: to spend from the appropriation DOD still needs approval from the authorizing committees of Congress. Sixthly, when supplementals are designated as dire emergency bills, they tend to escape the control of budget discipline, e.g., contribution to the national debt. Congress and the President have found the supplemental appropriation procedure to be such a convenient tool that the emergency designation has slipped into use in regular appropriation bills. Seventh, although small compared to normal appropriation bills, supplementals and especially those for defense, rarely if ever fund 100% of the need or costs of war. Eighth, wartime supplementals differ from defense humanitarian and peacekeeping activities: they are large, complex, increase the base in some cases and always affect the base in some way, decrease budget transparency, affect all services no matter who the primary responder is, and will have short, medium and long term consequences. Finally, supplementals have great symbolic political importance[11] for they show an immediate governmental response to a current year crisis.

We have drawn the distinction between peacekeeping supplementals and wartime supplementals. The normal 'peacekeeping' supplementals are usually small, passed quickly, and typically are not controversial. However, wartime supplementals have been large, frequent and complex and the usual distinctions between appropriations and supplemental have been blurred. Clearly supplemental activities have resulted in both increases and decreases in the defense budget base. These supplementals have also affected current and future years and are not simply payback for tasks already accomplished. As with disaster supplementals, they are more an estimate than a rigorous costing out, and it became clear at least through 2010, just as with disaster supplementals, more money would follow. For these reasons, war supplementals cannot be as transparent as the humanitarian and peacekeeping supplementals and this lack of transparency may hamper both DOD and congressional oversight of the main and supplemental appropriation bills because DOD and Congress may not be able to define or track accurately, or be able to project how supplementals and regular DOD appropriations fit together. In fact, this is a difficult and at times impossible task even for DOD comptrollers to accomplish with absolute certainty about accuracy in the short-term. This has resulted in both under-funding and over-funding parts of the defense budget base. Moreover, even though it may appear as if one service or military department is garnering the lion's share of a supplemental, it still may not be enough to properly fund that service for their mission. Also, since 2001 wartime supplementals have lead to longer term commitments, whereas the 1990s supplemental missions[12] were usually completed relatively quickly. It also is apparent that the war on terrorism will continue to impose requirements that all of the military services will face and modernization their weaponry while continuing to sustain readiness.

Politics always has been at play in the funding of wartime supplementals as it is in the regular budget process. DOD has put some costs in the base budget that should have been there (Army troop strength) and Congress has made cuts in the main DOD appropriation bill that it later has funded in supplementals. For example, in 2005 funds were shifted out of the defense appropriations area into non-defense domestic areas and the resulting shortages in defense were then made up by taking money from the Iraq Freedom Fund, passed as a supplemental. Thus, the result was funding of continuing DOD operations out of a supplemental. This mixing of the base and the supplemental is something DOD and Congress will have to sort out the consequences once a President no longer supports and Congress ceases to pass war supplementals. Given the obvious usefulness of supplementals and considering their complexity, it would seem that supplemental appropriations deserve more attention than in the past as a useful tool in the budgetary process.

Finally, we want to make one more point about the financing of supplemental appropriations. Throughout the nation's history wars, and war operations in non-declared war conflicts, have been paid for using debt financing. This is the case with the war on terrorism since 9/11/2001 and will continue to be the practice for the foreseeable future. In fact, debt is a legitimate way to finance the extraordinary costs of fighting major war operations over long periods of time. Critics and Congress should bear this fact in mind, along with an understanding of the long history of war financing, when they try to blame defense for increasing the size of the national debt or when they consider ways to reduce federal government spending in the period 2011–2020.

APPENDIX A
Table 6.2: Supplemental Appropriations 2000–2010

CBO Data on Supplemental Budget Authority for the 2000's (in millions of dollars)

FTN Bill Number Public Law	Informal Title	Date Enacted		Request			Enacted				
				Discretionary	Mandatory	Total	Discretionary Supplemental	Rescissions	Total	Mandatory	Total
2000											
HR-4425 106-246	Military Construction, 2001	July 13, 2000	Def	2,289	0	2,289	7,360	-322	7,036	0	7,038
			Nondef	3,988	35	4,023	8,248	-148	8,100	35	8,135
			Total	6,277	35	6,312	15,608	-470	15,138	35	15,173
HR-4576 106-259	Defense, 2001	August 9, 2000	Def	0	0	0	1,779	0	1,779	0	1,779
			Nondef	0	0	0	0	0	0	0	0
			Total	0	0	0	1,779	0	1,779	0	1,779
		Total, 2000	Def	2,289	0	2,289	9,139	-322	8,817	0	8,817
			Nondef	3,988	35	4,023	8,248	-148	8,100	35	8,135
			Total	6,277	35	6,312	17,387	-470	16,917	35	16,952
2001											
HR-2216 107-20	Emergency Supplemental and Rescissions, 2001	July 24, 2001	Def	5,841	0	5,841	6,915	-1,081	5,834	0	5,834
			Nondef	702	936	1,638	2,064	-1,355	709	936	1,645
			Total	6,543	936	7,479	8,979	-2,436	6,543	936	7,479
HR-2888 107-38	Recovery and Response to Terrorist Acts	Sept. 18, 2001	Def	13,744	0	13,744	13,744	0	13,744	0	13,744
			Nondef	6,256	0	6,256	6,256	0	6,256	0	6,256
			Total	20,000	0	20,000	20,000	0	20,000	0	20,000
		Total, 2001	Def	19,585	0	19,585	20,659	-1,081	19,578	0	19,578
			Nondef	6,958	936	7,894	8,320	-1,355	6,965	936	7,901
			Total	26,543	936	27,479	28,979	-2,436	26,543	936	27,479
2002											
HR-3338 107-117	Defense Appropriations Act, 2002 (2nd $20 billion - 9/11 attacks)	January 10, 2002	Def	7,468	0	7,468	3,840	0	3,840	0	3,840
			Nondef	12,533	0	12,533	16,160	0	16,160	0	16,160
			Total	20,001	0	20,001	20,000	0	20,000	0	20,000
HR-4775 107-206	Emergency Supplemental and Rescissions, 2002	August 2, 2002	Def	14,048	0	14,048	13,884	-616	13,268	0	13,268
			Nondef	13,515	1,100	14,715	12,670	-1,721	10,949	1,100	12,049
			Total	27,563	1,100	28,763	26,554	-2,337	24,217	1,100	25,317
		Total, 2002	Def	21,516	0	21,516	17,724	-616	17,108	0	17,108
			Nondef	26,148	1,100	27,248	28,830	-1,721	27,109	1,100	28,209
			Total	47,664	1,100	48,764	46,554	-2,337	44,217	1,100	45,317

Source: Congressional Budget Office, 2010b: 1.

NOTES

1. It is an interesting irony that a supplemental can result in a decrement for someone when a supplemental is not designated as a dire emergency. In this case, some departments and programs will be used as billpayers and will find their programs decreased, not supplemented. If the money comes from an under executing procurement or mandatory account and goes into permanent salaries as a pay raise, not only will the salary account have to be increased in the following years, but the original funds will probably have to be restored to the procurement or mandatory program account. The net effect in future years may well be an increase all around.

2. Note that in 2005 the defense appropriation bill was the last one passed and it broke the discretionary spending caps and resulted in a 1% across the board cut. However the supplemental part of the bill was designated a dire emergency and sheltered from the cut. Then, in 2006, as is discussed later, Congress cut from the defense appropriation bill and used the defense cuts to fund their domestic priorities, while making up the defense cuts in the dire emergency supplemental part of the defense appropriation bill. This kind of behavior is sometimes called smoke and mirrors because Congress is using procedural magic to evade the discipline of agreed upon budget process mechanisms.

3. See for example, Defense Aid Supplemental Appropriation Act, 1941, Mar. 27, 1941, ch. 30, 55 Stat. 53; Defense Aid Supplemental Appropriation Act, 1942,Oct. 28, 1941, ch. 460, title I, 55 Stat. 745; Defense Aid Supplemental Appropriation Act, 1943, June 14, 1943, ch. 122, 57 Stat. 151.

4. The part of the appropriation bill that would contain the supplemental was labeled a 'dire emergency' and thus put that part of the DOD appropriation bill outside of any discretionary spending caps. This would become common practice with the later bridge supplementals.

5. At first glance it may seem odd that Army did not change more quickly at the end of the Cold War, but remember three things: Army and DOD did draw down about 33% in personnel levels in the early 1990's, DOD planners had to be prepared for the eventuality that the Russian experiment might not work out and central Europe could again become a battleground, and Desert Storm/Desert Shield in 1991 reinforced cold war planning and tactics since it was basically won by air power and the M1A1 tank and its gun, which allowed US tanks to stand our of range and destroy Iraqi tanks. Army and DOD did move to a new base force profile, but stopped well short of the transformation urged later by Secretary Rumsfeld.

6. Note here that this bill was passed on time. Also note that it is common for appropriation bills to fund the next fiscal year; the fact that the supplemental was passed in the current year, but funded activities at least through the next six months of the next fiscal year is a departure from our model because it is funding future costs, not past costs.

7. For further explanation, see FY2007 Justification Materials. This complexity also manifested itself in the budget execution process. One USMC officer observed that while in a financial management billet he had about 300 accounts

to monitor in the execution of these supplementals for Iraq and any account that showed a 10% variation from the previous month was flagged and had to be explained. His estimate was that meant about 10% of the accounts had to be explained each month, irrespective if the variance were up or down.

8. In fact, in 2005 Congress took the initiative in providing money to uparmor Humvees and for body armor which the Pentagon had not requested.
9. Diamond, as quoted in Friedman, 2007.
10. Remember also that one of the unintended outcomes of Vietnam was the elimination of the draft and the change to an all-volunteer force.
11. On a cautionary note, the huge supplementals passed quickly after 9/11/2001 and after Hurricane Katrina in September 2005 appear to have led to wasteful, inefficient, and even fraudulent fiscal behavior. The continuous deployment of huge sums of money in Iraq and Afghanistan for contractors in support of the military effort and for reconstruction seems also to have led to misuse of money. When it comes to money, good intentions are not enough; proper controls must be put in place for appropriate execution. If not, the cost of symbolic actions will be higher than anticipated, although the return on the investment may still be positive. In 2007 a GAO audit discovered more than 22000 cases of fraud stemming from Katrina relief efforts. See Cohen, 2007.
12. Efforts in Bosnia, Kosovo, and the No-Fly zone enforcement in Iraq are examples of missions that lasted longer than a year, but their fiscal consequences were relatively low and the mission requirements could be predicted and easily absorbed within the current DOD capability envelope, thus we chose not to treat them as wartime supplementals. After FY2001, these efforts continued at a steady pace and in FY2002 were directly appropriated into the annual budget for each military service in the O&M and Personnel accounts, according to Evans. (Evans, 2006: 15). See also Office of the Secretary of Defense, 2003. Justification for FY2004 Component Contingency Operations and the OCOTF, February.

ACKNOWLEDGEMENT

Some data for the analysis provided in this chapter has been drawn from annual volumes of the Congressional Quarterly Almanac, vols. 30–54, 1974–1999 and Table 4.1, Supplemental Appropriations, 1974–1999 in Godek, 2000.

CHAPTER 7

BUDGET EXECUTION IN THE FEDERAL GOVERNMENT AND THE DEPARTMENT OF DEFENSE

INTRODUCTION

Most of what is written about budgeting focuses on the factors that determine resource allocation decisions. Significantly less has been written about financial occurrences after that decision is made and approved. Budget formulation is about the processes of identifying requirements, making policy choices among requirements, estimating the costs to deliver those policies, and having them legitimized through the formal legislative appropriation and authorization processes. In contrast, budget execution is about the processes associated with managing the budget authority once appropriated: how it is distributed through the agency, the locus of decision and approval rights for actual spending decisions, how the status of budget authority is measured and reported, and how the fiduciary role of tracking budget authority intersects other business processes such as program management, payroll and contracting.

Execution is that stage in the budget process where matters of policy shift from being tentative and prospective to being real and the agency struggles with implementation. At this stage, budgetary and programmatic

Financing National Defense: Policy and Process, pages 235–278
Copyright © 2012 by Information Age Publishing
All rights of reproduction in any form reserved.

requests have been fulfilled, changed or denied. Winners and losers of the policy disputes have been determined and the task at hand is to do the winners' work. Thus, whereas the budget was a plan and authorization was the approval of the plan, execution works the plan. And whereas the budget requested authority and the appropriation granted authority, execution is the exercise of authority. *Exercising authority in support of a plan*—that is the essence of budget execution in the Department of Defense.

In most governments execution includes a very broad range of activities that span both the revenue and expenditure sides of government finance. Execution at the state and municipal levels of government is often defined by the task of balancing forecasted and collected revenue with planned and real expenditure of those funds. Amounts available to spend can be uncertain and contingent on factors such as unemployment rates and local economic activity. It is particularly challenging to balance mandatory spending programs in times of economic contraction when revenues fall and demands rise. Execution is defined at the state and local level to include functional areas like human resources, contracting, physical plant maintenance. It also includes cash flow management which involves things like tax collection, borrowing and debt service, capital expenditure planning, intra-governmental transfers, and investments.

In comparison, budget execution in the Defense Department is much simpler because it encompasses only a subset of those activities. DOD is a spending machine with little regard for balancing its appetite with the revenue stream (taxes) that pays for it. Budget execution is often done in offices separate from offices that perform HR and contracting tasks, although there is necessary interaction. The financing of capital projects is largely resolved in the budget formulation phase when full and incremental funding decisions are made. Still, despite a relatively narrow focus, budget execution is a critical phase and there is no shortage of challenges confronting a DOD program manager, budget analyst, or comptroller.

Complex public organizations usually seek to achieve multiple goals, some of which conflict. That demands tradeoffs and compromise across the goals. With respect to budget execution those goals normally include:

- complying with law and regulation
- ensuring rapid and accurate processing of administrative tasks
- controlling the activities of the organization
- preserving flexibility over resources in order to adapt to contingencies
- mitigating the effects of funding shortfalls or unexpected mandates
- managing information flows
- identifying the locus of decision rights for financial matters
- measuring the cost and performance of the organization

- reducing the burden of overhead functions
- bridging the goals and analytical frameworks of line managers, budget staff, and program evaluators/auditors (Joyce, 2003; McCaffery and Mutty, 2003)

In this chapter we address the topic of budget execution in the Defense Department by suggesting these goals fall along a continuum where one end is anchored by strict compliance with rules of appropriations and tight operational control of financial activities and the other end is anchored by a strong sense of mission and program performance (sometimes in spite of the rules!). The organizational activities associated with budget execution, of course, are positioned along this continuum and when all goes well missions are accomplished while following all the rules. The chapter begins with an outline of defense budget execution fiduciary responsibilities: key aspects of appropriation law and the fundamental tasks associated with budget execution. This opening section is viewed from the fiscal control end of the spectrum and suggests the view of a comptroller. The section after considers the perspective of a program manager or operational commander for whom program success and mission support are paramount. There the chapter covers budget execution activities that are designed to support organizational goals more than compliance goals. The section after that addresses problems and special cases—those matters that seem to impede both good budget practices and good program management. The chapter concludes with some general observations about the nature of defense budget execution that draw from the entire chapter and other portions of the text.

OBJECTIVES

The objectives of this chapter are to provide the following understandings:

1. To distinguish the operational and management control aspects of budget execution; and distinguish budget execution from program execution
2. To explain the legal framework within which the DOD executes it budgets
3. To describe the flow of funds from appropriation to obligation
4. To be able to discuss the importance of measuring budget execution
5. To understand the primary activities that define the execution phase of budgeting to include allocation, creating slack, measuring and reporting, mid-year review, reprogramming, and the end of year dynamic

6. To relate implications of choices regarding the methods of allocating budget authority
7. To identify special cases that disrupt the routine of budget execution
8. To describe the revolving fund (working capital fund) model and distinguish the key differences between it and traditional mission funding of defense programs.

FIDUCIARY RESPONSIBILITIES

The introduction to this chapter implies that budget execution consists of a set of activities. When one "does budget execution" what is one doing? First is the flow of funds to the organization in the first place. Once an appropriation is enacted, the spending authority must pass to the leaders of the organization who are responsible for managing those funds and making the ultimate spending decisions. They, in turn, will normally allocate portions of those funds throughout their organization following one or more methods. The spending itself—which we refer to as entering into obligations (the government obliges itself to make a payment from the Treasury in exchange for goods or services received)—is done through the organization's business processes such as human resources for payroll, facilities management for rents and utilities, and contracting for goods and services. Because departmental spending is the exercise of legal authorities provided through appropriations, money is not as fungible as it is in the corporate sector. There are restrictions that limit what the department may buy and when. Throughout this process, measurement and reporting systems track the status of the funds, how much has been obligated or expended, by whom, for what. Such tracking ensures compliance with the restrictions and, because budgets are financial plans to achieve organizational objectives, managers review progress toward those objectives and suggest reallocations of current funds and seek to influence the formulation of future years' budgets. In the sections that follow, these activities and characteristics are described in detail.

Operational control

All budgeting systems, to some extent, serve three purposes for an organization: they support *planning at the strategic level* and are concerned with alignment of resources for the organization, they support *management control at the goal or project level* and are concerned about efficiency and effectiveness at the sub-organizational level, and they support *operational control at the task level* and are concerned about the accountability of routine processes (Schick, 1966). Budget formulation tends to bridge the two higher order layers where decisions are prospective; budget execution tends to bridge the two lower layers where decisions are immediate and involve implementation of plans.

Section 4 of this chapter covers the management control layer. At the operational control layer, effective budget execution means that the organization as a whole, and managers and staff individually, comply with all specific terms and conditions of appropriations, principles of relevant fiscal law, and agency rules and regulations. It means that all the transactional processes that perform the activities of budget execution (described in the next few sections) are accurate, repeatable, and auditable so that the information systems that capture those transactions provide high quality management information. These transactional processes include business functions like awarding and administering contracts, managing human resources, allocating and tracking funds, but also the primary line functions of the organization to the extent those functions consume financial resources. These may be operational (flying aircraft), support (maintenance or testing), overhead (headquarters administration or training), or services (health care). Finally, operational controls should limit staff discretion to a range of choices that are designed to support both mission accomplishment and legal compliance

Terminology, Appropriations, Fiscal law

It is important to note that most of this chapter is about the flow and use of legal authority, not the flow or use of funds. With the exception of the occasional disbursing officer in the field (who actually serves as an agent of the Treasury Department in that role), there is almost no money in the DOD. The money comes into and out of the Treasury based on information provided by the DOD. Defense officials very rarely "spend money." More precisely, they exercise the *legal authority under an appropriations act to bind the government to make a payment from the Treasury*. We call this budget authority. Because the Constitution says that "no money shall be drawn from the Treasury, but in Consequence of Appropriations made by Law" we necessarily begin with some terminology about appropriations and the process of obligating the government. After that, basic concepts of fiscal law are presented since these rules constrain the activities of defense managers.

Budget authority allows agencies to acquire goods and services, hire people, make loans, provide grants and guarantees, all of which may result in outlays from the Treasury. Congress holds the "power of the purse" which means they authorize programs to exist and appropriate budget authority to implement and sustain them. Through appropriations, Congress grants government agencies the authority to bind, or oblige, the government. When a government official signs a contract, hires and employee, enters into a support agreement with another agency, or provides a grant, they have formally and legally bound the government to make a payment in exchange for the successful completion of those tasks. Such binding actions are called obligations.

An obligation is a *legal reservation of funds.* It signifies that the government has entered into an agreement with another party and has taken that agreement the point where, should the government take no further action, but the based on sufficient actions of the other party, a payment from the Treasury is required. A commitment is an *administrative reservation of funds* that signifies a tentative agreement that has not yet "ripened" into an obligation. For example, budget authority set aside for a potential contract award fee would be recorded as a commitment; it would be an obligation once it was clear that an award fee had been earned and was payable. An expenditure, in traditional federal government terminology, an outlay, is the actual *payment from the Treasury* made to satisfy or liquidate an obligation. Expenditure is nearly synonymous with disbursement, outlay, or liquidated obligation. Examples include a government employee paycheck or a paid contractor's invoice. Expenditures may occur at the time of obligation, such as with a cash purchase; quickly thereafter, such as with payroll or a credit card purchase; or they may occur years later, such as with the construction of a building or ship.

Most defense appropriations are definite, meaning they set an upper limit on the amount of obligations. They are also definite in the time period during which obligations can be made. Expense-type appropriations like O&M and MilPers generally have a one-year obligation availability period meaning that new obligations may only be created during the fiscal year of the appropriation that will fund the liability. Investment-type appropriations generally have multiple year obligation availability periods to deal with the complexity and long lead times to acquire a vehicle, aircraft or ship.

Figure 7.1 shows a timeline for the life of an open appropriation. From the time of enactment or the first day of a fiscal year (whichever is later) the appropriation is open and new obligations can be made citing that budget authority. At the end of the specified obligation period, the appropriation expires, meaning no new obligations may be created. Existing obligations may be modified or cancelled, though. Expenditures can be made at any time after the obligation is made until five years after the expiration date, at which time the appropriation closes, or lapses. It is commonly said that at the time the appropriation expires, any unused budget authority is returned to the Treasury. This is an unfortunate expression because no funds had left the Treasury. An appropriation is not a bag of currency extracted from the Treasury; it is merely permission to bind the Treasury. What ex-

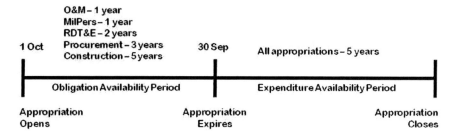

Figure 7.1 Appropriation Timeline.

pires is the authority to bind and if that authority is never exercised, the Treasury never moves a cent.

The term "appropriation" derives from the Latin *appropriare* meaning "to set aside" or "to make one's own". Fiscal law reinforces these notions. A fundamental principle of fiscal law is that appropriations are made for particular uses, to the exclusion of all other uses; there is a nearly 1:1 match between any object of expenditure and the proper appropriation to fund it. Nearly because in a small number of cases a particular object of expense could be said to fall equally well into two appropriations and the department has some discretion to choose which to use.

All appropriations have three characteristics that bind the actions of government managers. For an appropriation to be available for a legal expenditure of funds, all three of the following must be observed: (1) the purpose of the obligation or expenditure must be authorized; (2) the obligation must occur within the time limits prescribed by the Congress; (3) the obligation and expenditure must be within the amounts prescribed by the Congress (GAO, 2004). Purpose, time, and amount are the three primary limits placed on an agency. In short, Congress determines what an agency will do, when they will do it, and to what level of effort. "For almost all budget managers this body of law comprises a set of rules that guides their daily actions. For example, they know that they are not to commit funds before they have been appropriated, that they may not spend in excess of an appropriation, and they may only spend the appropriation on items for which the appropriation is made" (Jones and McCaffery, 2008, 322).

The purpose doctrine found in 31 USC §1301 states: "Appropriations shall be applied only to the objects for which the appropriations were made except as otherwise provided by law." This is a positive authority meaning that an authorization to spend and an actual outlay are valid only if authorized in the appropriation. It is not the case that any expenditure is authorized unless prohibited (a negative authority). The courts[1] have held "The established rule is that the expenditure of public funds is proper only when authorized by Congress, *not* that public funds may be expended unless prohibited by Congress." Some appropriations are quite narrowly written where both the desired ends and means to those ends are defined. For example, the fiscal year 2006 Defense Appropriations Act included this language:

> "SEC 8105. (b) From amounts made available in title II of this Act under the heading "Operation and Maintenance, Marine Corps", the Secretary of the Navy shall make a grant in the amount of $4,800,000, notwithstanding any other provision of law, to the City of Twenty-nine Palms, California, for the widening of off-base Adobe Road [...]" (Defense Appropriations Act, 2006: 2322)

This language is so specific that there is no discretion; the USMC must use part of its O&M,MC appropriation to widen this city street. Other appropriations are rather vague, allowing managers great discretion in determining how best to fulfill the mission. An extreme example occurred on September 18, 2001, when an appropriation was enacted (P.L. 107-38) that provided $40 billion, with no expiration, to the President who could transfer those funds in any amount to any agency, to spend on anything related to recovering from and responding to the terrorist attacks of a week earlier. Funds could be used for things ranging from cleaning "ground zero," to deploying forces to fight terrorism, to increased security of the nation's transportation system, to prosecuting offenders.

But even in the more vague cases, the discretion is limited by a legal rule called the "necessary expense doctrine." An early statement of the rule states that "when an appropriation is made for a particular object, by implication there is authority to incur expenses which are necessary, proper or incident, for execution of the object, unless there is another appropriation which is more specific about a provision, or if the purchase is unlawful."[2] A necessary expense is one that is vital to the success of the authorized program; a proper expense is one that supports, but may not be critical to, the program; incidental expenses are only tangentially related, but are somehow tied to an otherwise necessary or proper expense. There are degrees of necessity. And one should not impute on the necessary expense rule a requirement for efficiency or economy. One is not limited to the most efficient or best alternative under the necessary expense rule. However, a logical connection cannot be construed too broadly. The tastes and desires of a senior official, a best practice from private industry, or the importance of the item do not, in and of themselves, constitute a logical connection (Candreva, 2008).

A significant restriction on the use of appropriations derives from the structure of the budget request. If the DOD submits a budget requesting a particular dollar amount and justifies that with several pages of budget exhibits, and Congress appropriates that exact amount in a single line item identifying the name of the program, the legal presumption is that Congress intended what the budget requested and the program manager is expected to do what the budget described. For the majority of defense programs, this is precisely what happens each year.

The time restriction, as noted earlier, relates principally to that period during which obligations can be made. Appropriation acts give two basic authorities; authority to incur obligations and the authority to make expenditures. Regarding time, GAO states: "A time-limited appropriation is available to incur an obligation only during the period for which it is made. However, it remains available beyond that period, within limits, to make adjustments to the amount of such obligation and to make payments to liquidate such obligations" (GAO, 2004: 5-4). Federal law (31 U.S.C. §1502) states, "The balance of a fixed-term appropriation is available only for payment of expenses properly incurred during the period of availability or to complete contacts properly made within that period of availability." Expenditures may occur prior to the expiration of the appropriation, but per Figure 7.1, the department has five years to liquidate the obligations after the appropriation expires.

With respect to time limits, there are single-year, multi-year, and no-year appropriations. All appropriations are assumed to be a single-year annual appropriation unless they are otherwise stated, thus they are available for new obligations between October 1st and September 30th, the federal fiscal year. In the DOD, O&M and MilPers appropriations are single-year, but RDT&E is a 2-year appropriation, Procurement is a 3-year appropriation, Construction is 5-year, BRAC and working capital funds are no-year appropriations. Questions about the propriety of the timing of an obligation are resolved by applying the *bona fide needs test*: a fiscal year appropriation may be obligated to meet a legitimate, or bona fide, need arising in (or, in some cases, arising prior to but continuing to exist in) the fiscal year for which the appropriation was made. Bona fide need questions address the concern whether an obligation bears sufficient relationship to the needs of the agency in the period during which the appropriation was available. An example: is an item ordered on September 10th that is not expected to be delivered until October 20th (next fiscal year), a bona fide need of the prior year or the subsequent year? If the need actually arose in September, "yes." But if this purchase is merely using up expiring budget authority and there is only an expectation it might be needed in October, "no." The key is when the need arises, not when the need is satisfied.

In some cases, special authorities can be requested to get around the limitations imposed by the bona fide rule. Advance procurement appropriations permit the acquisition of long lead time items in acquisition programs in advance of the funds provided to complete the purchase. Multi-year procurement authorities allow the DOD to contract for several years of production to garner cost savings.

Restrictions on amount take several forms and are not as simple as a prohibition against spending two million dollars when an agency has only been appropriated one million. The restrictions are concerned more generally with spending money one does not (yet) have. The Antideficiency Act is the principal statute that addresses the amount characteristic. The Antideficiency Act prohibits:

- Making or authorizing an obligation or expenditure from any appropriation (or apportionment) in excess of the amount available in the appropriation (or apportionment). 31 U.S.C. § 1341(a)(1) (A) and 31 U.S.C. § 1517(a).
- Involving the government in any contract or other obligation for the payment of money for any purpose in advance of appropriations made for such purpose. 31 U.S.C. § 1341(a)(1)(B).
- Accepting voluntary services for the United States, or employing personal services in excess of that authorized by law, except in cases of emergency involving the safety of human life or the protection of property. 31 U.S.C. § 1342. (GAO, 2004: 6-36 and 6-37)

The first is the common definition and prohibits spending more than is provided in the appropriation or a legal subset of that appropriation, such as an apportionment or allotment. The second prohibits writing contracts or hiring personnel before the appropriation is made. The third prevents the employee or contractor from working "for free" pending that appropriation if they intend (or have the right) to eventually bill for their efforts.[3]

To round out a discussion of the legal framework for execution, it is important to note that the recording statute (31 USC 1501) requires that accounting records accurately portray the status of an appropriation. ". . . an amount shall be recorded as an obligation of the United States Government only when supported by documentary evidence." Both the under-recording of obligations (e.g., to avoid the detection of an Antideficiency Act violation) and the over-recording of obligations (e.g., to boost the apparent rate of spending) are illegal. Recording obligations only evidences them, it does not create them. An obligation exists when the action of the agency creates it, and then it must be documented and recorded.

Flow of funds

Once an appropriation bill is enacted and after the start of the fiscal year, the Treasury Department issues an appropriation warrant to the Office of

Management and Budget (OMB). A treasury warrant is a financial control document. The warrant establishes the amount of funds authorized to be withdrawn from Treasury accounts for each appropriation title. When an appropriation covers different agencies, OMB must apportion the correct amount to each agency covered by the act, as for example to the Departments of State, Commerce, and Justice from the appropriation act covering all three departments (Jones and McCaffery 1997: 18).

Given the warrant, OMB may then apportion funds to the agency. Apportionment is the distribution of an amount available for obligation in an appropriation for specified time periods, activities, projects or combinations thereof as approved by OMB and the Office of Under Secretary of Defense (Comptroller). The purpose of apportionment "... is to ensure that agencies spend at a rate that will keep them within limits imposed by their annual appropriations" (Lee, Johnson and Joyce, 2004: 328). O&M funds are apportioned by calendar quarter by the OMB under the authority of Title 31 Section 1513 of the U.S. Code. The apportionments are generally available on a cumulative basis. Department Secretaries are responsible for enacting regulations to administratively control and divide the apportionment. The system is designed to limit obligations to the amount apportioned, to fix responsibility for violations of the apportionments, and to provide a simple way to administratively divide the appropriation between commands. Apportionments are an effective tool for preventing overly rapid obligation of total funds at any point in the fiscal year. However, because funds are available on a cumulative basis, apportionments are better able to prevent obligation surges at the beginning of the fiscal year than at the end of the fiscal year (Jones and McCaffery, 1997: 39–40). Apportionments are requested by the agency based upon the spending plans created when the budgets were formulated. As part of the budget formulation process, major commands submit phasing plans (or obligation plans) that describe the expected patterns of spending based on projected contract award dates, manning levels for payroll, and scheduled activities. Normally individuals commands are apportioned funding based upon the plan they submitted. They are later held accountable for deviations from that plan, as we discuss in the next two sections.

Once the agency receives its apportionment they allot funds to subordinate organizations. The Undersecretary of Defense (Comptroller) allots to the military departments their shares of the defense appropriation. In turn, the Assistant Secretaries (Financial Management & Comptroller) of each service allot to their major commands the authority to incur a specific amount of obligations in specific accounts. For example, in the Navy the ASN(FM&C) will allot Aircraft Procurement, Navy (APN) budget authority to the Naval Air Systems Command. Each major command will further distribute budget authority to subordinate organizations, program offices, installations, etc. For expense-type accounts (O&M, MilPers) these are often referred to as operating budgets; for investment-type accounts (procurement, construction) these are often referred to as allocations. At the operational unit level (e.g., a ship or squadron), the term operating target (OPTAR) is used.

As authority is allocated down the chain of command so is the responsibility to meet the provisions of the Antideficiency Act (31 U.S Code 1341 and 1517) and not overspend. This strict legal responsibility normally extends to the major command or one of its major divisions. This office is said to hold operating budget responsibility and the offices below it are guided by planning targets, such as operating targets, allowances, and expense limitations. Those who hold legal liability are often staffed with formally trained and senior level comptrollers, lawyers, accountants, and other staff to ensure such strict liability is maintained. Quite properly, the Military Departments have flexibility to determine the appropriate level for the Antideficiency Act responsibility in order to see that small budget holders are not subject to over-control or held to account for spending patterns not within their administrative control (Jones and McCaffery 1997: 19). Thus, we find a ship or squadron may have a budget officer, but no comptroller or lawyer.

Creating Obligations

At the end of the process of allocation commitments are made, obligations incurred, and eventually outlays paid when the work or service is completed or the equipment delivered. Commitments, as noted earlier, signify that the process for obligating funds has begun. Obligations generally come in one of three forms: Contracts with private industry, intragovernmental reimbursable transactions (one government activity does work for another and is reimbursed for that), and payroll. Payroll is rather straightforward and is normally obligated on a monthly basis using real-time data about the number of employees and cost of their salaries and benefits. In many cases, the payroll is met by obligating the funds appropriated for the mission of the organization. In some cases, if the person is doing work to benefit another organization, the cost of the person's salary and benefits may be obligated against the benefiting organization's funding.

Over 25% of the defense budget formally changes hands inside the government before a dollar leaves the treasury in a payroll check or payment to a vendor. No command is entirely self-sufficient and there are commands that function solely in a support role. Anytime a ship or squadron requisitions a spare part, an acquisition program office hires a warfare center for engineering support, or when a tenant command enters into a support agreement with a host command, there is an exchange of budget authority for goods and services, not unlike a contract with the private sector. The department uses Military Interdepartmental Purchase Requests (MIPRS) in lieu of contracts and they permit the supplying command to use the budget authority of the customer command. Entering into the agreement is an obligation for the buyer and an increase in budget authority for the supplier. As the supplier incurs costs to do the work, they obligate the budget authority that was provided under the MIPR.

Contracts represent the largest dollar amount of obligations. A contract is a bilateral agreement between the government and a commercial vendor to provide a specific good or service to the government. There are extensive rules governing contracts which can be found in the Federal Acquisition Regulation (FAR). At the time a contract is signed, the dollar value of the terms and conditions that are firm must be obligated. Contingent costs, such as growth on a cost-reimbursement contract, award fees, penalties, options, or change orders are obligated at the time they occur or, in some cases, may be obligated based on a most likely scenario given past performance and adjusted later.

Measuring and managing obligations and expenditures

To this point, we have summarized the distribution of budget authority, defined what it means to exercise that authority and described the actions that create obligations. Because appropriations are limited in time and amount, it is both necessary and important to measure spending. It is necessary to ensure compliance with fiscal law and it is important to ensure that resources available to the department are used effectively. If Congress appropriates $x to the DOD to protect the United States and advance its interests, they expect to get $x of value. The commonly accepted evidence of full value is full use of the resource. If the DOD is fortunate or well-managed and can deliver $x worth of security for under $x, the expectation is that even more security should be provided. The legislature does not want the balance to go unused; in their collective mind the investment is made. The nature of defense budgeting is that requirements exceed resources, so institutional norms inside defense are that all the resources should be consumed. Remember from the budgeting lesson: budgets create expectations and serve as social and legal contracts between branches of government. The expectation is that the U.S. will receive as much national security as $x will buy. Thus, the DOD feels compelled to spend $x, and thus needs to measure progress toward that goal.

The most common measure of budget execution is the obligation rate. The obligation rate is simply the amount of budget authority that has been obligated divided by the total budget authority. It is expressed as a percentage. Similarly, an expenditure rate is also calculated using total budget authority as the denominator. Assume a command or program office expects $25 million in budget authority. As of the end of April and they have obligated $17 million. Their obligation rate would be $17 / $25 = 68%. Is 68% a good number? April is seven months into the fiscal year and 7/12 is 58% so their obligations are higher than a straight line projection. But is a straight line projection good? Perhaps, if obligations represented work accomplished, but they do only some of the time. If most of the obligations support government labor, a straight line may be appropriate, but if much of the work is contracted, an accelerated obligation rate would be better. Why? Work cannot begin until funds are obligated. Assume a contract for a year's worth of support services. That contract would normally be signed at the beginning of the year with the full year's worth of funding obligated at the time. So obligations normally occur at an accelerated pace, but this still does not tell us if 68% is good.

During the budget formulation stage, each budget submitting office prepares a phasing plan. A phasing plan is a month-by-month projection of the rate at which funds are expected to be obligated. The phasing plan is built by considering factors such as payroll, contract award dates, and the timing of projected intragovernmental purchases. At the department level, summing all the phasing plans for all the commands provides the basis for the quarterly apportionment requests. Commands normally receive funds at approximately the pace they projected in their budgets that they would obligate those funds. In our example 68% is considered good if it matches the phasing plan submitted by the command in the budget. We see the budget performance of our example in Figure 7.2. They received about $8 million in their first quarter apportionment, another $9 million in the second quarter and now have about $23 million in April. Their phasing plan projected that at the end of April, they would have obligated about $18 million (72%), their obligation rate is currently 68%, so one could say they are under executing: they are obligating at a pace below their stated plan.

The example above is at the individual command or program office level. At the department level, such data are aggregated and similarly measured. Tables 7.1 and 7.2 show the results of a study of 14 fiscal years of obligation data for operation and maintenance spending. As expected, O&M account managers obligated over 99% of their funds before the appropriation expired at the end of the year. Rates of obligation varied among the military departments but tended to follow a very consistent pattern. The pattern is most clearly seen in Table 7.1 and the graph that follows it, Figure 7.3.

In general, the *first and last* months of the fiscal year are the months with the highest obligation rates. For reasons noted above, October generally has a high rate of obligation because annual contracts are being awarded and any pent up, unfunded needs from the prior year may be acquired.

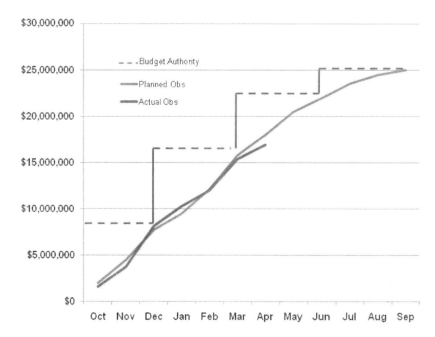

Figure 7.2 Sample Phasing Plan vs. Actual Obligations. *Source*: Authors, 2011.

TABLE 7.1 Average DOD O&M Monthly Obligation Rates with High-Low Ranges (Percent) 1977–1990

Month	Average	High	Low
October**	11.235	12.3	9.83
November	8.018	9.73	7.03
December	7.346	8.18	6.47
January*	10.048	11.42	8.11
February	7.165	8.35	6.1
March	7.223	7.96	5.65
April*	9.083	10.61	8.3
May	6.708	7.61	6.09
June	6.726	7.49	5.89
July*	8.778	10.57	7.43
August	6.887	7.39	6.17
September	10.616	12.05	9.78
Total	**99.833**		

Note: High = highest October and low = lowest October in 14 year period
** = start of fiscal year; * = first month of quarter.
Source: Jones and McCaffery, 2008: 344

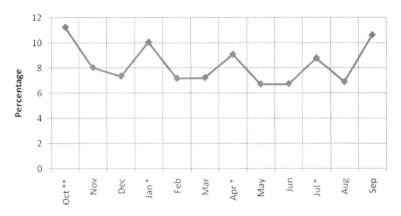

Figure 7.3 1977–1990 O&M Monthly Obligations
** = Start of Fiscal Year, * = first month of quarter
Source: Jones & McCaffery 2008, 346.

The summer months tend to have the slowest obligation rates as much of the funding is already "in place" on contracts and intragovernmental purchases, and because fund managers are positioning the organization for the end of year dynamic. Heading into September, funds on expiring projects are examined to ensure there is no excess or shortage, funds held back for contingencies are released, and plans are beginning for the transition to the next year. Because of the expectation to spend all budget authority coupled with Antideficiency Act concerns about overspending, September is normally characterized by an end of year spending surge. Whereas August records the lowest obligation rate, September and October have the two highest.

   Also to be noted is a quarterly pattern. In a "normal" year the first months of the second through fourth fiscal quarters (January, April, and July) experience the third, fourth and fifth highest rates of obligation. Every quarter begins with a high rate of obligation, followed by two months of successively lower rates. This pattern is due to the quarterly apportionment of funds to the department. New budget authority is quickly obligated to ensure work continues and, in years when funding is particularly tight, pent up demand. The last quarter of the year is the notable exception to the pattern.

   While our analysis has focused on O&M spending, other appropriation titles have similar expectations and patterns of obligation. Multi-year appropriations such as procurement and RDT&E do not have the pronounced quarterly patterns seen in O&M, but there are expectations for rates of obligation and expenditure. Table 7.2 reflects the standards that were published by DOD for fiscal year 2010. One should note that although RDT&E appropriations are available for two years, the expectation is that 93% of

TABLE 7.2   Obligation and Expenditure Rate Norms for Other Procurement, Navy and Research, Development, Test and Evaluation, Navy Appropriations

APPN	FY	Month	%Oblig	%Expen
OPN	FY10	October	7.17%	2.08%
OPN	FY10	November	14.33%	4.17%
OPN	FY10	December	21.50%	6.25%
OPN	FY10	January	28.67%	8.33%
OPN	FY10	February	35.83%	10.42%
OPN	FY10	March	43.00%	12.50%
OPN	FY10	April	50.17%	14.58%
OPN	FY10	May	57.33%	16.67%
OPN	FY10	June	64.50%	18.75%
OPN	FY10	July	71.67%	20.83%
OPN	FY10	August	78.83%	22.92%
OPN	FY10	September	86.00%	25.00%

APPN	FY	Month	%Oblig	%Expen
RDTEN	FY10	October	7.75%	4.58%
RDTEN	FY10	November	15.50%	9.17%
RDTEN	FY10	December	23.25%	13.75%
RDTEN	FY10	January	31.00%	18.33%
RDTEN	FY10	February	38.75%	22.92%
RDTEN	FY10	March	46.50%	27.50%
RDTEN	FY10	April	54.25%	32.08%
RDTEN	FY10	May	62.00%	36.67%
RDTEN	FY10	June	69.75%	41.25%
RDTEN	FY10	July	77.50%	45.83%
RDTEN	FY10	August	85.25%	50.42%
RDTEN	FY10	September	93.00%	55.00%
RDTEN	FY09	October	93.58%	58.00%
RDTEN	FY09	November	94.17%	61.00%
RDTEN	FY09	December	94.75%	64.00%
RDTEN	FY09	January	95.33%	67.00%
RDTEN	FY09	February	95.92%	70.00%
RDTEN	FY09	March	96.50%	73.00%
RDTEN	FY09	April	97.08%	76.00%
RDTEN	FY09	May	97.67%	79.00%
RDTEN	FY09	June	98.25%	82.00%
RDTEN	FY09	July	98.83%	85.00%
RDTEN	FY09	August	99.42%	88.00%
RDTEN	FY09	September	100.00%	91.00%

the appropriation be obligated in the first year. The logic is that RDT&E is inherently risky and uncertain so the second year provides flexibility, but because new funds are appropriated each year, no program should plan to obligate in the second year. If the funds are not needed until the second year, they should be budgeted in the latter year. Likewise, Other Procurement, Navy—a miscellaneous procurement account—is mostly (86%) obligated in the first year, even though it is available for three years. The longer time frame recognizes that these are capital purchases that may take more than a year to negotiate and contract. Because these are capital purchases, often with a long lead time before they are fulfilled, the expected expenditure rate as of the end of the first year for procurement is much lower than it is for RDT&E (55%) or OMN (75%).

Creating Budget Slack

The budget formulation and review process is designed to buy as much defense program as possible within resource constraints. The process is designed to scrutinize budget requests and to ensure they are funded as economically as possible. Spending categories that appear discretionary and non-specific, such as "management reserve" or "change orders," are routinely cut and funds allocated to seemingly more productive use. The benefit of doing so is to buy more programs; the danger is that all programs may be underfunded or there is insufficient slack in the budget to deal with the inevitable contingencies. Project management best practices acknowledge the need for a management reserve to deal with the inevitable contingency (Frame, 2002). Likewise, governments should have some slack to ensure effective implementation of policy initiatives (Schick, 2009). However, excess slack is wasteful and reduces the incentive on managers to be effective and creative. (Merchant, 1981, Mann, 1988, Yuen, 2004) Many budgeters learn how to hide slack in their budget, but assuming such slack is stripped away during the budget formulation phase, it is incumbent on the budgeter to create slack during execution.

"A substantial portion of budget execution is driven by the necessity of rescuing careful plans from unforeseen events and emergencies and unknowable contingencies." (McCaffery and Mutty, 2003: 77) What can the budgeter do to create slack in the budget to respond to these events? As funds are allocated to subordinate units, a small percentage may be held back in a discretionary account. While this results in the organizational units receiving less than they planned for, it creates an incentive for them to be creative to live within tighter means. Absent this, the incentive is to spend all of it regardless of whether the mission could be fulfilled for less. Some percentage of the organization will successfully execute the mission with this smaller share. Some other percentage will not; they will become the recipients of the discretionary funds. These reallocations are often identified during the mid-year review, discussed in a following section of this chapter.

Budget slack may also be created by managing the shape of the workforce. If there is a high rate of employee turnover, the organization may choose to operate part of the year understaffed to save on payroll expenses to free those funds for other uses. Another technique is to "sell" excess capacity to other organizations that may be able to offset operating costs and some fixed costs through the reimbursement process. Favorable price or foreign currency exchange rate fluctuations may create windfalls in the budget that can be held to cover contingencies. And there are scores of other management techniques that may be used to save funds, ranging from more favorable contract terms to lean process designs.

One concern for the budget office is that slack in the budget may be revealed through lower obligation rates. There is a temptation to overstate some obligations with the goal of "banking" excess budget authority. When those funds are required, they are de-obligated from the original order and then used on the new requirement. While such practices may be effective at managing the obligation rate and creating slack, they violate the recording statute and result in an inaccurate financial status for the organization that also impedes decision making at higher levels of the department.

End of year dynamic

The end of the fiscal year is an eventful time as annual funding must be fully obligated before it expires. There are several key elements in managing this process: understanding and adjusting the state of current obligations, estimating final payroll figures, compiling a valid list of unfunded requirements, gathering and centrally controlling all remaining budget authority, and preparing for the subsequent fiscal year. With good planning, the end of the fiscal year need not be wasteful, but studies show that it often is. As noted previously, because this is not money, but expiring legal authority, the incentive is to obligate the entire available budget because there is no way to "save" it. Thus, the quality of spending goes down (Liebman and Mahoney, 2010). "Moreover, any under execution may cause reductions in the following fiscal year. At the very least, under execution will result in a budget cut (mark) that the command will have to respond to get the funds back" (Jones and McCaffery, 2008: 335) Wastefully spending is considered less bad than not spending at all. How do organizations get in the position of a year end spending surge?

As noted in the last section on budget slack, there is a tendency to hold reserves throughout the year as a way of creating flexibility to deal with contingencies and unexpected events. Given these "rainy day" accounts, if it does not rain the funding is released late in the year. If one assumes four layers in the chain of command and each layer holds back 3%, that means the lowest layer only received 88.5% of the funding and the remaining 11.5% will come cascading down late in the year. In some cases, that last unit—an installation, squadron, or program office may see 10–15% of its annual budget authority appear in the last few weeks of the year. It might also be the case the funds delegated internally have not been used by all local managers and are swept up into a central account. The flow of funds into the budget office late in the year flow back out to those managers who are capable of spending it. At the end of the year, the money does not usually go to those with the most important need, but to those who are simply able to spend it quickly. One way to spend it quickly is to increase the pace of training or operations. One naval aviator at a training squadron in Virginia, commenting about the glut of flying time in September, said, "While I don't mind flying to Key West for lunch, it seems kind of wasteful."

Recalling the section on fiscal law, the bona fide needs rule becomes a concern at the end of the year. Spending must be related to the current year and it is improper to forward fund items for the following year. Thus, one cannot sign a contract in late September for grounds keeping for the next fiscal year. However, one can "top off" inventory levels of materials that are routinely kept on hand. It may have been the case that inventory levels were allowed to drop in order to create budget slack; now that the year is drawing to close remaining funds can be used to restore those levels.

Before spending decisions are made, existing obligations are validated. Some intragovernmental orders are valid only for the current year and funds cannot carry over into the next year. A host-tenant agreement is in this category; so is the purchase of support labor. It is incumbent on program managers and comptrollers to assess the level of funding on these MIPRs and to pull back any excess (or fill any shortages). The comptroller must also estimate carefully the final payroll and set aside sufficient funds or pull back any excess from that budget category. Ideally, line managers should be updating lists of unfunded requirements—those new needs that arose during the year for which funds were not budgeted, or lesser important things that did not make it into the budget at all. Having such lists ready to go, prices estimated, and contractual vehicles identified makes end of year spending more effective and productive. If the organization has several offices capable of creating new obligations, those authorities need to be temporarily curtailed otherwise they could lose control. Centralized accounting and spending in the closing weeks and days of the fiscal year is critical for avoiding under or overspending.

As we have observed thus far, throughout the fiscal year there is a great deal of emphasis on operational control of budgets. The fact that appropriations provide legal permission to bind the government, and that permission has strict boundaries, requires a degree of control over the budget, spending and accounting processes. These requirements, embedded in law, must be well understood and complied with. Budget authority flows through the department from those with executive responsibility to those who execute the day to day business of the department. In some cases, that authority flows horizontally across the department as one organization supports another on a reimbursable basis. It is incumbent upon everyone involved in these processes to ensure they understand the status of the budget authority at all times: who holds it, whether it is committed, obligated or expended, for a legitimate purpose. This budget authority not only enables the implementation of defense policy, the rate at which funds are spent bears on decisions for future budgets. Thus, the department actively manages obligation and expenditure rates to ensure all funds are used in a timely fashion. The legal strictures create incentives to behave in certain ways to ensure the rules are not violated. These behaviors may at times be uneconomical and may not be in the best interests of the department and the military missions it is assigned. Now we turn our attention away from managing the money to managing the organization.

MISSION SUPPORT

Within the Department of Defense, budget execution is primarily governed by the Government Accountability Office's Principles of Federal Appropriation Law (GAO, 2004) and the DOD Financial Management Regulations (FMR) (DOD, 2011). The DOD FMR chapter on budget execution covers topics such as apportionment, transfers and reprogramming, standards for recording commitments and obligations, status of prior year accounts, and special circumstances such as unmatched disbursements. The perspective one gets from the FMR is that budget execution consists solely of the tracking of budget authority as it flows through the system: where it originated, to whom it was passed, and its status. Both regulations exist to ensure compliance with the law and the threshold levels of fiduciary responsibility, but there is little to no mention of terms like *efficient* or *effective* use of funds; there is little mention of the programmatic aims for which the funds were provided. The requirements for recording information only deal with financial transactions, not whether the goals and objectives of the program were achieved.

While that is helpful for comptrollers, the line managers in the department care more about the goals and objectives. They have their own sets of rules and information systems that keep track of everything from the readiness of operating forces to ammunition supply levels to hospital bed utilization rates and acquisition milestone progress. Some of these data are used in the budget formulation phase, but in budget execution there can often be a difference between the goals of the general managers of the organization and the goals of the financial managers who support them. Compliance with fiscal law parameters on an appropriation may mean that the comptroller must tell a commander that his good idea cannot be lawfully implemented, or the comptroller pressures a command to award a contract quickly to keep obligation rates high when more time at the negotiating table could result in a better structured contract. This section of the chapter addresses the management aims of the department and how budget execution serves those aims.

Management control

As noted, all budgeting systems, to some extent, serve three purposes for an organization: they support *planning at the strategic level* and are concerned with alignment of resources for the organization, they support *management control at the goal or project level* and are concerned about efficiency and effectiveness at the sub-organizational level, and they support *operational control at the task level* and are concerned about the accountability of routine processes (Schick, 1966). As noted, budget execution tends to bridge the two lower layers and we already have discussed operational control.

At the management control layer, effective budget execution means that project prioritization, staffing, and resourcing support organizational goals and strategies. It means that a control environment has been established by senior managers as evidenced by control systems such as clearly documented processes, checks and audits, segregation of fiduciary duties, robust information systems, and well-trained staffs. It means that those control systems are designed in a fashion that mitigates the greatest risks to the organization and the public while retaining sufficient flexibility to respond to contingencies. It means that accounting and other information systems generate timely, relevant, and reliable information flows between line and staff managers. That is, line managers have the financial information they need to effectively direct the organization and staff managers understand programmatic needs sufficiently well to recommend financial actions to ensure success.

Allocation/reallocation

A fundamental financial management design decision for an organization is determining the basis on which budget authority is allocated. Since a critical aspect of budget execution is to provide useful information for management decision-making, the underlying data that is converted to information must be obtained and maintained in a suitable manner. Raw budget data is in the form of transaction descriptions such as the commitment or obligation of funds on a particular contract. Meta data must be identified and the transactional data coded effectively to support managerial decision-making. With respect to budgeting, there are four basic approaches to allocating budgets (and, conversely, aggregating data), and each supports different management or reporting objectives.

The first is an organizational structure approach. Here, budget authority is allocated based on the design of the organization: departments receive a share which they break down into divisions or teams which may break down into work centers or individuals, which may even break down into days or hours. In a typical functional bureaucracy, such an allocation scheme may be appropriate as it aggregates information according to the bureaucratic design. Other organizations may find their work is more project oriented and a project management approach is a better fit. Here, budget authority is allocated to a portfolio of projects and sub-allocated by project, sub-project, task and sub-task. The organization is less important than the project and people from multiple organizations may be working on the task. This allocation scheme allows the aggregation of spending information by project. In a third case, the main concern is not who is doing the spending or why, but what the funds are buying. Here, allocations are made by an object of expense approach: budget allocations are made to rent, utilities, salaries, facilities, supplies, and so forth. The earliest public budgets were (and many household budgets are) organized this way and remnants of the system still exist in the defense and federal budget today. Lastly, an organization may choose to allocate budget authority according to the method of expense approach: by contract or intergovernmental purchase order. If an organization's mission is supported by a few large contracts, funds may be allocated accordingly: by contract, line item, task, job order or individual expense element. These four methods are not only used at the individual organization level, the pattern scales so that one could look at one command or the entire Navy and see these four basic approaches.

In reality, most organizations use a blended approach in which portions of the budget are allocated by one method and other portions by another method. Imagine a divisional organization that has a few large centrally managed expenses, such as the rent on the building they occupy. This fictional organization has a workforce that is a combination of government and contractor employees and the commander of the organization has embarked on a large project to which each division has contributed a few personnel. The comptroller may choose to set aside the rent and utility money and pay that centrally; likewise the contracts for the support labor may be managed centrally and each division augments its staff with contractors. The special project has its own budget under the direction of the project manager. The rest of the funding is distributed to the division directors. Figure 7.4 is suggestive of such an organization. While such an allocation scheme may reduce administrative burdens and transaction costs (e.g., one allocation for rent instead of having each division responsible for their own), it may also limit the analytical capability of the organization and impede future budget decisions. For example, the organization may not be able to accurately say what portion of contracts A & B support project Σ. They may not know how much overhead Division 3 drives compared to Division 4. And, if during the budget planning phase, the functions performed by Division 2 or Project Σ are considered for a potential cut, does the organization know how much funding that represents? Absent a robust cost accounting system that can cross-walk these different methods of allocation, answering such questions can be challenging and may impede the type of analysis most helpful for running the organization efficiently or for making trade-offs in the PPBE process.

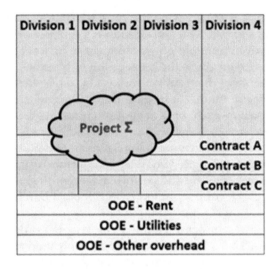

Figure 7.4 Suggestion of Allocation Scheme. *Source*: Authors, 2011.

Sometimes allocations are fixed if a firm decision is made in the PPBE process that a program element is immune to further cuts; these generally occur only if the service chief or secretary declares it so. Decisions made at lower levels are often revisited. Sometimes Congress fences a program from cuts that may befall other, similar programs; they do this by specifying that "no funds may be transferred" out of that line. Congress may also specify minimum levels of funding for a program by using "no less than" type language in the appropriation. But most allocations, once made, are not fixed. There are processes available to the department to re-allocate funds when doing so is necessary to best align resources with demands. That process of realignment, known as reprogramming, is frequently preceded by a thorough review of requirements and resources. This leads us to the next two sections.

Mid-Year Review

When looking at allocations of funds and the rates at which they are used, it is beneficial for the manager to do so in light of the organization's strategic plan. Public entities need to know what they are supposed to accomplish and have clarity of tasks and purpose. Strategic planning is an important focus of not just budgeting but the measurement of how funds are executed; "it establishes the context in which performance and cost information is considered. In order for any organization to evaluate either its performance or its use of resources in pursuit of that performance, it must first know what it intends to do." (Joyce 2003: 16) With proper planning and accurate measures of budget execution, managers can evaluate the execution of the budget and make effective adjustments. These adjustments can occur anytime, but normally in the middle of the year there is a formal comprehensive review.

Around the halfway point in the fiscal year, the military department budget offices conduct a mid-year review with the major commands. This review monitors both program and budget performance. Program performance is measured against goals identified in the non-financial information systems, such as readiness rates, supply fill rates, hospital bed utilization rates, system development progress, or earned value. All areas and organizations within the department are assessed for their progress toward objectives and their ability to meet prescribed standards of performance. Those exceeding expectations, as well as those lagging behind, are identified. "Fact of life" changes are presented by program offices, operational units and support organizations. These include contingent events like unexpected maintenance requirements, special provisions of law, a contractor breach, operational tempo changes, fuel price spikes, unfavorable test results, and the like. Those who desire additional funding to cope with such events propose "unfunded requirements lists" or UFRs.

At the same time, budget performance is measured by comparing obligation and expenditure rates against the obligation and outlay phasing plans that were developed during the budget formulation process. Any deviation from the plan invites scrutiny. For those who are underexecuting (spending less than forecast) questions arise: did they over budget? Are they falling behind schedule? Is there a technical problem that is keeping them from progressing? Is there a management problem? Can the excess funds be taken without undue risk to the rest of the year's program? For those who are overexecuting (spending more than available) similar questions arise: are they ahead of schedule? Is there a technical problem they are throwing money at? Is there a management problem? Do they need additional funds or can they make it the rest of the year with what remains?

In all cases, the central budget office considers any implications on future years' budgets. The mid-year review occurs in the late winter months, just prior to the critical period for programming the service budget in the current POM cycle. This detailed review of current program and budget status is considered as future year allocation decisions are made.

At the end of the mid-year review process, a set of reallocations are processed. Funds may be taken from accounts that are under-executing and given to accounts that have run into some unexpected difficulty and need more money. Generally, only urgent needs are funded such as operational contingencies, readiness-impacting matters, critical acquisition program issues that could affect a milestone, or items with intense media or political scrutiny. Mid-year review is the time when reserves held by commands begin to be released to fund unfunded requirements. Holding reserves is a habit with commands and comptrollers; sometimes they admit to it up front, because it is only prudent to pull off a reserve to help solve unanticipated problems that arise. Sometimes they argue that they do not have a reserve and they mean it. What they do not say is that they know very well which accounts to go to if they need additional money. No major command comptroller could function well within DOD without such knowledge (Jones and McCaffery, 2008: 334).

Essentially, a midyear review is conducted to ensure current funding levels are adequate. Those under executing may face the possibility of losing some funding. Those with compelling unfunded requirements may see additional funds flow their way. As with the application of any control or accountability tool, many managers can become nervous as the mid-year review approaches, especially if their spending deviates from the plan. Having a compelling reason is vital to ensure that a program or organization does not become a "bill payer" for someone else. The process is also sometimes considered unfair as the better performing organizations often lose funding to prop up those who have not performed well. It can feel like the bad manager is being rewarded and the good manager is being punished. Such an attitude is consistent with the theory of bureaucracies and the notion that the size of one's budget is an indicator of importance or value. A healthier attitude would recognize that the military services, and the nation, are better served by the reallocation.

While the mid-year review serves as a control mechanism. It is also a significant tool for using the flexibilities provided in the appropriations. When Congress appropriates funding to the DOD, it builds into that authorization the ability to move some funding between accounts. They recognize that the budget is merely a plan and the plan is over a year old by the time the funds are executed. Some flexibility is required to meet the nation's needs. Until the mid-year review, such changes in need can be met through the apportionment process, but by the middle of the year, a give-and-take process is necessary. A significant tool for effecting those changes is the ability to transfer, or reprogram, funds from one purpose to another.

Reprogramming

Reprogramming, simply put, is the conversion of some amount of budget authority from its current purpose to a new purpose. This commonly occurs late in the fiscal year when the DOD is ensuring just the right amount of funding is in each pay account. It is common to see the department late in the fiscal year, for example, reduce Military Pay, Navy (officer pay) account and increase Military Pay, Air Force (enlisted pay) if there are fewer naval officers and more airmen than originally forecast in the budget.

The idea of reprogramming is as old as the nation. An 1809 act, for instance, gave the President the authority to transfer funds when congress was not in session and such authorities have ebbed and flowed ever since (Fisher, 1975). Recent changes (in the last few decades) to reprogramming rules have redefined the congressional review process. Under current rules, the four defense committees (House and Senate Appropriations Committees and House and Senate Armed Services Committees) review reprogramming actions, and each committee may approve, deny, reduce or increase the amounts requested. (Roum, 2007)

There are four types of reprogramming actions: prior approval, internal, notification letter, and below threshold. The first three are commonly and collectively referred to as "above threshold" reprogramming. Prior approval reprogramming, as the name implies, requires congressional approval before the funds are transferred. Certain criteria trigger the use of a prior approval type action: increasing the number of units of a major end item, starting a new program or terminating a program; if the action affects funding for a congressional special interest item; if general transfer authority is used; or if the value exceeds specified thresholds. Typically, each year, following the mid-year review process, a large omnibus reprogramming action is sent to congress. In fiscal year 2011, for example, the Undersecretary of Defense (Comptroller) forwarded to Congress a prior approval reprogramming request that was 91 pages long and totaled over $5.1 billion in changes to several hundred programs after the mid-year review.

Internal reprogramming is used when the criteria for prior approval reprogramming are not met and the purpose for which the change is made is not significant. These are often used to make adjustments to foreign currency exchange rate fluctuation accounts, for example. Or they may be used to move money from transfer accounts into regular accounts. A transfer account is budget authority that exists for a particular event, but must be moved to a regular appropriation account (such as O&M or Mil Pay) before it can be obligated. An example is the Overseas Contingency Operations Transfer Fund (OCOTF) that exists for unexpected military operations.

A letter notification is used when there are adjustments to existing programs that fall below the thresholds for prior approval actions. Thus, a small change to a procurement program that does not alter the quantity may be affected and congress is simply notified after the fact. Finally, a below threshold reprogramming is a cumulative authority for very modest changes. The DOD comptroller keeps a running tally of actions taken and sends semi-annual reports to congress.

How often does reprogramming occur, in what accounts, and what dollar amounts? Roum (2007) examined eight years of reprogramming actions, and their individual transactions. In his work, one action consists of two or more transactions (at least one account gains budget authority and at least one account is reduced; in most cases there are multiple accounts affected). He found that a typical fiscal year is characterized by dozens of small internal actions that consist of 8.0 transactions on average, about two dozen prior approval actions that consist of 4.4 transactions on average, plus there is normally one large omnibus request of about 100 transactions. Added together those actions constitute about 4–5% of the total DOD budget authority for each year. Reprogramming most often occurs in the March–May time frame, just after the mid-year review, and in August–September as the fiscal year winds to a close. In years that were affected by continuing resolutions, more reprogramming actions occur. Eighty-five percent of the time, Congress did not alter the request, but let it proceed as requested. Reprogramming actions primarily involve current year budget authority, but may involve prior year authority as well. In fact, unobligated balances of expired appropriations are frequently reprogrammed into accounts that make those balances currently available. In descending order of dollar amount, O&M is most affected, fol-

TABLE 7.3 Complexity of Internal Reprogramming Actions

Reprogramming year	# of IR Actions	# of IR Transactions	Transactions per IR Action
1999	73	687	9.4
2000	86	597	6.9
2001	80	582	7.3
2002	69	566	8.2
2003	71	651	9.2
2004	106	739	7.0
2005	116	839	7.2
2006	58	612	10.6
Total	659	5,273	8.0

TABLE 7.4 Complexity of Prior Approval Reprogramming Actions

Reprogramming year	# of PA Actions	# of PA Transactions	Transactions per PA Action	Omnibus Transactions
1999	13	42	3.2	68
2000	20	68	3.4	112
2001	32	98	3.1	118
2002	14	59	4.2	56
2003	26	94	3.6	76
2004	29	161	5.6	110
2005	46	173	3.8	128
2006	28	223	8.0	167
Total	208	918	4.4	835

TABLE 7.5   Typical Dollar Values of Reprogramming Actions

	Reprogramming Year							
	1999	2000	2001	2002	2003	2004	2005	2006
IR Actions								
Min	148	502	350	680	600	511	445	400
Mean	193,081	140,207	153,298	239,985	251,911	153,562	123,235	238,609
Median	43,294	14,421	11,609	12,224	29,650	30,445	21,750	51,260
Max	2,250,296	2,206,030	2,840,923	6,700,000	4,734,462	2,163,538	1,810,000	3,048,686
1st Quartile	8,339	4,731	3,932	5,786	9,000	7,175	6,755	11,657
2nd Quartile	43,294	14,421	11,609	12,224	29,650	30,445	21,750	51,260
3rd Quartile	193,600	57,550	51,662	83,260	155,227	125,044	75,841	219,030
4th Quartile	2,250,296	2,206,030	2,840,923	6,700,000	4,734,462	2,163,538	1,810,000	3,048,686
Total IR	14,094,913	12,057,779	12,263,833	16,558,985	17,885,655	16,277,540	14,295,278	13,839,313
% of Total	93%	93%	79%	91%	85%	78%	59%	55%
PA Actions								
Min	1,900	1,436	1,927	6	500	2,000	924	2,500
Mean	25,046	21,685	84,779	114,677	115,211	69,431	182,435	301,883
Median	16,450	17,000	14,700	36,329	28,800	25,600	42,000	80,000
Max	61,872	87,500	1,265,465	661,695	1,333,402	379,933	1,539,848	1,650,000
1st Quartile	11,779	3,148	7,397	24,500	5,700	10,174	13,760	44,100
2nd Quartile	16,450	17,000	14,700	36,329	28,800	25,600	42,000	80,000
3rd Quartile	36,221	25,824	27,582	87,250	45,919	80,732	150,000	360,150
4th Quartile	61,872	87,500	1,265,465	661,695	1,333,402	379,933	1,539,848	1,650,000
Total PA	300,557	412,015	2,628,144	1,376,120	2,880,279	1,944,054	8,209,584	8,150,839
Omnibus	775,813	469,028	633,193	283,834	289,578	2,686,197	1,602,789	3,153,234
Grand Total	15,171,283	12,938,822	15,525,170	18,218,939	21,055,512	20,907,791	24,107,651	25,143,386
DoD BA	278,420,000	290,339,000	318,678,000	344,904,000	437,714,000	470,933,000	483,864,000	593,780,000
% of DoD BA	5.4%	4.5%	4.9%	5.3%	4.8%	4.4%	5.0%	4.2%

TABLE 7.6 Transactions by Fiscal Year

	Reprogramming Year							
	2006	2005	2004	2003	2002	2001	2000	1999
2006	829							
2005	89	870						
2004	47	123	764					
2003	7	74	73	655				
2002	8	15	67	64	582			
2001	11	12	15	37	41	676		
2000	8	13	13	9	33	51	673	
1999	1	23	18	7	2	30	59	707
1998	1	2	21	11	3	6	30	47
1997	1	2	11	10	5		8	25
1996		4	11	6	1	8	1	4
1995	1	2	17	4	1		1	2
1994				17		4	1	4
1993					7	4		2
1992					1	8	1	1
1991						5	1	1
1990					5	5		4
1989							2	
1988						1		
Total	1003	1140	1010	820	681	798	777	797

(*Fiscal Year of Funds* — row axis label)

TABLE 7.7 Number of Transactions by Appropriation and Fiscal Year

	Reprogramming Year								
	1999	2000	2001	2002	2003	2004	2005	2006	*Total*
O & M	328	290	272	221	261	278	332	297	2279
MILPERS	108	89	88	119	101	113	133	137	888
PROCUREMENT	138	170	202	123	151	227	269	252	1532
RDT&E	96	111	130	78	115	89	172	134	925
MILCON/HOUSING	63	43	36	54	98	166	108	100	668
Revolving and Management Funds	17	18	17	17	27	38	32	12	178
Contingency Operations and Other Transfer Funds	20	27	16	21	20	45	48	21	218
Other DoD Programs	27	29	37	48	47	54	46	50	338
Total	797	777	798	681	820	1010	1140	1003	7026

TABLE 7.8 Funds Reprogrammed into and out of Appropriations

Appropriation	Amount To	Amount From	Difference
O&M	181,338,633	17,908,049	163,430,584
MILPERS	19,260,327	14,051,967	5,208,360
RDT&E	9,041,206	5,637,313	3,403,893
Procurement	28,414,763	10,896,502	17,518,261
MILCON/Housing	3,153,012	1,899,543	1,253,469
Revolving and Management Funds	11,531,061	22,506,788	(10,975,727)
Contingency Operations and Other Transfer Funds	2,197,362	163,703,731	(161,506,369)
Other DoD Programs	3,544,256	21,246,124	(17,701,868)

Source: Roum, 2007: 29–35.

lowed by procurement, military personnel, and research and development. During wartime in the 21st century, O&M is less affected as O&M needs have been met through supplemental appropriations instead. Summary statistics from Roum (2007) are provided in Tables 7.3–7.8:

Revolving Funds

The working capital funds (WCF) are a type of revolving fund that have been used in the Navy in one form or another since the late 1800s. In a revolving fund *all income is derived from the activity's operations and is available to finance continuing operations without a fiscal year limitation.* Simply stated, a revolving fund activity accepts an order from a customer, finances the costs of operation using its working capital, and then bills the customer who reimburses the fund.

In general, the WCF is a financial strategy that employs private sector business-like techniques for resource management, cost accounting, and cost allocation in clearly established customer-provider relationships that mimic contracts. Characteristics of the fund's operations include:

- Identification of the full cost of delivering goods and services, including all direct, indirect, depreciated capital investment, military personnel, and allocated general and administrative (G&A) costs.
- Recovery of the full cost by the WCF activity through its rate structure so that they do not need to receive appropriations from Congress.
- Passing the full cost of delivering the goods and services to the ultimate customer who drives the demand.
- Stabilized rates set in the budgeting (PPBE) process and fixed during the year of execution (except in extreme cases); the WCF activity absorbs price fluctuations rather than the ultimate operational customer.

There are several advantages of using the WCF model. The process identifies the total cost of goods and services to the buyers, sellers, and oversight bodies, thereby promoting more transparent allocation and utilization of resources. It provides managers with the financial authority and flexibility to procure and use labor, materials, and other resources more effectively. In some cases, procurement can be made in anticipation of customer demand to ensure responsiveness. The WCF places customers in the position of critically evaluating purchase prices and the quality of goods and services ordered; in some cases, permitting the "invisible hand" of competitive market forces to drive economies. In less competitive markets, it provides information for buyers to hold providers accountable. The system allows for financial flexibility within the WCF activity as there are no fiscal year limitations on the WCF corpus of funds and moderate gains and losses are acceptable. It allows for financial stability for the appropriated fund customer since it establishes standard prices, or stabilized rates, enabling customers to budget and to execute their plans with less uncertainty. Lastly, it shifts the management focus from acquire and spend ("use or it or lose it") to cost and cash management.

To be included in the WCF financial structure, a proposed business area must meet four criteria. First, they need outputs (i.e., goods produced or services provided) that can be clearly identified and measured. Second, they must identify organizations (i.e., customers) that need and will order those products or services. Third, they need to have an approved cost accounting system. Finally, the organization had to have evaluated the advantages and disadvantages of establishing such a buyer/seller relationship and made the business case. A chartering process is used to formally establish WCF business areas and to identify their organizational structure, as well as their assets and liabilities.

The cost accounting system is necessary in order to allocate all costs associated with the activity across the products and services sold. Recall, the objective is full cost recovery. The WCF received its initial 'working capi-tal" through an appropriation and/or a transfer of resources from exist-ing appropriations, and uses those resources to finance the initial cost of operations. The initial appropriation, or working capital, is called the cor-pus. The primary financial goal of the WCF activity's management team is to manage their business such that the long term operating result nets to zero—the organization does not make money nor lose money. That long term operating result is referred to as the Accumulated Operating Result (AOR) which is the net sum of each fiscal year's calculation of Net Operat-ing Result (NOR), the difference between revenue generated (sales) and costs incurred. NOR is similar to a for-profit business' earnings or losses. The corpus remains indefinitely and "revolves" as sales revenue flows in and costs of production flow out.

Certainly there are times when conditions result in a gain or loss. In pe-riods of high inflation for commodities such as fuel in 2005–2010, a WCF activity may lose money since its stabilized rate was based on a lower cost assumption. On the other hand, in periods of increased demand, such as that which IT support activities experienced responding to the Y2K prob-lem or TRANSCOM experiences during a rapid mobilization, they will re-cover excess amounts and show a gain. In those cases, the annual NOR is negative (loss) or positive (gain) which may lead to a positive or negative AOR. There are two responses to this, one focused on the near term (pre-dominantly cost) and one focused on the budget (both cost and revenue). In the near-term, the activity can attempt to control costs through conser-vation efforts or hiring freezes. They might also attempt to affect revenue by marketing their services to additional customers. In the budget process, the WCF activity will always drive the AOR back to zero since the goal is full cost recovery, no more, no less. So a positive AOR represents past recovery of more than the full cost and therefore rates will be lowered in the budget to intentionally lose money to reset the AOR to zero. Conversely, a nega-tive AOR represents less than full cost recovery in the past and budgeted rates will be increased to recover the remaining cost. If AOR is at zero, the budget will reflect a rate that is equal to the projected full cost of the item sold. One should note that WCF rates are not a commercially competitive price, it is merely a reflection of the sum of direct cost and the allocation of indirect and G&A costs.

To set the rate, the WCF activity gets an estimate of the total volume and mix of sales for the budget year from its customers. It then adds the full cost of providing those goods and services and divides by the volume of activity to compute a Unit Cost Goal (UCG). The full cost includes all pay and ben-efits for all the employees, overhead expenses, material, maintenance to capital equipment as well as the eventual replacement of that equipment.[4]

The UCG is the "should cost" figure to be used in the year the budget is executed to compare against actual costs.

$$\text{UCG} = (\text{total estimated cost}) / (\text{total estimated output})$$

For many WCF activities the volume of activity is measured in labor hours or work-days. If the AOR is other than zero, the numerator is adjusted accordingly to earn (lose) the negative (positive) AOR balance. This revised cost per unit becomes the rate charged to the customers.

$$\text{Rate} = (\text{total estimated cost} \pm \text{AOR adjustment}) / (\text{total estimated output})$$

If the WCF activity sells precisely the volume and mix estimated and contains their costs to precisely what was budgeted, they will recover the full costs and adjust the AOR. Since variation is inherent in the process, differences occur and the cycle repeats with the next budget continuously aiming to bring AOR to zero.

What has just been described is the model of the industrial fund type of working capital funds used at depots and R&D centers. Industrial funds may incur expenses to benefit a customer only after the customer has submitted—and the working capital fund activity has accepted—a funded reimbursable work order. There is another form of working capital fund, commonly referred to as the stock fund activities. Examples of stock fund activities are the Defense Logistics Agency and the Navy's Inventory Control Points. These working capital fund activities operate under the same philosophy of full cost recovery and stabilized rates, but they need not wait until a customer order is received to begin incurring costs. Given their mission to maintain ready stocks of material, these command have contract authority to incur obligations in anticipation of customer demand based on engineering, life cycle management and fleet usage data. These commands sell consumable items and repair parts at the cost of goods plus a surcharge to cover their operating costs. The surcharge is computed during the rate setting process and fixed during the year of execution.

PROBLEMS AND SPECIAL CASES

Thus far we have described and defined some of the normal events that define the execution phase of budgeting: the legal context, the flow of funds to the end user, criteria for allocating resources, measuring execution, control and review of execution, and flexibilities inherent in the system. But not all fiscal years are normal, not all events are typical. There are situations that demand additional attention because they interrupt the routine. And these situations have implications on both operational and management control of the department. Here, we discuss four of the most common situations: late annual appropriations and continuing resolutions, congressional additions to the budget, "taxes" and withheld budget authority, and contingency operations. Each disrupts the programmatic and budgetary execution in its own way.

Continuing resolutions

The new fiscal year begins October 1, and by this date Congress should have passed the annual defense appropriations bill. Routinely, this is not the case. Since the constitution states that no funds will leave the treasury but in consequence of an appropriation made by law, *if the regular appropriation act is not finished, a temporary appropriation is necessary.* These temporary appropriations are called continuing appropriation acts (CRAs), or more commonly continuing resolutions. How often does this occur? According to the Congressional Research Service, "From FY1977 to FY2011, after the start of the fiscal year was moved to October 1, all of the regular appropriations acts were enacted on time in only four instances (FY1977, FY1989, FY1995, and FY1997). No continuing resolutions were enacted for three of these fiscal years, but continuing resolutions were enacted for FY1977 to fund certain unauthorized programs whose funding had been dropped from the regular appropriations acts." (CRS, 2011: 3). Thus, in 32 of those 35 years, at least one CRA was enacted. And in the 25 years prior to 1977, in no year were all appropriations passed on time (CRS, 2011). CRAs may be of very short duration if the passage of the appropriations bill is imminent; in other cases, agencies have been funded for the entire fiscal year under a CRA if there is significant debate over programmatic or funding matters. Figure 7.5 from the Congressional Research Service displays the number and duration of CRAs from 1998 to 2011. On average there are six CRAs per year with an average duration of nearly a month, but in 1999 and 2001, they averaged less than 4 days each. This does not mean the entire federal government is on a CRA on average for six months of the year. Of the 12 or 13 regular appropriations bills, some are late while others are enacted on time; thus some agencies fall under a CRA and others do not.

A CRA normally does not specify a dollar amount, but rather "usually funds activities under a formula-type approach that provides spending at a restricted level, such as 'at a rate for operations not exceeding the current rate'" (CRS, 2011: 2). Each CRA will have a time limit stated in the resolution as the maximum time during which funds appropriated by the CRA are available for obligation. If the regular appropriations act is not complete by the expiration date of the CRA, a subsequent CRA is issued.

This is why there are multiple CRAs in most years. The rate for operations specified in the resolution is an annual amount and can be in terms of an appropriation act which has not yet become law, a budget estimate, or the current rate (the year under execution). If it is in terms of pending law, generally the rate is set at the level of the lowest rate of the bill passed in either the House or Senate. Ultimately, it is Congresses' choice (Jones and McCaffery, 2008).

It should be observed that budget execution is made much more difficult when Congress funds DOD and much of the federal government on a series of Continuing Resolutions. For example, for FY 2011 budget comptrollers had to plan and execute seven separate CRAs before the regular defense appropriation was approved by Congress. This required a huge amount of civilian staff overtime (with commensurate increases in costs of execution) and caused a high level of stress to military and civilians in the comptroller and operating communities. Having to pay the costs of warfighting from a combination of CRAs, supplemental and base budget funding was a challenging experience for DOD budget executors. The lesson for budget execution is that when Congress fails to pass regular and, when needed, supplemental war related appropriations on time, or even roughly

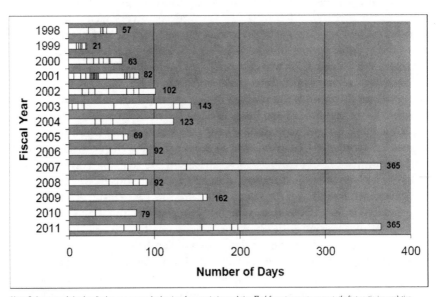

Note: Each segment of a bar for a fiscal year represents the duration of one continuing resolution. The left-most segment represents the first continuing resolution, effective beginning on October 1 (the start of the fiscal year). Duration is measured, in the case of the initial continuing resolution for a fiscal year, from the first day of the year through the expiration date. For subsequent continuing resolutions for a fiscal year, duration is measured from the expiration date of the preceding continuing resolution.

Figure 7.5   Duration of Continuing Resolutions, FY1998–FY2011. *Source*: CRS, 2011.

on time, DOD staff time is wasted and the risk that funding will not flow to the best purposes and highest needs is greatly increased.

In DOD a CRA generally means that the spending will occur at the pace it did in the fiscal year just concluded. Any growth in the budget is not provided in the CRA. No new programs may start under a CRA and no new purchases or new employees should be hired unless congress has said otherwise. Programs that were set to end, may be forced to continue. For the majority of administrative and operational functions in DOD, it is business as usual and many people will not notice any difference (Jones and McCaffery, 2008: 336). For some, though, it may give an unwanted hardship like those who may be trying to sign a new contract or advance a program past a key milestone.

At the operational level, comptrollers and commands must forecast accurately temporary requirements for funding: the amount needed for 7 days, 10 days, 14 days, one month, etc. and pass those requirements up the chain of command. It is vital to understand when key contracts must be let, whether they can be deferred or partially funded, the date of payrolls, and the like. At the management level, the plans articulated in the budget may be delayed; transitions from development to production may be affected; new programs are delayed and programs reaching the end of their usefulness may be extended. Should a CRA continue long enough, plans may be affected enough to impact future budgets.

Congressional Add-Ons

Congress plays a large role in budget execution through both the restrictions imposed in the authorization and appropriation processes and in their oversight of budget execution. Managers seek flexibility in order to make trade-offs to effectively manage their programs or respond to contingencies, but Congress and higher level managers feel a need to maintain control over the budget process to ensure the department stays on course and in order to make a public accounting for monies spent. Jones and McCaffery said, "Where managers might prefer more flexibility, the Congress insists that public monies be safeguarded, even if this results in more control than might be necessary to meet policy ends efficiently. We also have to remember that to some extent Congress is more concerned with where and how money is spent than it is interested in establishing incentives that stimulate efficiency" (Jones and McCaffery, 2008: 378).

Congress takes seriously its constitutional prerogative to "raise and support armies" and "provide and maintain a navy" and every year will make some adjustments to the budget presented by the DOD, adjusting some line items, eliminating others, and adding things that were not requested. Once the authorization and appropriation bills are enacted, it is incumbent upon the department to implement what was enacted, not what was requested. In those cases where congress added a program or a specific requirement to an existing program, not only does the responsibility to execute that program need to be assigned, the budgetary impact must be assessed. Within the Department of the Navy, each congressional addition is met with a set of questions to understand its origin, the programmatic and budgetary implications, and the appropriate actions the department must take. Those questions include:

- Provide a description of what this item and the proposed add is or does.
- What contractor(s) is/are involved (indicate percent of contract and associated personnel employed) and in which States?
- Is funding for this item already contained in the current budget? If so, identify the line item and how does the add differ from the budget submitted.
- Is funding for this project contained anywhere in the FYDP? Describe.
- If Congress were to provide the additional amount indicated, how much more additional funding would be required in the current fiscal year, and subsequent years, to complete the project? How much of that is already in the FYDP?
- Did congress add funds last year for this? How much?
- How much has been invested in this program to date?
- Does a written and validated requirement exist for this item? Explain.
- Do you assess that the line item has no/low, some/medium, or high military value?
- Does funding for the proposed add-on interfere with your plans to competitively develop or procure such equipment?
- Do you know of any reason that you could not or would not execute additional funds for this item in Fiscal Year 2009?
- Why are (additional) funds for this item not in your budget request?
- If the CNO has submitted an Unfunded Program Requirements List, is this item on the priority list?

We understand that these questions address issues of both operational control (ensuring the budget accurately represents the financial impacts) and managerial control (the alignment of this add to the overall defense strategy). Our PPBE chapter describes the manner in which the DOD seeks to balance the defense program across the services, appropriations, major force programs, capabilities, and the FYDP. A congressional add on disrupts that balance and the implications of any add on must be considered in future budgets. It is also important to make sure that if congress adds $x for an item that $x is actually an amount that will result in the desired capability. It is possible that sum may be excessive or insufficient and there are other programs that will be affected.

In addition to adding new programs, Congress will sometimes put restrictions into existing programs. For example, the FY2011 defense appropriations act states:

> Sec. 8023. (a) Of the funds made available in this Act, not less than $30,374,000 shall be available for the Civil Air Patrol Corporation, of which,
>
>> (1) $27,048,000 shall be available from "Operation and Maintenance, Air Force" to support Civil Air Patrol Corporation operation and maintenance, readiness, counter-drug activities, and drug demand reduction activities involving youth programs. (Department of Defense, 2011b)

Presuming congress appropriated the full amount requested in the O&M, Air Force budget, all the other air force programs (flying hours, depot maintenance, flight training, etc.) would need to be decremented to make available $27 million for the Civil Air Patrol. Although it may seem as if a program is fully funded in an appropriation, there are often general provisions like this that erode the amount. This requires some adjustment of plans by the various O&M, AF program managers, and leads us to the next topic.

Taxes and Withholds

Previously, we discussed the issue of creating budget slack and some techniques for doing so. The other side of that coin occurs when higher levels of an organization create slack by holding back a percentage of the budget and the lower levels of the organization suffer what is called a "tax" or a "withhold." Generally a "tax" is funding that is taken from a program or organization budget with no expectation of it being returned; it is assumed to be a permanent reduction to the program budget. A withhold is a temporary reduction that is more likely to be released later in the fiscal year. Both are done for legitimate management reasons, but result in necessary programmatic responses by the affected parties.

From where do taxes and withholds originate? The one already mentioned occurs when each layer in the hierarchy of the flow of funds "shaves a little off the top" to create budget slack. Another common example derives from the implementation of general provisions in the appropriations act. Congress may appropriate 100% of the budgeted request for a program, but later in the appropriation there is often language like this example from the FY2010 Defense Appropriations Act.

> SEC. 8097. Notwithstanding any other provision of this Act, to reflect savings from revised economic assumptions, the total amount appropriated in title II [Operation & Maintenance] of this Act 4 is hereby reduced by $194,000,000, the total amount appropriated in title III [Procurement] of this Act is hereby reduced by $322,000,000, the total amount appropriated in title IV [Research, Development, Test & Evaluation] of this Act is hereby reduced by $336,000,000, and the total amount appropriated in title V [Revolving and Management Funds] of this Act is hereby reduced by $9,000,000: Provided, That the Secretary of Defense shall allocate this reduction proportionally to each budget activity, activity group, subactivity group, and each program, project, and activity, within each appropriation account. (Department of Defense, 2010)

In this situation, we see reductions to various appropriations that equate to a decrease of about 0.3%. The act says this is to be distributed proportionally. Thus, every program will begin with 99.7% and go down from there, depending on other provisions and taxes and withholds. The phrase "revised economic assumptions" normally means that the latest economic forecast at the time the appropriation was written assumed slightly lower inflation than was assumed when the budget was drafted months earlier.

Withholds occur when it is deemed necessary to hold back the distribution of funding temporarily, often to confirm compliance with some provision of law. Sometimes the provisions of the authorization act and appropriation act differ in quantity or amount; all or part of the funds may be withheld pending confirmation from congress regarding their intentions. In another common situation, a program may be scheduled to transition from development to production; both RDT&E and procurement funds are appropriated, but the procurement funds are withheld pending the milestone decision to advance into production. In a third case, a portion of government contracting is required to go to small or disadvantaged businesses. In order to comply with those provisions, the service comptrollers may hold back a portion of funding, this is especially true in RDT&E and procurement accounts, until such thresholds are met, and then any remaining balances may be released to the programs.

In the FY2010 defense appropriations act, we see yet another example.

> SEC. 8037. Of the funds appropriated to the Department of Defense under the heading "Operation and Maintenance, Defense-Wide", not less than $12,000,000 shall be made available only for the mitigation of environmental impacts, including training and technical assistance to tribes, related administrative support, the gathering of information, documenting of environmental damage, and developing a system for prioritization of mitigation and cost to

complete estimates for mitigation, on Indian lands resulting from Department of Defense activities. (Department of Defense, 2010)

In this case, at least $12 million in O&M funding would likely be withheld by the DOD comptroller until a plan was in place to ensure compliance with this provision. Only then would it be released. And later in the fiscal year, the comptroller would follow up with that program office to ensure "no less than $12 million" had been made available for this program. Since that $12 million also came from "of the funds appropriated" if it was not already budgeted these funds would result in a "tax" to one or more other O&M,DW programs.

One can see that the actions of the managers at each level of the hierarchy, coupled with the provisions of the appropriation and permanent law, result in some percentage of funding being permanently or temporarily withheld by the department and not forwarded to the program office, installation or operating unit. Managers at the lower level of the department must learn to plan to operate on less than the amount fully budgeted.

Contingency Operations Funding

The DOD Financial Management Regulation plainly states, "DOD Components normally do not budget for contingency operations." (DOD, 2011: 23-3) This means that military contingencies that arise during the year of execution must be paid for from funds already appropriated for the operation of the department. Should funds need to be realigned to support the mission, the department will prepare a reprogramming request. Should costs exceed the level deemed affordable, the department may choose to submit a supplemental appropriation. Regardless, units that are tasked with supporting the mission will experience some budget stress while awaiting additional resources; other unit that are not tasked may see resources pulled from their operations and diverted to support the contingency.

"There are three different types of estimates that are developed and used during the course of an operation. The pre-deployment estimate is used to assess various operational assumptions and to inform the go/no-go decision making process; the budget estimate is used to define and defend requests for reprogramming or additional appropriations; and the working estimate is used during execution of the operation against which the Military Departments measure actual costs, and which can be used as the base for determining the changes in cost that would result from changes to the operational plan. All three types of estimates are important to ensure that senior leaders have the latest and most accurate information available for use in the resource allocation process." (DOD, 2011: 23-7)

To mitigate the financial effects of contingencies, the Overseas Contingency Operations Transfer Fund (OCOTF) was established to meet emergency operational requirements in support of emerging contingency operations without disrupting approved program execution or force readiness. The account has been used for things as varied as military missions, U.S. citizen evacuations, overseas humanitarian assistance operations, and support of civil authorities (e.g., firefighting or hurricane response). The use of the OCOTF enhances the DOD's flexibility for managing emerging costs resulting from the redeployment of troop and the rapid increase of forces.

As recently as 2011, the DOD requested funding of the Overseas Contingency Operations on the basis that, "The highly variable nature of the costs to rebalance United States Forces between Iraq and Afghanistan, the redeployment of U.S. troops from Iraq, and rapid increase of forces in Afghanistan to locations lacking infrastructure and established logistical support, make it difficult to plan for every requirement. These amounts will provide a minimal level of flexibility to transfer funds to appropriations incurring unexpected, unplanned for costs during the year of execution." (OSD, 2010: 3) Contingency Operation Funding allows for the agency flexibility to provide for unforeseen expenses without breaking rules that Congress has given.

CONCLUSIONS

The overarching structure of this chapter has been designed to demonstrate that budget execution is characterized by a duality. On one hand, there are legal and regulatory restrictions placed on the department with respect to the use of funding. Since funding comes in the form of legally prescribed budget authority, compliance with the law is a minimum requirement and controls must be placed on the financial transactions that support the operations of the department. On the other hand, the defense department is mission-driven and funding is but one of many resources that support those missions. Budget execution should be done in service to mission accomplishment, which can sometimes conflict with the fiduciary responsibilities. Underneath this overarching duality are other matters that should be apparent to a reader now, or will become apparent as he progresses through the text. The reader should understand that budget execution is characterized by these other considerations:

- There exists a perpetual tension between the need for resource flexibility and the need to maintain adequate control over those resources. In general, lower levels of the chain of command prefer more flexibility and those above prefer more control. "Authorizing and appropriation bills do not provide all the direction agencies require in order to operate, and the law does not anticipate all the circumstances that may arise in the course of managing federal programs" (Joyce, 2003: 29). Policymakers recognize this and draft appropriations with a degree of vagueness to allow managers to manage and they provide tools of flexibility such as reprogramming authority to deal with contingencies. In daily operation, budget execution is tightly controlled by comptrollers, lawyers, and resource sponsors, while line managers and program offices demand flexibility.

- The department strives to achieve multiple objectives simultaneously: to meet the schedule and performance goals of each program, to be prepared to respond effectively to contingencies, to use resources efficiently, to reduce administrative burdens, to report useful management information, and to meet the fiduciary responsibilities associated with appropriated budget authority. It is impossible to them all perfectly; there must be a prioritization.

- The department strives to effectively perform the activities of budget execution while accurately recording the financial effects of those activities so that managers, executives and policymakers have dependable information for decision-making. This requires a strong link between budgeting and accounting.

- The department strives to build mission-focused information technology systems that support the critical operational and support functions, but those systems must also interact with financial systems and be financially auditable. Unlike the for-profit sector where operational systems are designed to support a bottom-line financial goal; in the DOD, operational systems are designed to be superior operational systems first and financial considerations are secondary at best.

- The norms of budget execution—as evidenced by the way budget execution is managed, use-or-lose considerations, processes for dealing with special circumstances, and the overlapping phases of the budget process—means that managers who are responsible for executing today's budget and policy must be mindful of the effects on future years' budgets and policies. One cannot be effective by focusing solely on the current year issues.

- The department places a priority on the accomplishment of programs and missions that are directly funded and support those programs and missions through revolving funds. Paradoxically, the larger revolving funds become the more management attention they garner, taking attention, and possibly resources, from the direct missions.

- Every manager in the department finds he must reconcile the formal, written rules with the informal norms of the position they hold.

These characteristics explain much of the subtlety that exists in the execution of budgets and funding in the Department of Defense and to a more general extent the federal government.

NOTES

1. *United States v. MacCollom.* 1976. 426 U.S. 317.
2. Comptroller General. 1927: 6 Gen: 619, 621.
3. This is distinct from voluntary in the gratuitous sense, such as when a Boy Scout troop repaints the anchor at the entrance to the naval base as a public service project.
4. Capital expenditures are depreciated over the useful life of the item; the depreciation expense is included in the annual rate.

CHAPTER 8

BUDGET PROCESS PARTICIPANTS IN THE PENTAGON

INTRODUCTION

When critical observers conclude that the Secretary of Defense generally functions as a chairman of the board with power but limited ability to insure implementation of his will in the military departments and services, and that the Department of Defense is so large and complex that it is virtually 'unmanageable'—this perhaps *understates* the actual circumstance. This observation does not have the same impact as saying that the Secretary is confronted with three massive military departments whose attitude at any point in time may be characterized as competitive as well as cooperative. While other decision processes in the Pentagon also reveal competition, none is perhaps so well situated as the budget process to capture the efforts of military departments to get along with each other while, at the same time, getting a little more in resources than the others from the Secretary of Defense (SECDEF). Senior leaders in the Pentagon know that budgets are 'everything', i.e., that without sufficient funding little can be done programmatically in the appropriate way or at the required level of intensity. Lack of budget success in a particular year not only means waiting for next year; it also may set crucial departmental programs back five to ten years. As a result, senior leaders engage and negotiate at length over budget issues.

Financing National Defense: Policy and Process, pages 279–325
Copyright © 2012 by Information Age Publishing
All rights of reproduction in any form reserved.

In this chapter we describe how those who formulate and negotiate budgets in the Pentagon go about this task. First, we delve into the budgetary strategies employed by the military departments (MILDEPS) and the behavioral patterns of budget officials and analysts. Second, we describe how the budget offices of the military departments are organized and how they function. Third, we analyze the budget processes of the three military departments and the U.S. Marine Corps. In part of the chapter, we view budgeting in detail from the perspective of the Department of Navy and the Marine Corps to understand by example many of the complexities of deciding upon and representing military department resource requests in the DOD budget. Finally, we briefly examine the budget functions of the office of the Secretary of Defense.

MILITARY DEPARTMENT BUDGET STRATEGY AND BEHAVIOR

Within the Department of Defense, the budget submitting offices for claimants including the major commands and the systems commands joust for dollars with budget reviewers in central comptroller offices of the military departments and the Office of the Secretary of Defense (OSD). Competition for resources is intense and players earn reputations for their proficiency in the resource generation process. In terms of competitive strategy, entire military departments become characterized in terms of their approaches. For example, the military departments occasionally have been labeled by DOD level budget insiders as the 'dumb, the defiant and the devious.' None of these adjectives is complimentary but, as we will note, there are reasons to explain these characterizations.

Typically, the Army is cited as the 'dumb' because of an observed practice of unwillingness to submit full budgets to OSD. A member of the OSD Comptroller staff lamented, "We tell the (the Army) what we want but they can't produce it." (Secretary of Defense, 1990) However, based upon our research, a more likely explanation is strategic in nature. The Army has been known to fail to submit requested budget data to OSD. Consequently, at times OSD is forced to 'invent' the data for the Army in the DOD budget. This strategy permits the Army to deny authenticity and support for parts of the DOD budget when questioned in Congress. This is simply playing what our parents used to tell us was "dumb like a fox."

On the other hand, the Navy has been cast as the 'defiant'. In interviews with OSD comptroller staff we were told, "The Navy always has to do it their own way. We ask them to do something and they flat refuse ... period. This is always their first reaction. Then they [the Navy] give us what they want to do to comply with our request. This happens so much [that] we have come to expect it." (Secretary of Defense, 1990) What is accomplished by this approach we may ask? Interviews with Navy comptroller staff verified that the OSD view was correct. No one in the Navy comptroller office even denied this strategy. We were told, "What they [OSD] ask for often is wrong because they don't understand the Navy or our budget. We give them the best-scrubbed and most accurate budget of any service because we put our budget through a more thorough review process than the others [services]. Our numbers we can trust, but not theirs [OSD]." (Secretary of the Navy, 1990) In this case, when questioned about what is in the DOD budget before Congress, the Navy can offer up its own numbers to supplement or replace what OSD has provided when it so wishes.

Both the Army and Navy strategies as described here are typical in budgeting. They fall into the class of tactics referred to as the "end-run" (Wildavsky, 1964). Both approaches allow the respective military departments/services to support the DOD and President's budget, as they must once the Secretary of Defense has submitted his budget to the President's Office of Management and Budget, while providing opportunity for deviating from it before Congress. This leaves us with the third characterization, the 'devious'. This approach apparently is used often enough that it is expected by OSD budget analysts. "The Air Force always puts on the best show for us, but especially for Congress. They are the 'high tech' people. They put up all kinds of graphics and 'smoke and mirrors' and everybody in the room is 'wowed'. They [the USAF] get what they want, then when they leave we all ask, 'What did they say?' You can't figure it out. Their numbers are full of holes but the members and staffs [of congressional committees] don't care. The Air Force has their own strategy—dazzle them, get approval—and then do what you want." (Secretary of Defense, 1990)

The obvious advantage of the Air Force approach is first to win at budget competition and, secondly, to gain maximum flexibility in managing the funds they receive. While this flexibility has been abused in some notable cases (e.g., cost overruns and funds management nightmares with the B-1 bomber program), it probably works more often than not to give the USAF what it wants. Otherwise, why would the approach persist? Our point with respect to strategy is that the military departments have many different ways to try to get their share of funding in the annual budget wars with the Office of the Secretary of Defense (OSD) and Congress.

BEHAVIORAL STRATEGIES OF BUDGET SUBMITTING OFFICES AND ANALYSTS

In the complex world of budgeting, players often adopt strategies that simplify their perceptual costs of participation, increase their chances of winning or decrease the probability of losing. DON comptrollers and budget analysts have shared their perspectives on the budget process with us over the last decade. A fleet comptroller explained, "Everything in (Navy) financial management is by the book. If you miss a time deadline, you are no longer a player; you have given up control of your destiny." Since analysts up the line are looking for dollars to cut, by not submitting something on time, an agency invites a cut when it misses a deadline. The rule then is always to meet deadlines, with the best that can be done, even if it is not perfect. Secondly, this comptroller emphasized, "You must give the desired product. If you are asked to give five issues with their dollar offsets, and you provided 109, 104 will be thrown out. If the remaining 5 do not have offsets, they go too." Again, the reason is that everyone is looking for money for other programs and will use any pretext to find it. In this instance this is also a work management issue for the claimant budget analyst; time is always short so one should not waste time on things that will be thrown out when that time could be better used working up other issues.

Finally, this comptroller said that the process lent itself to some simplifying assumptions:

> I assume that you [program sponsors in BSOs] know your program and you will tell me when I make a mistake and you will do it within twenty-four hours...90% of the time people are not prepared to do this. If no appeal is made and made quickly, then the mark [cut] was a good mark. Our analysts think the proof of a good mark is when the claimant doesn't cut anything vital to its core function in response. If they don't cut flying or steaming hours in reaction to my mark, then it must have been a good mark.

One comptroller said he asked all his analysts to graph their accounts so that they could see the trend line in the recent past and the near future. Then the low point on the graph is taken as the starting point for a mark: "Whatever is the lowest point for your program in the past is a valid base, since you have survived, unless you can explain why it is too low." This comment is not applicable to the investment accounts, just to the O&M categories where readiness is the main perspective.

One FMB budget analyst explained why some uniform personnel did not look forward to a tour of duty in the pentagon: "Washington is different from the fleet. I just finished my executive officer (XO) tour. I had 300 people, all I had to say was "jump to it", and things would get done. Here it takes 16 chops (different signatures or clearances) on every memo I send. (Aboard the ship) I could really crank out the paperwork, sign things by direction of the Captain. Here I have responsibility for lots of money, but no authority... Without authority fighting battles takes more time; persuasion is your most important tool, but you do not have time to fight everything." This officer was functioning as a budget analyst in the Navy Budget office and his superiors and peers all thought he was doing an outstanding job. He continued, "Peons like me do the work...this hearing (a fleet budget presentation) was just for show. All the details get worked out between the analysts, people like me. We work all the technical things, then it goes to OSD and becomes a little political and then to Congress and it really gets political; people cut or change things because they do not like them, not on technical grounds." He added, "When political decisions are made, weenies like me have to get $50 million out of a program. Sometimes we only have a day, or a couple of hours to do it in, so we make the cut and say, "Tell us what impact this will have on your program." You have to be prepared to be the bad guy here (in the budget office)."

Working analysts from the Lieutenant Commander to Captain ranks in the Pentagon would often downplay their role in the resource allocation process; they see themselves as the "worker bees" implementing decisions that are made "above my pay grade." They would acknowledge that they had input to how the decision was shaped and how it was implemented; after all, that was the proper role of the worker bee.

THE PROGRAM VERSUS THE BUDGET

POM Glitches

Budget analysts typically have found problems with the Program Objectives Memorandum (POM) and programming process. Some of these objections have been addressed in the 2001–2003 reforms that have merged the programming and budget analysis processes. For example, one budget analyst remarked, "Programming is trade off time. The FYDP has a lot more program in it than we could possibly pay for. So everyone is looking to cut someone else's program. All the way into Congress, someone is looking to steal your money and put it into a better program." Another added, "Some programs are beautifully justified in the POM, but fall apart in translation into the budget. For example, suppose the money for military construction (MILCON) does not make it into the MILCON appropriation; now you can not build the building and the O&M analyst takes maintenance and operating dollars out of your activity budget and tells you to try again next year." Another observed, "The POM always bought back what you lost in the previous budget process.

The POM directive was to meet every requirement by putting something in every cup. On average, the POM may have been underfunded by 20 cents on the dollar," he said referring to the early 1990s." He added that he did not think this would work in an era of scarce resources and suggested that leaders ought to make vertical cuts in programs as soon as possible, rather than horizontal cuts across the board, so that whatever slack resulted from the vertical cuts could be put to use to prepare for the lean years. While this sounded like sensible advice, it did not seem to win many converts. Leaders preferred to keep programs alive across the board at diminished levels, like a golfer refusing to give up his short iron game because even at a low level he could keep the skill set alive and resurrect it quickly if it were needed.

One rookie claimant analyst was offered a deal his first year in Washington where he would get more money in the outyears (BY+4 and 5) of a POM for his program if he was willing to allow his program to be zeroed out in the middle year. This seems to be the equivalent of sending someone for a "left-handed monkey-wrench." A program that is zeroed out is cut; it is no longer in the POM and now must compete with all those other programs that did not make it into the POM previously to get back in. Chances of success are not good. Furthermore, everyone knows that the outyears never unroll as promised and that everyone is promised to get well in the outyears. It is a cheap promise to make since even the most credulous program manager knows that years four to five beyond the budget year are so remote from the current year that they are almost imponderable. Thus, experienced players will take almost anything in a current year as opposed to a rich promise in the outyears. The outyears could unroll as promised, but the chances are remote. To some extent, players are willing to judge the sophistication of other players by how they respond to claims about the outyears; seasoned players tend to express a bit of well-mannered disbelief. Our "rookie" friend refused the deal and kept his program alive.

Budget Process Glitches

If analysts are cynical about some aspects of the POM, they also recognize there are glitches in the budget process. A retired director of the FMB said, "It's the purpose of green eye shade types (budget analysts, comptrollers) to find loose money for things the CNO wants to do." To counteract this, he advised claimants to, "...have your numbers accurate and with good justification...so you do not come back empty-handed." He added, "When you talk to your boss about cost figures, do not wing an answer. Instead say, 'last time I checked it was X and as soon as the meeting is over I'll get back to you right away if that is wrong. Nothing is worse than having a reputation for 'bum dope.'"

Another FMB director advised claimants not, "...to put non-starters in your budget. You will lose the money and people will suspect the rest of your budget." A non-starter is a budget item that has insufficient support. A claimant must "ask around" about this kind of idea to see who will support it. For example, a big discretionary program enhancement in a lean year is a non-starter. He warned that some items always seem to be non-starters, e.g., staff travel or things that are nice, but not critical. When money is lost for non-starters, it is lost to another claimant and may threaten the rest of that program in the budget base. At the DON level, it could be money lost to another military department, thus increasing their budget base in the near future. The civilian world may allow for the expression of "nice to do" proposals in the budget process as consciousness raising items, military budgeteers are death on this tactic. To some extent, the arena for this kind of behavior is in the domain of the Authorization committees when they debate service roles and missions, force structure, nuclear weapons policy, and treaty provisions, controversial items that usually lead to intense discussion and late authorization bills without really threatening funding.

In the investment accounts, finding the correct cost estimate is a problem. One FMB leader noted, "Everyone has a number. The contractor, the program manager, the Navy Budget Office, OSD ... analysts will take the lowest number, so you have to be sure you can explain the difference between your number and the lowest number." Another analyst added, "You find out what numbers other participants in the decision process have by calling around, by "working the system" by "dropping in for a cup of coffee." A retired FMB Director commented that analysts should, "... know what the boss is thinking ... try to answer questions from his perspective." He advised preparing for the budget hearing by calling the officer who would brief the Admiral before the hearing, the pre-briefer. At a flag officer review, "Someone is going to tell him what he is going to hear. Make sure you talk to the pre-briefer. You might say, 'Can I help you get ready for your briefing of the boss? Then compare numbers and explain them. Try to avoid surprises."

One retired senior Admiral explained, "Nobody likes surprises, particularly when they cost more money. The field activity is often the first to see it ... to recognize program growth. They should take it up the chain of command. Unexpected program growth suddenly dropped on senior managers is very hard for them to handle." Since they have trouble accommodating it, their first response is to deny it and cut it. He also warned, "Little changes can mean a lot." He noted that a small change in some electronic gear for a helicopter was "a minor upgrade, but it eventually caused some very expensive program changes, in training and repair manuals, even for the airframe. This small change drove a huge change in program cost." Moreover, he warned that priorities change all the time and that when a claimant loses a program or item in his budget, it could be because of new technology elsewhere, or maybe something in the black budget has made the need for it obsolete. Additionally, treaties can introduce turbulence into operating budgets by changing roles and missions and their support costs, so claimants have to be prepared for a turbulent environment.

Even after the appropriation act is passed, claimants have to be careful. One analyst warned, "If you manage resources, you should know where your money appears, O&M, OPN, MILCON, because they all have different legal requirements. You should know if there are special requirements with it, say an Environmental Impact statement, of if Congress has written special provisions in the Conference report to guide how you spend it."

Analysts and claimants cope in this environment through prior training, by successive tours of duty in the budget world of increasing responsibility, by seeking budget or resource allocation tours in the pentagon, the center of all the action, and by cultivating good interpersonal skills that allow them to communicate effectively up and down the chain of command and laterally with other players who impact their budget process, for example, by taking money they might have won, or by losing money through unsophisticated budget behavior. They cope by being flexible and reacting to fact of life occurrences, by, "making lemonade out of lemons." To learn how this is done, and under what forms of organization it takes place, we examine the military department budget offices.

ORGANIZATION AND DUTIES OF MILITARY DEPARTMENT BUDGET OFFICES

The Department of the Navy Budget Office

The Navy Budget Office, officially titled Fiscal Management and Budget (FMB), is the central budget office of the Department of the Navy. It is responsible for the preparing both the Navy and Marine Corps Budgets. The goal of FMB is to fuse the strategic demands and requirements of both services with the strategic plans and guidance of the Secretary of Defense to produce a single Secretary of the Navy budget submission (BES) to the Office of the Secretary of Defense, represented by the OSD Comptroller.

The Navy budget office is part of the Office of the Assistant Secretary of the Navy, Financial Management and Comptroller (ASN/FM&C) who functions as the Navy Comptroller. The Assistant Secretary of the Navy, Financial Management (ASN/FM) is a civilian Presidential appointee. The Navy Comptroller was established initially in accordance with the provisions of Title IV of the National Security Act Amendments of 1949 that formalized control of DOD under the Secretary of Defense. The mission assigned to the Navy Office of the Comptroller is to implement principles, policies, procedures and systems to ensure the effective control over all financial matters within the Department of the Navy. The Navy comptroller is required by Congress through legislation enacted in the 1970s to prepare the budget. Congress required budgeting to be a civilian function in DOD so that the military department budgets are issued under the authority of the Military Department Secretaries rather than the military chiefs of staff (e.g., the Chief of Naval Operations– the CNO–in the Navy). The Comptroller has broad budgetary responsibilities performed almost exclusively by the Navy Budget Office (FMB). While a variety of information and support units are located in this office, the two main analytical divisions responsible for the budget are Operations, which handles personnel and their supporting expenses, and Investment and Procurement, which handles ship, aircraft and weapons procurement. The leader of the Navy budget office is called the Director and is typically a Rear Admiral. This position is unique in that the Director (N-82) reports directly to both the CNO and the Secretary of the Navy (i.e., this position is "double-hatted"). The professional staff is composed of approximately one-half military officers who rotate through a tour of duty as a budget analyst and one-half civilians who are career civil servants. The civilian executive staff of FMB are Senior Executive Service (SES) employees with special employment status that makes them accountable not only to the Secretaries of the Navy (SECNAV) and Defense, but also to Congress. (One year the House appropriations committee was displeased with the DON FMB SES people so it cut them out of the appropriation bill; the Senate restored them in conference work; being visible to Congress is sometimes a mixed blessing.)

The Navy budget guidance manual states, "The budget functions of the Comptroller of the Navy occur during all phases of the budget cycle, including formulation, presentation, and execution." (Secretary of the Navy, 2002) Some of the duties performed by the budget office include:

1. Establishment of the general principles, policies and procedures that control the preparation, presentation and administration of the Department of the Navy (DON) budget.
2. Establishment of the appropriation structure for preparation and justification of the budget.
3. Supervision of the analysis and review of DON budget estimates, and submission and negotiation of the budget with the Secretary of Defense, the Office of Management and Budget (OMB), and Congress.
4. Supervision of any reprogramming of funds by DOD or Congress.
5. Provision of information as principle point of contact for outside agencies and other military department budget offices in all DON budgetary matters.

The Navy is the only military department with two services (Navy and Marine Corps), each of which prepares separate budgets that must be melded into one. FMB has developed a unique structure to facilitate this process. As noted, the Navy budget office performs duties for the Comptroller (ASN/ FM&C) and the Chief of Naval Operations as the Fiscal Management Division (N82). Although the FMB and N82 offices are one in the same, the accountability structure for the two positions differs. As part of the CNO Operating Navy structure (OPNAV), the office is designated as N82 and has delegated budget preparation and execution responsibilities, although this duty is subordinate to the authority of the Secretary of the Navy. For the Comptroller, FMB has responsibilities to oversee preparation and execution of the Navy and Marine Corps unified budget. Historically, this office has functioned in managing "blue dollars", i.e., those that support the Navy only, "green dollars" that support only the Marine Corps, and blue-green dollars—Navy dollars to support combined Navy/Marine Corps functions. As noted subsequently, the Marines have their own budget and comptrollership function, somewhat analogous to the large Navy fleet claimants.

The Navy Budget Office (FMB) organization consists of six divisions: Appropriations Matters Office (FMBE); Operations Division (FMB1); Investment and Development Division (FMB2); Program/Budget Coordination Division (FMB3); Business and Civilian Resources Division (FMB4); Budget and Procedures Division (FMB5). Each division plays an essential and specific role in the development and implementation of the DON budget. The following is a brief synopsis of the responsibilities of each division as specified in the Budget Guidance Manual.

FMBE: Liaison Responsibilities

This division is responsible for maintaining liaison with the Congressional Appropriations Committees, the Office of Legislative Affairs, and the Congressional Liaison Offices of the Secretary of Defense, Secretary of the Army and Secretary of the Air Force for functions related to congressional hearings and congressional staff matters for all oversight committees. The FMBE coordinates all matters related to DON participation in hearings before the House and Senate Appropriations Committees and keeps FMB and all its fellow divisions advised on the current status of congressional action and appropriation requests. It provides the schedule of committee hearings, arranges for DON witnesses, coordinates the review of transcripts of hearings, coordinates responses to committee questions, arranges briefings for members of the committees, their staffs, and the members of the professional staffs, and provides any other coordination activities associated with these committees.

FMB1: Milpers and O&M

The FMB1 is responsible for reviewing, recommending, and revising estimates for the Military Personnel (Active and Reserve forces) and Operation and Maintenance (Active and Reserve) appropriations, and other funds of the Navy and Marine Corps. It looks for proper pricing of PM initiatives and their translation into the budget. It ensures that the primary readiness accounts are properly funded those that fund ship and aircraft tempo and depot maintenance, adequate funding of base support and appropriate funding of pay and allowances. It develops and uses operational cost models for programs such as the flying hour or steaming hour programs, uses average cost rates for personnel costs and checks the O&M accounts for the reasonableness of their estimates. The office is also responsible for assisting in the justification of estimates before OSD/OMB and the Congress, the continual review of program execution, and the recommending of adjustment to allocations, when required. For the OSD/OMB review, the FMB1 analysts act as the primary DON contact with the OSD/OMB staff analysts. They are responsible for publishing schedules of hearings, attending hearings, and coordinating and clearing all responses to requests for additional information. They also prepare or review reclamas (appeals) to Program Budget Decisions and issues for the Major Budget Issues meeting. Finally, they are responsible for ensuring the accuracy of the OSD decision recording system and for updating the DON tracking system. For the Congressional review, the FMB1 analysts are responsible for preparing or clearing budget material provided to Congress in support of Military Personnel and Operation and Maintenance appropriations. This may include budget justification material, statements, transcripts of hearings, answers to questions, backup or point papers, and appeals to authorization and appropriation reports. Representatives from FMB1 may attend hearings as backup or supporting witnesses.

FMB2: Investment and Development

FMB2 is responsible for reviewing, recommending, and revising estimates for the investment and development appropriations, including procurement, research and development, construction, family housing, and base closure and realignment. This office is also responsible for assisting in the justification of estimates before OSD/OMB and Congress, the continual review of program execution, the recommending of adjustments to allocations when needed, and the reporting of selected acquisition costs and data to the Congress. The focus is on most likely cost, including proper pricing and pricing consistent with the pricing of previous years. This process reviews for realistic phasing of projects to see that contract dates and delivery dates are consistent and it examines development and procurement milestones to see that they are properly phased. The office looks for a reasonable funding profile and an absence spikes or dips. They also check for prior year execution performance to see that contracts were executed on schedule and that obligations met obligation rate targets.

For example, in a ship construction project where only 10% of the appropriation was to be obligated in the prior year, it would check to see if all of the 10% were obligated. If only 7% were obligated, this could mean the

290 ■ Financing National Defense

program had a problem and imperil funding for the budget year. Critical indicators for these accounts involve time-phasing of production schedules, rates of production, lead-time, slippage in production schedules, and Congressional approval for production. For the investment account, the watchwords are most likely cost, realistically phased programs, with a reasonable funding profile, based on good execution performance.

For the OSD/OMB review, the FMB2 analysts act as the primary DON contact with the OSD/OMB staff analysts. They are responsible for publishing schedules of hearings, attending hearings, and coordinating and clearing all responses to requests for additional information. They also prepare or review reclamas to PBDs and issues for the major budget issues (MBI) meeting. Issues which cannot be solved between analysts or supervisors of analysts may rise to major budget issue status and be resolved by SECDEF. In any year 5 to 10 issues may do this; usually the military department has to bring a suggested solution with it, e.g., fund X by cutting Y, both within its own jurisdiction. It is not seen as fair to argue fund Navy X by cutting Army Y, although if SECDEF wanted to do this it would be seen as OK by Navy. Major budget issues generate lots of paperwork, sometimes filling several three-inch binders for each issue. Finally, FMB2 is responsible for ensuring the accuracy of the OSD decision recording system and for updating the DON tracking system.

For congressional review, the FMB2 analysts are responsible for preparing or clearing budget material provided to Congress in support of Investment and Development appropriations. This may include budget justification material, statements, transcripts of hearings, answers to questions, backup or point papers, and appeals to authorization and appropriation reports. Representatives from FMB2 may attend hearings as backup or supporting witnesses.

FMB3: Budget Guidance and Procedural Coordination

This division is responsible for the preparation of DON budget guidance and procedures; control and coordination of budget submissions; coordination of reclamas to SECDEF PBDs; preparation and/or clearance of all program and financing schedules included in the budget; coordination of DON participation in appeals to congressional action; development and operation of ADP systems in support of the budget formulation process at the DON headquarters level; administration of financial control systems and procedures for the apportionment, allocation of funds and the reprogramming process; and preparation of fund authorization documents for the appropriations under its cognizance.

This office also reviews and makes recommendations on Departmental budget issues and appraises the effectiveness of budget systems. Additionally, FMB3 prepares reports for the DON on the status of the OSD/OMB review, coordinates the DON participation in the Major Budget Issue meetings, monitors the OSD automated decision recording system, and operates the DON system for recording all decisions. FMB3 also coordinates the preparation of DON input into the President's budget, preparation of financial and other summary budget documents, and preparation of the Budget Officer's statement. This office is responsible for preparing budget material provided to Congress in support of Advisory and Assistance Services. FMB3 prepares bill digests and Congressional Action Tracking System tables pertaining to the DON budget at each stage of authorization and appropriation committee action on the President's Budget. It is also responsible for reviewing any issues that may arise during Congressional review concerning appropriation responsibility and all proposed legislative changes that affect the budget.

FMB4: Working Capital Funds

FMB4 is responsible for reviewing, recommending, and revising estimates for the Navy Working Capital Fund (NWCF) and civilian personnel for inclusion in the budget and the justification of these estimates to OSD/OMB and the Congress. This office also reviews and validates funding estimates in working capital fund activity budgets to ensure proper balance between NWCF "providers" and DON appropriated fund "customers." For the OSD/OMB review, the FMB4 analysts act as the primary DON contact with the OSD/OMB staff analysts. They are responsible for publishing schedules of hearings, attending hearings, and coordinating and clearing all responses to requests for additional information. They also prepare or review reclamas to PBDs and issues for the MBI meeting. Finally, they are responsible for ensuring the accuracy of the OSD decision recording system.

For congressional review, the FMB4 analysts are responsible for preparing or clearing budget material provided to Congress in support of NWCF activities and Civilian Personnel accounts. This may include budget justification material, statements, transcripts of hearings, answers to questions, backup or point papers, and appeals to authorization and appropriation reports. Representatives from FMB4 may attend hearings as backup or supporting witnesses.

FMB5: Budget Policy and Guidance

FMB5 is responsible for the development, coordination, and issuance of DON budget and funding policy and procedural guidance for all DON appropriations, funds, and organizations. This includes the promulgation of DON policy guidance required in the development of the budget; review and appraisal of budget policy and procedures and their implementation within the DON; development of improvements in organizational responsibilities and interfaces related to budgeting and funding; continuous appraisal of adequacy and effectiveness of financial management systems to ensure conformance with budget policy; resolution or adjudication of audit and inspection findings involving budget policy and procedures matters; analysis of implications of audit findings for financial management policy of the Department; review of budgetary policy impact of legislative proposals; identification and clarification of congressional direction concerning DON budget policy and procedures; and development of functional standards for and review of comptroller organizations. This division adjudicates such issues as when government funds may be used to provide cake for the U.S. Vice-President when he visits a ship to give a speech, e.g., it is legal when he awards a medal and praises the ship; it is not legal when he awards a medal and gives a campaign speech. The role of FMB5 is so important that often it is referred to as the Navy budget office.

Department of the Air Force Budget Office

The Air Force Budget Office was established to aid the Assistant Secretary of the Air Force (Financial Management and Comptroller—SAF/FM) in formulation of the Air Force BES (Department of the Air Force, 2002a). Its goal is to obtain funding to support the Air Force mission by translating program requirements into approved budget estimates. The Air Force Budget Office is composed of five directorates, including the Directorate of Budget Investment (FMBI), the Directorate of Budget and Appropriation Liaison (FMBL), the Directorate of Budget Management and Execution (FMBM), the Directorate of Budget Operations (FMBO), and the Directorate of Budget Programs (FMBP). The Air Force Budget Office depends on these directorates to help in the development and execution of the budget estimate. The following is a description of each directorate's mission

FMBI: Estimation and Execution

The FMBI directorate develops the budget estimate and tracks financial execution of Aircraft, Missile, Munitions, and Other Procurement. It also aids in the formulation and execution of the Research, Development, Test and Evaluation (RDT&E), Military Construction (MilCon), Military Family Housing (MFH), Base Realignment and Closure (BRAC) and Security Assistance Activities accounts. The directorate is organized into five divisions, namely Military Construction, Program Support, Security Assistance, Missiles, Munitions, Space & Other Procurement, and Aircraft and Technology

FMBL: Congressional Liaison

The FMBL is the Air Force liaison to the Congressional Budget and Appropriations Committees and the Congressional Budget Office. Its job is to develop and implement strategies to ensure Congress is aware of pertinent Air Force budget positions and issues. In addition, it is responsible for monitoring congressional activity and keeping Air Force officials informed of actions taken on the Air Force Budget. Furthermore, it acts as the Air Force point of contact for House Appropriations Committee Surveys and Investigations.

FMBM: Policy and Procedures

The FMBM directorate establishes financial policy and procedures. It provides oversight of Air Force Defense Business Operating Fund Activities and manages the Air Force's financial data systems, prior year financial adjustments and processes for appropriation distribution to subordinate activities. The FMBM reviews and validates all Air Force requests of the Department of the Treasury, and the Schedule of Apportionment and Reapportionment to the Office of Secretary of Defense (Comptroller).

FMBO: Milpers and O&M

The FMBO directorate is the Air Force focal point for all matters pertaining to planning, formulating, integrating, defending and executing of the Air Force's Operations and Maintenance and Military Personnel Appropriation budgets that support approved programs and mission priorities.

FMBP: PPBES and Budget Coordination

The FMBP directorate integrates the Air Force Budget within the Planning, Programming, Budgeting, and Execution System. It also coordinates the Air Force actions for the Budget Estimate Submission (BES) and Budget Review Process leading to the President's Budget Submission. The FMBP manages the Air Force database for the Force and Financial Plan and all fiscal control adjustments. In addition, it acts as the principal advisor to the Assistant Secretary of the Air Force for Financial Management and Comptroller, and the Deputy Assistant Secretary for Budget on Total Force Comptroller and Budget issues between the Air Force, Air Force Reserves and Air National Guard. These function are core to Air Force budgeting.

Department of the Army Budget Office

The Army Budget Office is the lead agency in the Department of the Army (DOA) responsible for the development and defense of the Army Budget. It directly assists the Assistant Secretary of the Army (Financial Management and Comptroller (ASA (FM&C)) by providing a link between the Secretary of the Army and the organizations that make up the DOA during the budget process. The Deputy Assistant Secretary of the Army

(DASA) for the Budget is the head of the Army Budget Office. The Army Budget Office is composed of four directorates, including the Management and Control Directorate (BUC), the Operations and Support Directorate (BUO), the Investment Directorate (BUI), and the Directorate for Business Resources (BUR).

In addition to these units, the Army Budget Office also has a Congressional Budget Liaison (CBL) office to assist in the handling of budget issues that occur on the Congressional level. Each of the directorates that make up the Army Budget Office has explicit tasks, and it is also implicit that each directorate must seek to ensure that the budget estimate presented by the Department of Army to the Secretary of Defense (SECDEF) speaks for every organization in the department from the lowest to the highest. The following is a brief synopsis of the responsibilities of each directorate taken from the ASA (FM&C) Organization and Functions Manual.

BUC: Budget Formulation and Execution Guidance

The BUC directorate is responsible for Army budget formulation and justification processes, issuing Army-wide budget formulation and execution guidance. In addition the BUC analyzes the impacts of changes to the Army's budget during the formulation, justification and execution phases of the DOA budget process. The BUC directorate is organized into three divisions, for Budget Formulation; Budget Execution, Policy, and Funds Control; and Budget Integration and Evaluation.

BUO: Milpers and O&M

The BUO directorate is responsible for formulating, presenting, defending, and managing the execution of the Operation and Maintenance, Army and Military Personnel, Army appropriations. The directorate coordinates budgeting of these appropriations from program development completion through budget execution completion. Also, this directorate participates in the program development process by membership on functional panels that interface with programs previously given resources in the budget cycle or being executed by the field. In addition, the BUO serves as the focal point for the Major Army Commands (MACOMs) to interface with Department of Army Headquarters (HQDA) on operating budget issues. The BUO directorate is comprised of three divisions, the Current Operations Division; the Military Personnel Division; and the Operating Forces Division.

BUI: RDT&E, Procurement and Milcon

The BUI directorate is responsible for financial management operations, budgeting, and execution for the Army's Procurement appropriations; Research, Development, Test, and Evaluation, Army appropriation;

Military Construction, Family Housing, and Chemical Agents and Munitions Destruction, Army appropriations; and for the Defense Department's Homeowners Assistance Program. The Director serves as an Assistant Secretary of the Army (Financial Management and Comptroller) (ASA (FM&C)) representative to the Army System Acquisition Review Committee. This directorate is the primary office for interfacing with the Office of the Under Secretary of Defense (Comptroller) on investment in Military Construction and multi-year appropriation matters. The Directorate is organized into four divisions: Weapons Systems; Acquisition and Integration; Facilities; Other Procurement, Army.

BUR: Working Capital Funds, FMS, IT and PPBES

This directorate is responsible for formulating, presenting, and defending the Army Working Capital Fund (AWCF), Foreign Military Sales (FMS), and Information Technology Systems Budget (ITSB) to OSD, OMB, and Congress. It develops and issues Army policy for business resources and manages the interface between the Army, other military services, DOD, and other non-DOD government agencies. The BUR advises the Deputy Assistant Secretary of the Army for the Budget DASA (B) and Assistant Secretary of the Army (Financial Management and Comptroller) (ASA (FM&C)) on issues relating to all other working capital funds and serves as the focal point for all aspects of the Planning, Programming, Budget, and Execution System (PPBES) for the Army's working capital fund, foreign military sales, and information technology budget. The BUR directorate is comprised of four divisions: Supply Management Division; Depot Maintenance/Ordnance/Information Services Division; Business Integration Division; and the Special Business Activities Division.

In concluding this section of the chapter, we may observe that each of the military departments have organized their budget offices differently, but commonalities exist in terms of functions performed and duties. All of these budget offices serve a "gatekeeper" function, as do budget offices in virtually all public organizations. To get funding and to use it, all parts of the organizations served must go through the budget offices to get their resources. All budget offices must organize to formulate, analyze, propose and execute budgets within department guidelines and according to established procedure. All budget offices have to defend their proposed budgets to OSD and Congress. Finally, all DOD budget offices have to be accountable for executing budgets and budget policy according to appropriation law so that, above all, what they do is legal and auditable.

Now that we have some understanding of how the military department budget offices are organized under their respective Department Secretariats, we turn to an analysis of the military department budget processes.

MILITARY DEPARTMENT BUDGET PROCESSES

Although the goals of budgeting in each of the military department are the same, the methods they use differ, as indicated in the following descriptions of the military department budget processes.

Department of the Navy Budget Process

To develop a budget, the Department of Navy (DON) depends on a decentralized budget formulation process driven by bottom-up responses to top-down controls. In the DON the budget process portion of the PPBS consists of four phases, including Office of Budget (FMB) Review; Office of the Secretary of Defense (OSD) and the Office of Management and Budget (OMB). Review; Congressional Review; and Appropriation Enactment and Execution

Navy Budget Office Review
During the first phase of the DON Budget process, Budget Submitting Offices (BSO) submit budget estimates for the organization they represent to the Office of the Budget (FMB). BSOs are also known and referred to as Major Claimants. In the DON, about twenty-four Major Claimants submit budgets, including the Atlantic and Pacific Fleets, the Bureau of Medicine, the Bureau of Naval Personnel, the Chief of Naval Education and Training, the Military Sealift Command, the Naval Air Systems Command, the Commander-In-Chief, Reserve Forces (NAVRESFOR) and so on. The Commandant of the Marine Corps (CMC) is also included as a major claimant.

The submissions by the Major Claimants are developed from the lowest level budget estimates usually referred to as Cost Center Estimates. These Cost Centers are at the lowest tier in the DON financial chain of command, for example, air stations in the Pacific fleet. They submit budget estimates to the Activity Comptrollers who in turn review, revise, and combine the Cost Center Estimates into an Activity Budget. For example, the type commander for Air Operations for the Pacific Fleet will assemble the budget for all air stations (and all the aircraft carriers too). This new consolidated budget is then forwarded to the Major Claimant who will ensure its reasonableness. The Major Claimants will then consolidate the Activity Budgets into one Major Claimant Budget Submission. As is discussed later, the Pacific Fleet Claimant will submit a budget that consolidates air, sea and undersea operations and their supporting expenses. This is then submitted to the Assistant Secretary of the Navy (Financial Management and Comptroller) (ASN (FM&C)) through the Navy Budget Office (FMB). The Marine Corps has its own internal PPB process and also submits its budget to the Assistant

Secretary of the Navy (ASN (FM&C)). Organizing the process like this allows the civilian comptroller position to adjudicate the needs of the Navy and the Marine Corps outside the military chain of command.

The process begins in late winter, when the budget call is issued and claimants prepare their budgets. These are based on control numbers from the previous year's budget and new local needs. During this process, claimants also include items from the POM if it is concluded in time; if not their budget is adjusted to include POM items during the summer in the Navy budget office. In June, these budgets are then submitted to the Navy Budget Office where the analysts review the package, including the official submission, summaries, and backup data. Analysts ask questions of the claimants and dialogues ensue, both over the phone and face to face as claimants try to sell budgets and analysts try to keep them within fiscal guidelines. Some questions may be sent in written form to the claimant, asking for a written statement in support or clarification of a position in the original submission.

Upon receiving the budget submissions from the Major Claimants, the budget office (FMB) will conduct a review of the estimates. If FMB officials think that certain estimates submitted need to be revised, they will issue a mark. A mark is basically an alert to the Major Claimant that their budget estimate submissions will be altered. At this point, a Major Claimant is permitted to submit an appeal (reclama) stating its position. It is only during this phase that the Major Claimants within the DON are provided with an opportunity to state their objectives and priorities for resources in the context of an executable budget. Beginning in June, major claimant budget hearings are held and issues are discussed. Subsequent to the hearings, 'marks' are distributed to the claimants in the name of the Comptroller of the Navy. These marks are normally cuts, made as adjustments, corrections, or denials of and to the requests for funds made by the claimants. A reason is provided for each mark to the claimant; usually the reasons have to do with incorrect pricing, a challenge to program executability during the fiscal year, hence a reduction of funds, or timing concerns where it is not believed that the funds will be needed during the year for which they have been requested.

The next step then is for the claimant to reclama or appeal the mark. In some years, some claimants appeal all marks; in other years, only a portion the marks are appealed. In general, the Navy Budget office considers a mark a final decision, unless new evidence can be brought to bear which gives the analyst new grounds for analysis. The FMB position is that the original mark is a studied and rational decision and not capricious or random, thus to appeal a mark is to question the analyst's best judgment. Hence, to make that appeal and still save face, a claimant would have to bring in new information that the analyst did not have at his disposal when he or she made the original decision. As one FMB Director said, "Reclama's are for money that is already lost. They have to be supported by new technical information." To get back 5 or 10% in reclama action then, is a good score, since the analyst fully intended that nothing be given back.

While claimants can hold back some information and appeal all marks, submitting new information at the later juncture, this ultimately would not be seen as 'fair.' Defense is a resource constrained arena and for one claimant to get more, someone else has to get less. A claimant could appeal many items one year, but in subsequent years, his credibility would suffer. The resolution of the reclama process occurs at the level of the Director of the Budget Office, between the Director, the claimants, and the analysts. Some issues may be of a high enough visibility that the claimant's community may appeal it to the Secretary of Navy level. Every year sees some of these, but most are settled at the analyst levels. Both claimants and analysts know where most of the trouble spots are in the budgets and the claimants usually begin to work these issues by alerting their analysts to potential needs long before the budget is submitted to FMB.

Budget analysts within the FMB view initial budget estimates with a preconceived downward bias. They assume some amount may always be cut from the claimant's initial request, if for no other reason than the claimant has had to begin his budget so much earlier in the year that accurate costs on certain items or programs or economic trends were not available. Consequently, analysts believe that program officials tend to over-estimate total program costs in order to compensate for this greater uncertainty. This logic is certainly typical of all budget reviewers and FMB analysts are no different. A reclama review is the final step in this process and provides a forum for resolution of adjustments. Its purpose is to guard against arbitrary or incorrect adjustments made to BSO budgets. Normally reclamas are due on the fourth day after the mark and the review session will be held subsequently to get final resolution on that issue.

The results of this process are designed to produce a budget that is timed correctly, and accounts for all known delays and disruptions (i.e., do not ask for 12 months of funding, if only 10 can be executed); contains current pricing of cost estimates based on execution experience (i.e., latest estimate of steaming hours) and latest cost factors (i.e., on local labor rates); is executable and avoids any mismatches between budgeted resources and program requirements.

Budgeting is often referred to as an exercise in timing, pricing, and executability so that program requirements and their dollars of support are matched to the fiscal year. Historically, some evidence exits about how these factors have been used in Navy budget account management. Smart and Shumaker studied Navy procurement accounts and found that executability problems (27%), pricing problems (19%), prior year performance flaws

(10%) and congressional action (22%) accounted for 78% of the marks analysts made in these accounts.

Marks studied marks to Navy O&M accounts from 1988 to 1989 and over the FYDP—six years, from 1988 through 1994 (Marks, 1989). In a detailed appraisal of four large claimants, Marks counted both the number of changes and dollars affected by the change. He found that 75% of the FMB marks involved pricing changes. Timing changes drove 8.5% of the cuts and programming decisions accounted for the rest. Marks found that programming changes drove almost 47% of the dollar changes, even though they were fewer in number than the pricing changes. Marks also added that in the reclama phase FMB gave back almost all it cut, but not necessarily to those from whom it was cut. He also indicated that the OSD review did not have large dollar consequences, averaging about 1.36% for the accounts he studied, and ranging from a cut of 11% to some accounts to an increase of 6% in others. In general, Marks found that aggressive claimants fared better, both in the original submission and in the appeal phase of the budget process. However, it is good to remember that these are incomplete studies (they only covered some accounts for some years) and dated. The current Navy budget manual says that marks may be made for various reasons, including pricing, congressional interest, program slippages and so on. At the end of the budget process, FMB works to construct a budget that has best pricing (most accurate, if not the lowest) to accomplish the mission, the best schedule, strong budget justification, clear dollar and manpower balance, timely execution plans, and a clear statement of funds needed during the fiscal year. The Navy budget guidance manual that indicates how budgets are to be prepared may be found online at http://dbweb.secnav.navy.mil/guidance/bgm.

In 2002, the budget process was changed with issue papers replacing the mark-reclama process. Proposed cuts were issued as issue papers on the FMB internet website, with email notification to other offices or claimants affected by the issue. If the issue affected them, they could write a response to the issue paper supporting or opposing the FMB position. Once FMB is satisfied with the DON budget estimates, they will then forward their version to the Secretary of the Navy for final approval. In 2003, the mark and reclama terms were dropped from the vocabulary of the Navy budget world. One budget officer said, "There are no marks anymore, just areas of interest and anyone can raise an issue for adjudication." (Secretary of the Navy, 2003) He suggested that the process was more collaborative, but still not an easy process. The jury is out on this. It may be a permanent change—or not.

The mark and reclama process had a long tradition in the Navy, a service given to cherishing its customs. Moreover, the reason budget offices are constituted is largely because no one else wants to raise the difficult little questions about inflation rates and workload, or botched procurement pro-

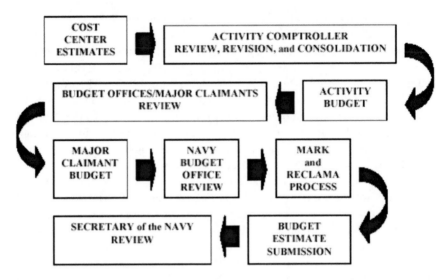

Figure 8.1 Navy Budget Process from Cost Center to Secretary of Navy. *Source*: Authors, 2011.

grams or badly sequenced ship repairs and exorbitant base repairs...all the little things that budget analysts get paid to raise questions about. Budget offices have been "bad guys" for a long time. Their job is to say "you can not do that because you do not have the money," or "Congress only gave us so much money for that," or "Congress did not give us any money for that." Budget offices enforce economizing discipline. In the next chapter, we describe the mid-year review process. Students of the budget review process realize that something has to replace the mark—reclama process simply because that process embodies a function that must be carried out. It is not clear that the budget review process can be delegated to others through a web-based system of issues and responses.

Phase II: Negotiation with OSD/OMB

Once approved by the Secretary of the Navy, the budget estimates are then submitted to the Office of the Secretary of Defense (OSD) and the Office of Management and Budget. This initiates the second phase of the DON budget process. OSD and OMB conduct a joint review of the DON budget to ensure the reasonableness of the DON budget estimate. During this phase, the Program Budget Decision (PBD) is developed. The PBD, like the marks issued by FMB in the first phase of the DON budget process, is a list of requested adjustments. These requests are submitted to each service in response to their budget estimates. Initially a draft PBD is issued in order to allow each service to reclama. Some PBDs are very simple, for example a change to an inflation or currency exchange rate; others can involve complex weapon procurement acquisition strategies spreading out a buy over several years or considering going from one source to multiple sources. Once all the reclamas to these PBDs have been reviewed and changes have been made, the finalized PBD is developed and issued to the services for action.

After making the changes required by the PBD, each service will then resubmit its BES to the OSD and OMB in order to create the DOD budget. The DOD budget will subsequently become the President's Budget and be submitted to Congress. This marks the end of the second phase in the DON budget Process.

Phase III: Negotiation with Congress

In the third phase of the DON budget process, the President's Budget is submitted to Congress for review. By the time the budget is ready for this congressional review, it has been reworked and revised many times to make it accurately reflect DON needs while staying within the DOD budgetary constraints. This is important since DOD budget will now compete with other departments for funding. Appeals may be made to Congress when a bill that has passed one or the other chamber has a provision that is detrimental to the needs of the Navy. No appeal may be made if identical bills are passed in each chamber. However, when bills differ, the Navy may appeal, for example, to support the Senate's position against the position taken in the House. Appeals are directed to the Conference Committee charged with resolving issues where the two versions of the bill differ.

Generally, appeals must stay at or under the higher number. Some analysts believe that successful appeals start by staying a little under the higher number. Conventional wisdom says that DON does not appeal all differences because it does not want to wear out its welcome and that it knows the bottom line at the White House so it must be judicious about appealing in order not to exceed that White House number. There are also cases where DON would like to appeal, but DOD (SECDEF) prefers that it not appeal. Usually the DON must support the position closest to that in the President's budget, but can negotiate with the DOD comptroller to diverge from this position. DON and the other services exercise this appeal process because the two houses of Congress often treat issues differently as a result of differing philosophies or constituency issues and the outcome may be injurious to Navy programs, for example, in reducing a multi-year procurement program to where the contractor can not perform to bid pricing specifications.

All things considered, the legislative process is even more complex than it seems. One would think that the Conference Committee proceedings would be a limited to legislators, but the administration can and does appeal bills even at this late stage. Moreover, sometimes how issues are finally settled in the Conference Committee has ramifications for the ongoing budget process in DOD. For example, in 2003, DON needed to know the outcome of several issues in the FY 2004 appropriation bill then under consideration, before it could finalize related issues in the FY 2005 President's Budget, being built in the Pentagon during the same time. No matter what DOD did, if it acted before the Conference Committee's actions on the FY 2004 appropriation bill, it would be offering up some programs for cuts or others for deeper cuts in the FY 2005 budget or even cutting programs itself that Congress might not have cut.

This kind of dependency is particularly annoying when the appropriation bill is late, which is the usual case. When this happens, DOD delays its decisions and makes them in the November-December period, after the final appropriation bill has passed. Usually, but not always, these issues are few in number and concern weapon systems purchases, either the absolute number or changing the sequence of purchases over the next several years. This is why these decisions can be held in abeyance; they affect the future more than the present or budget year.

Phase IV: Execution

The fourth and final phase of the DON budget process marks the enactment of appropriations by Congress in the DOD Appropriations Bill. This bill, once signed by the President, allows the DON to incur obligations and to make payments out of the Treasury. A detailed description and analysis of the Navy budget execution process is provided in chapter 9.

The above section emphasizes the similarities between military department budget processes. However, the Department of Navy is a study in contrast to the Army and Air Force in that it has a somewhat more complex budget process in part because it budgets for two services, the Navy and the Marine Corps. The Marine Corps has its own Financial Management office under the Commandant that prepares, examines, and executes the USMC budget. The USMC FM (civilian comptroller and director of the budget) has status comparable to the Navy ASN/FM. Because fiscal guidance is issued by OSD to the military department budget offices through their secretariats rather than to the military services, additional computations must be conducted within the Department of the Navy to divide Navy from Marine Corps funding. The Assistant Secretary of the Navy, Comptroller (civilian Presidential appointee) plays an important role in adjudication between Navy and the Marine Corps budget officials in close coordination with the USMC FM. In the section below we analyze aspects of the Marine Corps budget organization, budget process, and budget accounts with emphasis on the interdependence between Navy and Marine Corps accounts and management.

U.S. Marine Corps—Navy Budget Process

When determining how to divide money between the Navy and the Marine Corps, the Department of the Navy depends on a procedure known as the "blue-green split." This procedure is based on a formula established by the Navy and the Marine Corps in a letter of agreement more than 25 years ago. Although the division of funds is consistent, dollar amounts are not fixed and can be altered significantly in favor of one service or the other by the Secretary of the Navy in different fiscal years. On average, the Navy has allocated roughly 86 percent of its annual appropriated funding to the Navy and 14 percent to the Marine Corps (Williams, 2000: 84–85).

Marine Corps Appropriations

The Department of the Navy funds that are spent by or on behalf of the Marine Corps are concentrated in two appropriations clusters; the first is termed the "green" appropriations and the second the "blue in support of green" appropriations. The green appropriation consists of dollars controlled directly by the Commandant of the Marine Corps. These include Military Personnel, Marine Corps; Reserve Personnel, Marine Corps; Operations and Maintenance, Marine Corps; Operations and Maintenance, Marine Corps Reserve; and Procurement, Marine Corps.

Control is shared between the Navy and Marine Corps in the second group of appropriations, the blue in support of green. These include Military Construction; Military Construction, Reserve; Family Housing; Research, Development, Testing and Evaluation; and Procurement of Ammunition. These accounts are composed of funds provided by the Navy that provide "direct" and "indirect" support for the Marine Corps. The direct support funds are provided directly from the Navy budget. This category provides the money required to procure, operate, and maintain Marine Corps aircraft. The indirect support part of the "blue in support of green appropriations" is comprised of funds that the Navy would have to spend even if the Marine Corps did not exist. The blue in support of green appropriation budgets amphibious ships and their equipment, Naval Surface Fire Support, Corpsmen, and Chaplains. Figure 8.2 demonstrates the division of funds in the Department of the Navy.

Green Appropriations

Once the blue-green split is completed, the Marine Corps can begin to build their Program Objective Memorandum (POM), which in turn leads to the budget. The first thing Marine Corps financial officials do, once their amount of the Department of the Navy TOA is determined, is to "pay the bills" associated with the green appropriation. Funds must first be set aside for resources that have already been committed by the Marine Corps in pre-

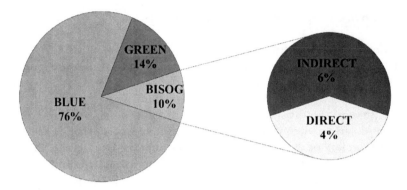

Figure 8.2 Blue-Green and Blue-in-Support of Green Dollars. *Source*: Authors, 2011.

vious years. These funds are referred to as the "core." The "core' is simply the summation of the previous funding decisions that the Marine Corps does not want or need to revisit, what non-DOD analysts would call the base (Burlingham, 2001: 60–64). Its purpose is to help the Marine Corps to:

- Fence entitlements,
- Maintain programmatic stability for well defined, executing programs,
- Recognize the cost of doing business,
- Establish a programmatic baseline, and
- Create a discretionary portion of program.

The core is developed from the minimum requirements of each account that makes up the green appropriation. In short, this category includes all the "must fund" fixed costs of the command (USMC Program Objective Memorandum 2004 Guide: November 2002). Budget analysis proceeds by examining the basic, minimum requirements for the core appropriation accounts in order to set core dollar number.

Green Dollar Core Review

In the Military Personnel and Reserve Personnel accounts, analysts determine the cost of maintaining the Marine Corps at its authorized end strength. In addition, these analysts price bonus plans, accession phasing, and all of the pieces that must fit together to ensure that the bills are covered. Since the Marine Corps is a people intensive service, it is no surprise that this account consumes the majority of the funds in the core. What funds should be set aside in Operation and Maintenance, for both active duty and reserve accounts is the most difficult to determine. The O&M account has been called an accountant's nightmare, because so many different people spend these funds in so many different ways (Williams, 2000). If there are any disagreements over the funds that make up the budget core, it is safe to assume that they will occur within this account.

In the Procurement and Procurement of Ammunition accounts, analysts determine minimum requirements based on two procedures. In Procurement, budget analysts meet minimum requirements by continuing to fund obligated programs. Analysts for the Procurement of Ammunition account determine minimum costs by looking at training and combat requirements. The training requirements are consistent from one year to the next, while the combat requirements are simply replenished when used.

In the Research, Development, Testing, and Evaluation account items that the Marine Corps intends to buy later, and those which they already started to invest in are funded. At the same time, the analysts for this account assess the minimum requirements to fund the Science and Technology facet of the Marine Corps. These dollars will be used to support the Marine Corps War fighting Lab and various advanced concept exploration programs.

The Military Construction and Family Housing accounts factor in both the current inventory and the resources needed to begin construction on the most urgent requirements of the Marine Corps when determining minimum required amounts. This account is the smallest one in the core.

Once the minimum requirements of the green appropriation are identified, the core is set. The Marine Corps will only use this money to fund the requirements determined in the core-setting process. The funds that remain are referred to as the discretionary funds and are available to the Commandant to satisfy all of the demands of the operating forces and supporting establishment of the Marine Corps.

Green Dollar Discretionary Funds

The discretionary funds of the Marine Corps are identified during the Planning phase of their Planning, Programming, and Budgeting System. These funds are only a small portion of the resources received during the blue-green split. In the Program Objective Memorandum for FY 2002-FY2007 for example, about 5 percent of Marine Corps funds were designated as discretionary funds—about $5 billion over the five year period.

The programming phase of the Marine Corps is a period of internal competition. Whereas the planning phase determines the discretionary fund amount, the programming phase decides to whom the fund should go. Requests for the discretionary funds are called "initiatives." In the Program Objective Memorandum for FY 2002-FY 2007, about 525 "initiatives" were submitted. In order to fund all of these, the Marine Corps would have had to spend over $17 billion more then they had estimated. As a result, over three fourths of the 525 initiatives were not funded. The goal of the

Marine Corps, as it is with the other three services, is to get the most for its money. In order to do this, Marine Corps leaders must ensure that the initiatives receiving the funds are the most beneficial. The core is not questioned, since the resources it funds have already been found to be beneficial in previous year reviews. The challenge is selecting from the numerous initiatives submitted, the ones that will most benefit the Marine Corps in the future and funding those with very limited resources.

Just as the other PPBES processes function, the Marine Corps uses a sophisticated committee review system to sift and winnow out the best initiatives. Initiatives are grouped into categories and compete against each other in the POM process. Winners then are funded in the budget process to the extent funds are available. In the POM process, initiatives are reviewed first by Program Evaluation Groups (PEGs). These committees are composed of Lieutenant Colonels, Majors, and civilian equivalents tasked with conducting the initial evaluation of the initiatives. For an initiative to receive resources in a POM, it must first compete successfully within its own PEG, e.g., investment, manpower, military construction and so on. The PEGs are not fiscally constrained. It is their job to hear briefings on selected initiatives that represent different Marine Corps missions or sponsors, judge priorities and relative benefit among the selected initiatives, and consider any objections. Each PEG prioritizes the initiatives in terms of its benefit to the overall mission of the Marine Corps, rather than by cost. Their ranking lists are then forwarded to the POM Working Groups (PWGs) for benefit/cost analysis and re-ranking.

The POM Working Group (PWG) process begins with the consolidation of the PEG lists into a single benefit-only list to create a merged list from the different PEGs so all initiatives may be rated against each other. The PWG then refines the list by taking the benefit value of each individual initiative and dividing it by its cost. This will readjust the order of the list by presenting one based on both benefit and cost rather then just cost. In addition to this refinement, the PWG makes further adjustments based on the professional knowledge, judgment and experience of the lieutenant colonels, majors, and civilians that make up the group. Once each of the initiatives is properly ranked the PWG initiates a process called "order to buy." In the order to buy process, the PWG begins at the top of the list of newly ranked initiatives and starts "spending" the discretionary funds. This process continues until all discretionary funds are spent.

The results of the PWG are submitted to the Program Review Group (PRG), a committee of the most senior officers in the Marine Corps, where they are combined with the core for final assessment. The PRG objective is to assess the war fighting capabilities, verify compliance with guidance, resolve intermediate issues, and make corresponding program adjustments. Once completed, the PRG will then form a single Marine Corps POM including both the fixed core and the discretionary initiatives, which is forwarded to the Commandant along with any major issues that need to be resolved. The Commandant may make final adjustments to this package.

When determining which initiative to support, members of both the PEGs and the PWG use various criteria. Although each analyst may have his own specific standards of determination, as a whole, they tend to look for the same things. The fist thing they look for when determining validity of an initiative is whether its sponsors have provided a concise, specific statement of fiscal need, based on sound funding estimates. They tend to favor initiatives that identify tradeoffs, offsets, and overlaps, and avoid blanket claims, slogans, and buzzwords. They are more likely to support initiatives that define their programs in simple terms and clearly explain the impact on the Marine Corps. In addition, they tend to reject claims of cost savings that cannot be identified by activity, amount, or year. The more quantifiable the supporting information, the more likely the initiative will be highly ranked. (Taylor, 2002: 59)

Department of the Air Force Budget Process

The Air Force Budget Office is the lead agent in the budgeting phase of the Planning, Programming, Budgeting and Execution System (PPBES). The Air Force budget process is described in "The Planning, Programming, and Budgeting System and the Air Force Corporate Structure Primer." (Department of the Air Force, 2002b) The budget office is the key player in this part of the PPBES process. Its objective is to formulate, execute, and control the allocation and use of resources based on requirements identified during the planning and programming phases of PPBES. In the Department of the Air Force the budget process consists of three phases, namely (1) Investment Budget and Operational Budget Review, (2) Budget Estimate Submission, (3) Budget Review.

Phase I: Scrub the Base
The Investment Budget Review (IBR) of the first phase begins with the review and evaluation of the execution and performance of programs funded with investment dollars within the Major Commands and System Centers. It is during this phase that analysts from the Air Force Budget Office determine the expected obligation and execution rates of each program. Their goal is to identify and adjust obligation and execution problems. If the Air Force does not identify and adjust these problems, the Office of the Secretary of Defense (OSD) will usually do so in the Budget Review. However, if the OSD adjusts a program, the savings do not automatically belong to the Air Force.

Based on the findings of the IBR, Air Force Budget Office analysts propose specific adjustments to specific investment accounts in selected programs. These proposals are then forwarded to the Investment Budget Review Committee (IBRC), which is chaired by the Director of the Directorate of Budget Investments (FMB-I). The IBRC will review each proposed adjustment in order to determine which will be sent to the Air Force Board (AFB), which is chaired by the Director of the Air Force Budget Office. This board reviews the IBR recommendations and decides which to keep, adjust, or delete. These results are then submitted to the Air Force Chief of Staff (CSAF) and the Secretary of the Air Force (SECAF) for final approval.

The second aspect of the first phase of the Air Force Budget process is referred to as the Operational Budget Review. It progresses similarly to the IBR except that it focuses on the Operations and Maintenance Accounts. The Operational Budget Review Group (OBRG), chaired by the Director of the Directorate of Budget Operations (FMB-O), reviews the proposals and briefs them to the Air Force Board. As in the IBR, this board evaluates the proposals and submits their findings to the CSAF and the SECAF for consensus.

The main objective of the Investment and Operational budget reviews is to prevent the OSD from adjusting the Air Force Total Obligation Authority. By correcting funding issues within the department, the Air Force is able to make changes that will result in net savings. The recommendations developed during these reviews, once approved by the Air Force Chief of Staff and the Secretary of the Air Force will then be used in the development of the Air Force Budget Estimate.

Phase II: Adjust Scrubbed Base to POM

The beginning of the Air Force Budget Estimate development marks the commencement of the second phase in Air Force Budgeting. It is during this phase that, the approved recommendations of the IBR and OBR are merged with the guidance provided in the Program Decision Memorandum by the Air Force Budget Office in an attempt to readjust the Program Objective Memorandum (POM). The Budget Estimate Submission (BES) is developed from the newly adjusted POM.

The BES is like a bill and it is the job of the budgeters and programmers within the Department of the Air Force to find offsets to lessen its cost. Once the BES is determined, it is briefed by the AFB to the Air Force Council (AFC), CSAF, and SECAF. Once approved the BES is submitted to the OSD thus concluding the second phase of Air Force budgeting process.

Phase III: Negotiate with OSD

Once OSD has received the Air Force Budget Estimate, the third and final phase of the Air Force budgeting process begins. During this phase, the

OSD and Office of Management and Budget (OMB) conduct a joint Budget Review of the Air Force BES. It is their objective to identify more cost effective pricing or programming alternatives. These alternatives are presented in Program Budget Decision Memorandums (PBD). Initially the OSD budget analysts will prepare PBD drafts in order to alert service representatives of the impending marks (cuts) certain accounts will receive. It is imperative that Air Force representatives be proactive and involved in the hearings during this process. Often OSD budget analysts will not write a draft PBD (make a cut) if the Air Force can explain away the analyst's concerns. If however a disagreement emerges, then the Air Force has an opportunity to challenge a draft PBD in the form of comments or reclama (appeal).

Once a reclama is initiated, it is the job of specific representatives, appointed by the Air Force Budget Office, to defend the programs involved in the reclama. This entire process takes only a few days, but its effects can be lasting. After this process has been completed, the Director of the Air Force Budget Office signs a memorandum of either acceptance or rebuttal, in part or completely, to the OSD comptroller. If a rebuttal is initiated, the program in question is elevated to the status of a Major Budget Issue (MBI) and is negotiated between the Air Force Chief of Staff, the Secretary of the Air Force and OSD, represented by the Defense Review Board (DRB).

Once all marks and reclamas are final, the Air Force PBD is used to adjust the department BES that is delivered to the OSD for action. The Air Force BES will then be merged with the BESs of the other services into the Department of Defense budget. This final action will conclude the Department of the Air Force budget process. The Air Force budget process is more centralized than that of the two other military departments.

Department of the Army Budget Process

The Army budget process is managed by the Assistant Secretary of the Army (FM&C) through the Deputy Assistant Secretary of the Army (Budget). It is the Deputy Assistant Secretary who takes charge in the budgeting and execution phase of the Army's Planning, Programming, Budgeting, and Execution System (PPBES) through the Army Budget Office. The Army Budget Office divides the budget process into three parts, namely formulation, justification, and execution.

Budget Formulation

The formulation phase of the budgeting process begins with the development and approval of the Army Budget Estimate Submissions (BES). It is during this phase that the first two years of the programs in the Program Objective Memorandum are converted into the department budget estimate submission. The Army Budget Office supervises the entire formulation process. Major Army Commands (MACOMs) and installations, such as airfields, barracks, camps, depots, and other facilities, aid in the development of the BES by providing a budget request by means of their Command Budget Estimates (CBEs). In the Department of the Army, about sixteen major army commands, ranging from Europe, to the Pacific and Korea, and including such commands as the Corps of Engineers, Medical, Traffic Management, and Special Operations submit Command Budget Estimates.

These estimates developed by the installations are summarized in the Major Army Command Current Budget Estimate, which is next reviewed by the Program Budget Committee and the Army Resource Board. Then it is merged with the program revisions submitted by the Director, Program Analysis, and Evaluation who works directly for the Army Chief of Staff. Next, it is sent to the Secretary of the Army and the Chief of Staff, Army by the ASA (FM&C) for final approval. Decisions are made on the budget at all these stages. Once approved, it is the job of the Army Budget Office to forward the Department of the Army Budget Estimate Submission to the Office of the Secretary of Defense (OSD) and Office of Management and Budget (OMB).

OSD and OMB review the submission to ensure reasonableness and Army estimates are either approved or adjusted. Usually disagreements with adjustments are handled at the lowest level; however, if the dispute persists they are labeled Major Budget Issues and forwarded to the Defense Review Board and Joint Chiefs of Staff for examination. Once the adjustments have been finalized, the results are combined with the other services in order to form the Program Budget Decision (PBD) document, which is then sent to the Deputy Secretary of Defense for final approval. The PBD list the required changes that each service must make to their Budget Estimates. After these changes have been made the services budget estimates can then be included into the Department of Defense Budget and eventually the President's Budget that marks the end of the Department of Army budget estimate formulation phase.

Justification in Congress

The second phase of the DOA budget process, budget justification, initially begins after the President's Budget is submitted to Congress for review. The House Budget Committee and Senate Budget Committee begin the congressional review. Their goal is to ensure that the President's Budget is within the discretionary spending caps of the Omnibus Budget Reconciliation Act of 1990 as updated. During this period that the SECDEF and representatives of each service within the DOD testify before the House National Security Committee, Senate Armed Services Committees, and Appropriation Committees in order to justify to Congress the DOD Budget

Estimates. To support the SECDEF and the DOA representatives, the Army Budget Office provides detailed budget justification books to the authorizing and appropriations committees as well as any other assistance that might be needed to prevent a congressional adjustment to the DOD Budget Estimate.

Execution

The third and final phase of the DOA budget process is Execution. Here it is the job of the ASA (FM&C) to supervise and direct the financial execution of the funds appropriated by Congress. By utilizing the Command Budget Estimates submitted by the major army commands and installations for guidance the Army Budget Office acting on behalf of the ASA (FM&C) distributes all funds approved by the budget and monitors their execution during the budget year. This picture of the Army budget process is drawn from "Budget Office of the Deputy Assistant Secretary of the Army, [http://www.asafm.army.mil/budget/budget.asp] August 2002. Notice that the Army description emphasizes Congress more than the Air Force does.

DOD BUDGET ACCOUNTS: WHAT DO THE DOLLARS BUY?

Previous sections described and analyzed budget office organization and budget processes, i.e., who does budgeting and how the process works. Now we examine military department budget accounts using a case study of the Department of Navy budget to explain how the funds structure is organized and what is purchase by account, as well as some of the issues that analysts face in making budget decisions. The Department of the Navy has a total of five major appropriation accounts that fund both Navy and Marine Corps activities. These appropriations include Operations and Maintenance (O&M); Military Personnel (MilPers); Procurement; Research, Development, Testing, and Evaluation (RDT&E); and Military Construction (MILCON). The majority of the funds go to O&M, Military Personnel, and Procurement as can be seen below. In the following section, we discuss what is in each of these accounts and how they are treated in the budget process.

Military Personnel

The Department of the Navy Military Personnel (MilPers) appropriation is divided into two categories, active and reserve. The active MilPers is composed of Military Personnel, Navy (MPN) and Military Personnel, Marine Corps, (MPMC). The reserve MilPers consists of Reserve Personnel, Navy (RPN), and Reserve Personnel, Marine Corps (RPMC).

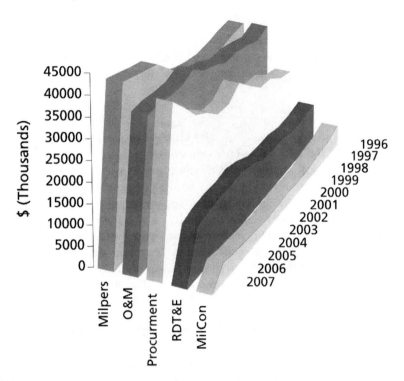

Figure 8.3   DON Appropriated Funds 1996–2007. *Source*: Authors, 2007.

The budget activities funded by the MPN and MPMC include pay and allowances of officers, pay and allowances of enlisted, pay and allowances of midshipman, subsistence of enlisted personnel, permanent change of station travel, and other military personnel costs. The budget activities funded by the RPN and RPMC appropriations include unit and individual training and other training and support. The following exhibit (table 8.1) shows each MilPers appropriation and its budget activities.

When determining the MilPers budget estimate, analyst use an average cost basis. The numbers of people promoted, departing, arriving and already serving are all factors that affect the level of the MilPers budget estimate. The most important factor is the determination of the average cost rates utilized by the Navy and Marine Corps analysts in the formulation of their Military Personnel (MILPERS) budget estimates. Although the rates for military pay and allowances are established by law, the average rate of base pay must be established from these estimates. The base pay average is the largest contributor to the MILPERS account.

Other estimates that affect the level of the MILPERS account include: the number of promotions expected, the number of personnel gains or losses,

TABLE 8.1 Components of the Navy Military Personnel Account

MPN	MPMC
• Pay Allowances Officers	• Pay Allowances Officers
• Pay Allowances Enlisted	• Pay Allowances Enlisted
• Pay Allowances Midshipmen	• Pay Allowances Midshipmen
• Subsistence of Enlisted Personnel	• Subsistence of Enlisted Personnel
• Permanent Change of Station Travel	• Permanent Change of Station Travel
• Other Military Personnel Costs	• Other Military Personnel Costs
RPN	**RPMC**
• Unit and Individual Training	• Unit and Individual Training
• Other Training and Support	• Other Training and Support

Source: Authors, 2008.

and the longevity raises which will accrue during the budget year. Utilizing the above estimates, an average cost rate is computed for each paygrade from the lowest recruit to the highest officer. The same procedure is followed to establish the average cost rate for basic allowances for quarters. Turnover rates are very important for analysts, because turnover rates affect the cost of allowances for changes in duty station, reenlistment bonuses, and clothing for new recruits, and separations. The personnel accounts are not as complex as the other accounts in that they proceed from known factors like specified salary and bonus costs and as well as the number of authorized personnel and use historical averages to develop average costs for other factors; however, it is a large job that takes a lot of brute computational power.

The MPN account may not be as complicated as others, but its issues are important and subtle. In a volunteer force, the innate capability of personnel is of great importance. Over the last decade, CNO guidance to MPN managers and others in the resource allocation process has stressed that people are of paramount importance. MPN managers have recruiting accession and standards goals as well as retention targets. The MPN account has support for recruiters, bonuses to stimulate reenlistment and retention and to fill out communities where expertise is always in demand, as well as support for such things as advertising and change of station travel. All of these matters can be important. For example, when change-of-station dollars get short, moves are often pushed into the next fiscal year. What this does is extend time on station for both officers and enlisted personnel often for another three to four months. This can mean extended sea duty, with a resulting decline in re-enlistment, billets ashore gapped or left vacant for months, and a slowed career progression, directly affecting morale as well as again threatening retention and force structure goals.

Some of the costs in this account are historically based on standards; others are driven inversely by the economy. When the economy is good, recruiting tends to decline and thus more money for recruiters and advertising is needed. Basic salary is also a concern. While the employment cost index, a standard measure of wage cost in the economy, increased by 6.4% from 1989 to 2000, enlisted pay for the same period was increased 2.8% and officer pay only 1.3%. Personnel managers know that you get what you pay for, no matter what is said about making personnel a high priority. The pay and pension reform in FY2000 budget were a significant step forward, but like many issues, their effect will only be felt in the long run. In the meantime, analysts look to historical averages, pay structure, the health of the economy, competing labor markets (for pilots, for example), and decide on dollars for advertising, recruiter activity and bonuses. Information technology also appears to be a large unmet bill in the MPN account because the systems that deal with personnel include many unrelated systems, with many requiring frequent manual data entry. Adopting modern corporate business systems practices could help manage this account.

Operation and Maintenance

The DON Operation and Maintenance (O&M) appropriation, like the MilPers appropriation is divided into active and reserve. The active category includes Operation and Maintenance, Navy (OMN) and Operation and Maintenance, Marine Corps (OMMC). The reserve category includes Operation and Maintenance, Navy Reserve (OMNR) and Operation and Maintenance, Marine Corps Reserve (OMMCR). The budget activities funded by the OMN and the OMNR appropriation include operating forces, mobilization, training and recruiting, and administration and service wide support. The budget activities funded by the OMMC appropriation are similar to the OMN and OMNR appropriation, but do not include mobilization. The OMMCR appropriation funds operating forces and administration and service wide support.

The O&M account is very diverse and includes everything from civilian pay to fuel, paper, and ammunition-all the items, which the DON consumes as an operating entity and most of the expenses it incurs to keep operating. It is easier to state what is not in the account then what is in it. The following exhibit shows the budget activities under which all the items of the O&M account fall.

The O&M account is the most inclusive of the five major appropriations in the DON. It includes funding for items such as the flying hour program, costs for operating the fleets, civilian salaries, base maintenance, ammunition, yellow tablets and pencils, other administrative expenses, and temporary active duty travel for the military and civilians, all the items that the Navy consumes as an operating entity and most of the expenses it incurs to

TABLE 8.2   Components of the Navy O&M Account

OMN	OMMC
• Operating Forces	• Operating Forces
• Mobilization	• Training and Recruiting
• Training and Recruiting	• Administration and Service-wide Support
• Administration/Service-wide Support	
OMNR	**OMMCR**
• Operating Forces	• Operating Forces
• Mobilization	• Administration and Service-wide Support
• Training and Recruiting	
• Administration/Service-wide Support	

Source: Authors, 2008.

keep operating. O&M budget requests are determined either by formula or by a historical cost average. Formulas determine such budget categories as the cost of steaming hours for the Atlantic fleet. Historical costing procedures determine budget amounts by making an estimate based on cost already incurred, such as the average cost for temporary employees in the Washington D.C. area or office supplies or base maintenance over the last three years.

When determining the reasonableness of these estimated amounts, analysts depend on experience, work measurement standards, cost accounting information, employment trends, price level changes, and prior budget execution performance. Selecting the best measurement technique however depends on which program is being estimated. For example, cost and work measurement data are best used when examining ship and aircraft overhauls, fleet operations, flight observations, medical care, supply distribution, and real property maintenance. The flying hour program and ship steaming hours are examples of formula driven categories. Due to its variety, no standard methodology can be described for review of this account.

Procurement

The DON Procurement appropriation encompasses a number of appropriations. Aircraft Procurement, Navy (APN), Weapons Procurement, Navy (WPN), Shipbuilding and Conversion, Navy (SCN), Other Procurement, Navy (OPN) Procurement, Marine Corps (PMC), and Procurement of Ammunition, Navy and Marine Corps (PANMC) are all included under the Procurement appropriation as is shown below.

The planning of pricing and milestone schedules in the acquisition cycle of DON programs is dependent on accurate procurement appropriation

TABLE 8.3 Components of the Procurement Account

APN
- Combat Aircraft
- Airlift Aircraft
- Trainer Aircraft
- Other Aircraft
- Modification of Aircraft
- Aircraft Spare and Repair Parts
- Aircraft Support Equipment and Facilities

WPN
- Ballistic Missiles
- Other Missiles
- Other Weapons
- Torpedoes and Related Equipment
- Ammunition
- Spares and Repair Kits

OPN
- Ships Support Equipment
- Communication and Electronics Equipment
- Aviation Support Equipment
- Ordnance Support Equipment
- Civil Engineering Support Equipment
- Supply Support Equipment
- Personnel and Command Support Equipment
- Spares and Repair Parts

PMC
- Ammunition
- Weapons and Combat Vehicles
- Guided Missiles and Equipment
- Comm. and Electronic Equipment
- Support Vehicles
- Engineer and Other Equipment
- Spares and Repair Parts

SCN
- Fleet Ballistic Missile Ships
- Other Warships
- Amphibious Ships
- Mine Warfare and Patrol Ships
- Auxiliaries, Craft and Prior Year Program Costs

PANMC
- Ammunition, Navy
- Ammunition, Marine Corps

Source: Authors, 2008.

funding. Production schedule, inventory requirements, spare part philosophies, and lead-time are all taken into account when determining how much of the budget should be allotted to this appropriation. It is, however, the determination of an accurate cost per unit estimates that is most important to analysts. These cost per unit estimates are determined in two ways, depending on whether the item is newly acquired or already in development. The cost estimates for existing items is a historical cost supplied by the accounting system, while cost estimation for new items is developed using engineering cost estimates.

In addition, cost per unit estimates for newly acquired items also use factors such as amount of inventory on hand, projected consumption rate, requirement for spare parts, status of Research, Development, Testing and Evaluation programs, production time schedules, slippage of production schedules, required lead time, mobilization base and approval for production to aid in the determination of the most accurate cost per unit. These procurement accounts are a particularly interesting part of the Navy budget in that they represent the purchase of capital goods items, whose mission is often unique to warfare (but not always-search and research and communications hardware have obvious civilian uses) where DOD specifies an end use and a private corporation attempts to estimate what it would cost to build that item and how much profit can be made. The account is also interesting because defense corporations have large amounts of money at stake; loss of a big contract as an old contract phases out could threaten the very existence of a corporation. Congressmen and Senators in whose jurisdictions these corporations are located have an obvious stake in these events. Whether new systems are being considered or whether the question is to buy a few more aircraft or ships, each session of Congress sees a lot of lobbying focused on the Armed Services and Appropriations Committees to ensure that DOD keeps buying weapon systems.

Budget folklore has it that members of Congress are particularly interested in these committees because they can use them to get defense spending for their states and districts. Members of Congress from states rich in defense spending have to get on these committees to protect the investment in their state in its human and physical capital base, and to ensure that they get their appropriate share of defense dollars.

The mechanics of the weapons account are also interesting. First, the budget for a weapon system is developed by a program manager who is a DOD employee, usually a uniformed expert in the military system being purchased. It is his job to oversee it, although few program managers ever oversee a program from inception through RDT&E to full production through the close out of the production line, a process that could take 10 to 15 years or longer. Thus, the program manager enters into a program, somewhere in midstream, and takes over from someone else. His job is to guide the program over the hurdles that will occur "on his watch"—like new requirements, or unexpected glitches, or fiscal slowdowns. He works closely with the corporation which produces the system and when it comes time to budget, he and his business manager sit down with their corporate counterparts and estimate what the cost will be to produce x number of missiles for the next budget year or years. It is a co-operative, yet adversarial relationship over the cost to produce hardware where no one knows precisely how much the item will cost until it is built.

The central budget office analyst exercises a review function, just as he or she would were the ship or aircraft or weapons accounts the budget for temporary labor to paint the walls, something which can be understood more easily and for which a multitude of real world reference points exist for comparison. Analysis in these accounts often proceeds from a series of assumptions about what happens to cost when experience building a system begins to accumulate. The central budget analyst will expect to see unit costs in a program decrease under certain circumstances. Some of these include:

1. As experience accumulates building a weapon—this is called the learning curve.
2. As quantity increases: this is a cost versus quantity relationship that notes efficiencies of scale should lead to decreased unit costs
3. With repetition: repeated construction of the same system should drive down the unit cost of the system. This is the learning curve in another guise, but it is often broken when new requirements are put into the systems and modifications made.
4. With dual sourcing: competition between or among contract bidders is expected to drive down the price that a corporation can ask, hence apparently driving down cost. How far down is often a source of contention between the program manager and the central budget office analyst.

Costs may also change when a corporation decides to take less in the way of profit from a program for reasons it may not readily disclose, e.g., to keep its plant fully occupied, to maintain core skills while it waits on the outcome of another contract, or perhaps because it knows that once the buy has started it can build back its profit margins as change orders come in to modify the weapons system over its lifetime. This is likely to be proprietary information that government analysts may be able to estimate some years later, but it is not likely to be available during the annual budget process.

These attempts to guess which way cost will go and how large the increment of cost change will be are important because these weapons are bought with appropriated funds that have to be requested from Congress. Since dollars are always in short supply, a weapon system that can be made to appear to cost less looks better than one that costs more, if their capabilities are approximately equal. It is no wonder that weapon systems incline toward a low-ball cost strategy when the contractor is anxious to get the contract, elected representatives are anxious to have him get it, program managers are focused on fielding a good weapon system and improving their chances for promotion, and budget analysts at several levels are interested in seeing unit costs decline.

Another phenomenon experienced observers detect in the weapons procurement area is often described as the bathtub curve. It describes the attempt by a corporation to keep its cash flow, and hopefully profits, up as it is transitioning out of an old program and production lines are shutting down, while the new weapons contract that will replace it is just beginning to phase in. This results in a gap which occurs in the corporate profit stream when one system is phased down and another phased in. Lobbying efforts are often focused on the size of the gap, in an attempt to decrease it, and keep the cash flow stream to the corporation going, by extending the construction from the current weapon system or beginning a new system before the old production line has closed down, thus advancing the new profit stream before the old stream has dried up.

The task for analysts in these accounts is quite complex. An investment review checklist suggests that the basic question for these accounts was to ask if it made sense to program the item in question at the requested quantity. Analysts were urged to review the document that provided the basis for the requirement, find the inventory objective and how it was derived, and review any munitions, spare parts or overhaul or rework schedules. Then they could move to a budget scrub, concentrating on such questions as: is the item budgeted to most likely cost? Is the program realistically phased? Is the total program funding profile reasonable? and Is the current program being executed on schedule? For program pricing help, they might request a unit cost track between fiscal years to find any fluctuations, request identification of non-recurring costs to make sure they do not show up in future years, verify a learning curve was used in program pricing, ensure that only OSD/OMB inflation rates were used for cost escalation, and request a contract status report to analyze it for cost and schedule deviations compared to the budget request.

In terms of program phrasing, the analysts might ask if the production build-up is too rapid, review the equipment delivery schedule to ensure that the funded delivery period does not exceed 12 months and examine the factors limiting production ramp-up such as test equipment, personnel, and raw materials. In the funding profile, the analyst is cautioned to watch for and avoid funding spikes as well as programs where the profile is too stretched out to be effective, investigate the possibility of multiyear contracting and Acquisition Improvement Program initiatives like competition and economic production rates.

Investment account analysts have access to multiple budget schedules that allowed them to cross check programs, numbers, and dollars. They can look to see if the out-year profile made sense, what system was suggested for replacement, how it would interface with other systems and if so, how it was funded. For example, if an item were related to military construction, was the military construction item funded? Certain basic themes persist in these questions, none more important than ascertaining if the future profile is reasonable given the past record. "What is the basis for the estimate?" and "how does it compare to the last negotiated cost?" followed by "what are the elements of change from the last negotiated cost?" are questions that are asked by those who are responsible for budgeting these accounts. However, each type of account has its own intricacies, from airframes, to engines, to ships, to missiles, to ammunition, and so on.

Research, Development, Testing, and Evaluation

The DON Research, Development, Testing, and Evaluation appropriation includes basic research, applied research, advanced technology development, demonstration and validation, engineering and manufacturing. Table 8.4 shows these budget activities more clearly.

The budget estimate for the RDT&E appropriation tends to be fixed across the DOD, despite some annual fluctuations depending on variations in budget climate. Insiders say that when the account is larger than 10% of total budget, the risk is that not all projects started can be put in the field. When the account is below 10% the risk is of limited innovation, that not enough new projects are being started. Additionally insiders know that some systems will have teething problems while others may never develop as expected. Running this account at a bare minimum ignores these fact of life situations.

Budget insiders believe that historical logic suggests that an investment of under 10% in this account indicates that the DON is not investing in enough weapons development to keep up with potential competitors in the long run while investments of over 10% of the [DON] budget raises concern about the ability of the organization to successfully man, deploy, and maintain the range of weaponry under development. Although this metric may have been based on the cold war world, it tends to persist in the conversation of knowledgeable budget insiders.

An examination of past trends computed from Figure 8.4 of the DOD Comptroller's Green Book indicates that total DOD RDT&E has been about 12.6% from FY 1981 through 2003, lower that than during the Reagan build-up—which was primarily a procurement build-up—and higher than that during the down years of the 1990s as DOD fought to protect its technological edge. With the procurement holiday of the 1990s, aging systems and a new foe, DOD plans to allocated 15.3% of its budget each year

---

TABLE 8.4   Components of the RDT&E Account

RDTEN

- Basic Research
- Applied Research
- Advanced Technology Development
- Demonstration and Validation
- Engineering and Manufacturing Development
- RDTE Management Support
- Operational Systems Development

Source: Jones and McCaffery, 2008: 467.

DOD RDT&E

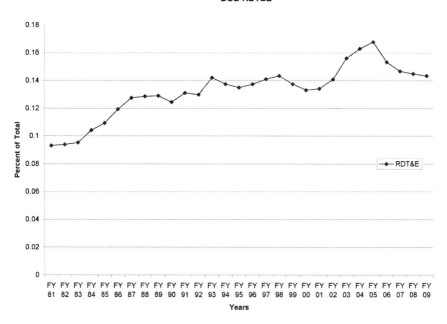

Figure 8.4 RDT&E History (RDT&E as a Percent of Total DOD Budget Authority, Average by Fiscal Year). *Source*: Jones and McCaffery, 2008: 483.
Averages by Selected Period:
  RDT&E BA (Actuals): 1981–2003 = 12.6%
  RDT&E BA: FY81–85 (Reagan Buildup) = 9.9%
  RDT&E BA: FY 91–98 (Drawdown years) = 13.7%
  RDT&E BA: FY2004–09 (DOD FYDP Plan years) = 15.3%

on RDT&E over the course of the Five Year Defense Plan. History would indicate that this is a somewhat optimistic figure.

Whatever the truth to the 10% rule, analysts for the account have some set routines. They annually review each program's financial balance in order to determine the status of unobligated and unexpended balances. Unexpended and over-obligated funds are automatic warnings to analyst that a program needs to be reviewed. For example, a program which is under-obligating and has high balances may have run into substantive trouble (under-obligating because contractor can not make some part of system work and thus can meet milestones and thus not get paid) and ought to be reviewed more deeply before any additional money is committed. Conversely, a program that is in danger of over-obligation may also be in trouble since it is costing too much and also ought to be reviewed. Perhaps the original cost estimates were wrong and the engineering cannot be done for the price that was estimated.

The budget office also reviews the account for balance with the other accounts, for total level of expenditure, for areas in which further research ought to be pursued or curtailed, and the availability of scientific personnel and research facilities. Analysts may look for unrealistic growth planned over a prior year, no recognition of schedule delays, unobtainable milestones, and failure to consider changes in limiting factors like personnel or test equipment, and proposed changes in technical scope. Analysts also have to be sensitive to congressional action that may restrict progress or funding, pending submission of data to Congress. Analysts also look for duplication of effort where other services or government agencies are developing a similar item or where other programs satisfy a similar requirement and could lead to a cut against this program. Analysts examine the budget estimate to ensure that only those funds needed to carry the program over the 12-month budget period are requested. They may examine contractor and service data to see when the work will be performed and they may examine outlay data to see that the program is on schedule thus far. Schedules and milestones will be examined to make sure the program is still on track and no slippage from the approved plan is evident.

THE BUDGET FUNCTION IN THE OFFICE OF THE SECRETARY OF DEFENSE

The budget functions of the Office of the Secretary of Defense are performed by staff of the Deputy Secretary of Defense, Comptroller. The title of the DOD Comptroller has changed over time. Presently it is Principal Deputy Undersecretary of Defense, (Comptroller), and Deputy Undersecretary of Defense for Financial Management Reform. All of the tasks associated with budgeting are managed under the auspices of the DOD Comptroller, not to mention some of the tasks related to programming not elaborated here. Briefly, the DOD Comptroller budget related responsibilities include issuance of guidelines for budget preparation and review (including preparation of exhibits), review of military department budget submissions, preparation of budget marks to these budgets and responses to military department reclamas, issuance of Program Budget Decisions (PBDs), compilation of military department and defense agency budgets into the budget of the Secretary of Defense and the President's budget. Preparation and issuance of the official DOD budget document, the National Defense Budget Estimates or "Green Book" for each fiscal year is an important responsibility of the DOD Comptroller staff.

When the defense budget is reviewed by the committees of Congress, DOD Comptroller staff support testimony by various defense department officials including the DOD Comptroller, review proposed and enacted legislation for fiscal impact, and provide a myriad of reports required and requested by members of Congress and their staffs. They also conduct various special studies on behalf of the Secretary of Defense and develop estimates of savings projected to result from implementation, e.g., management efficiency initiatives.

As explained in previous chapters, within the PPBES process the development of the Department of Defense budget begins with the dissemination by the Secretary of Defense of the Defense Fiscal Guidance that specifies the "top-line" budget targets for each military department and agency. This comprises roughly the amount of money targeted as not to be exceeded when the military departments set their respective resource requirements. Once the fiscal constraints are issued, the Department of the Army, Navy and the Air Force may begin formulating their budgets.

In budget execution, the DOD Comptroller and staff have responsibility for submitting the military department and defense agency apportionment requests to the Office of Management and Budget, allocating regular and supplemental appropriations into appropriate DOD accounts, and internal allotment of the budget to what DOD analysts refer to as the military components (military departments and services). DOD Comptroller responsibilities also include establishing and distributing spending targets and procedures when Congress makes regular and Continuing Resolution appropriations, monitoring spending patterns across DOD accounts, monitoring and reporting to the Secretary of Defense and other DOD officials on spending trends and patterns. In budget formulation and execution, DOD (C) officials review the control of accounts and spending activity to comply with the law and congressional requirements relative to their authority to do so, responding to congressional budget initiatives including those originating in the Congressional Budget Office, and from committee and subcommittee staffs. They submit DOD reprogramming requests and monitor the reprogramming process, submission of requests and distribution of supplemental appropriations. They maintain coordination of end of year accounting and reconciliation to appropriations requirements, responding to internal (Inspectors General) and external (e.g., GAO) audit findings and a number of other post-spending year tasks. Among these other tasks is monitoring the working capital funds operated and managed by the military departments, and provision of all types of information to the Secretary of Defense, and to other DOD officials including data to be issued to the public and news media through press releases. This is only a synopsis of tasks performed by DOD Comptroller staff.

The DOD budget process cycle revolves around the schedule established by the DOD Comptroller. The Comptroller also coordinates with the other DOD level secretariats (e.g., the Under Secretary of Defense for Acquisition, Technology and Logistics) to provide information essential for administration of the responsibilities of these offices. This is especially important for the acquisition budgeting process that operates semi-autonomously as described later in this book. This summary of the budget responsibilities of the DOD Office of the Comptroller is illustrative and not comprehensive. More specific detail on the roles of the DOD Comptroller and of OSD staff are provided in earlier chapters of this book, particularly chapter 4 on PPBES and chapter 7 on budget execution.

CONCLUSIONS

In this chapter we have demonstrated the complexity of the budget processes that operate within the military departments and services and in the Office of Secretary of Defense in the Pentagon. We have reviewed how each department budget office is organized and functions within its budget processes to produce submissions to the Office of the Secretary of Defense. We have explained that the PPBES process has undergone some recent modifications. Since 2003, a combined POM and budget review performed by the military department budget and programming offices and the Comptroller staff of OSD have attempted to determine the desired force structure and fiscal estimates.

In this chapter we have not explored in detail what happens once the military department budget submissions to OSD are reviewed and approved by SECDEF and staff. The budget process as it operates outside of DOD has been explained thoroughly in earlier chapters. As noted, OSD does not have a separate budget office that operates identically to those of the military departments. The staff of the OSD Office of the Comptroller performs the function of budgeting and analysis for the Secretary of Defense. In analysis of the military department budget processes we analyzed the PBD and reclama cycle that eventually leads to translation of the DOD budget into the formats required by the President's Office of Management and Budget, and then the appropriation format employed to by Congress. We have traced the origin of the Pentagon budget process beginning with preparation of future year estimates shown in the Five Year Defense Plan (FYDP), and the Quadrennial Defense Review. We have noted that these data are not incorporated directly into budgets because they are generally formulated prior to the beginning of the budget process. We have explained how great care is taken to make the budget year estimate precise and accurate.

Each of the military departments maintains a central budget office to coordinate, facilitate, and oversee this process. It is the job of these offices to review the budget for conformance with the directives issued by Office of the Secretary of Defense and DOD Comptroller at the start of the budget process and through any subsequent changes to those directives. Ultimately, the military department budget offices must produce budgets on behalf of their Secretaries and military chiefs that integrate the guidance set forth by the Secretary of Defense with the requests of the organizations within their respective departments. As we have shown in this chapter, the budget offices and the budget processes of each military departments and OSD are designed to accomplish a very complex task of rationing and coordination. We have shown that each military department and the budget officials and analysts that work in them employ their own unique strategies and tactics to obtain and protect their budgets and programs.

CHAPTER 9

FINANCIAL MANAGEMENT AND DEFENSE BUSINESS PROCESSES

INTRODUCTION

It is axiomatic to observe that to do anything programmatically in government you need money. Everywhere that money is authorized for expenditure, spent, and an accounting provided, one finds financial management knowledge, skills and abilities performing the functions to ensure that fiscal law and administrative rules are abided by and enforced. The public as well as the legislative and executive branches of government rely on financial managers to be the guardians of public fiscal affairs to see that financial functions are performed to the letter and intent of the law and to provide the means for assuring accountability in spending of public money. Financial management in DOD is, as a consequence, a very important function, in fact one that is crucial to the achievement of national security and defense missions and objectives.

Financial management in the Department of Defense can be viewed broadly or narrowly. Broadly, financial management encompasses a diversity of managerial functions including the full PPBES process, comptroller, accounting and reporting, auditing (financial, management and performance), treasury functions (from revenue generation to investment), operations and maintenance, capital asset planning, acquisition and procure-

Financing National Defense: Policy and Process, pages 327–366
Copyright © 2012 by Information Age Publishing
All rights of reproduction in any form reserved.

ment, logistics management including inventory management and control, contracting, personnel and human resource management, risk analysis and management, program assessment, evaluation and policy analysis, and retirement account management and fiduciary stewardship over such accounts. Financial management touches nearly every function of DOD, especially the business-like functions. More narrowly, one may view financial management as a set of practices distinct from budgeting, where financial management encompasses only those financial processes associated with the more technical fiduciary responsibilities rather than the more general resource allocation decision-making. The broad view is covered by the entirety of this text; this chapter takes the more narrow view.

Taking the narrow view that budgeting is distinct from financial management, one should note that "[b]udgets are undeniably more important than financial statements in local as well as national governments." (Smith and Chen, 2006: 16) The process of budgeting is where policy disputes are waged, priorities are established, resources are allocated, and winners and losers determined. Budgets are also used as control devices during execution to keep public managers focused on the policy priorities. Financial management is often regarded by managers as a "housekeeping" task that does not influence decision-making. Most comptroller's offices within DOD are separated into two groups and "[a]t times, the communications between budget offices and accounting offices resemble a visit to a foreign country where the language spoken is different from the visitor's" (Hoge & Martin, 2006: 137). Line managers rarely communicate with the accountants yet frequently interact with the budget side. The National Academy for Public Administration found that despite a lot of progress since the passage of the CFO Act in 1990, "[t]he key functions of financial management—strategic planning, budgeting, accounting and financial reporting—have not always been properly integrated [...] financial data and reporting are not always user-friendly to decision-makers, including policy makers and program managers [and] cost accounting has not been widely implemented" (NAPA, 2006: 7–8). However, most experts agree there has been substantial improvement over the past 20 years integrating financial management processes, practices, data systems and strategic decision-making processes. This chapter explores these issues in detail.

Before the discussion goes further, it is important to note that the scope of financial management in the Defense Department is unparalleled in any corporation or other government agency. There are approximately 50,000 financial managers in DOD (Hale, 2011). In fiscal year 2009, the Defense Finance and Accounting Service (DFAS) processed over 180 million pay transactions, paid nearly 8 million travel payments and 13 million commercial invoices, accounted for 956 active appropriations, and maintained over 175 million general ledger accounts (DFAS, 2010). DOD financial statements for fiscal year 2009 identify $1.8 trillion in assets, accounted for $1.18 trillion in budgetary resources, and report the net cost of operations for the year at $652 billion. (Department of Defense, 2009)

OBJECTIVES OF CHAPTER

The purpose of this chapter is to develop understanding in the following areas:

1. To define financial management in the public sector and explain its relationship to budgeting and core business processes
2. To outline the legislative requirements for financial management in the DOD
3. To differentiate among the three bases of accounting used in the DOD, highlighting the differences in information content and the needs of the users of the data
4. To identify the basic financial statements, relate them to financial statements produced in the for-profit sector, and contrast the differences
5. To summarize and evaluate the arguments for and against auditable financial statements for the DOD
6. To indicate the major defense management reform initiatives of the past 20 years, identifying their successes and failures, and drawing inferences about their consequences
7. To explain the persistent challenges faced by DOD leaders as they attempt to reform financial management practices
8. To evaluate and discuss the merits of proposed reform initiatives in light of lessons of the past and the nature of defense management.
9. To assess whether the costs of implementing the Chief Financial Officer's Act of 1990 are worth the benefits produced by doing so.

FINANCIAL MANAGEMENT IN THE DEFENSE DEPARTMENT

Financial Management Defined

Adopting the narrow view of financial management, and in recognition that this text has chapters dedicated to budgeting and budget execution, we may define financial management in the DOD as *those functions and supporting processes that encompass the department's ability to give a public accounting for performance of programs and the associated budget authority entrusted to it, including the financial impacts of the department's activities.* Let us examine that definition a little more closely.

Giving a public accounting includes the formal formulation and issuance of the department's quarterly and annual financial statements, accompanying statements of assurance, and performance information, but it also includes less formal responses to inquiries from inside or outside the department about the status of appropriations, trust funds, special interest projects, cost of contingency operations, liabilities, foreign currency fluctuations, cost overruns for acquisition programs, and the like. The definition includes not only the processes for formulating those reports, but also the underlying practices that ensure those reports are timely, accurate, and repeatable.

An accounting for the budget authority entrusted to it implies a strong relationship between budgeting, program management, and financial management. In the chapters on PPBE and budget execution, we discussed how the department determines what programs should be included in the budget plan and the obligation and expenditure of funds in pursuit of the programmatic goals. Financial management provides a snapshot of the financial condition of the department which can be used by decision makers when making resource allocation decisions. To the extent that past budgetary and program performance affects future budgets, financial management provides the evidence of past performance. Later in the budget cycle, financial management intersects the execution phase. Recall that appropriations do not provide money to the DOD, but rather the authority to obligate the government to expend funds from the Treasury in support of DOD, and that such authority is bound to specific purposes, time frames and amounts. It is imperative that the department have in place processes, tools, and expertise to account for the status of that authority to ensure none of the three bounds are exceeded. Such accounting of budget authority is an element of financial management.

Finally, the financial impacts of the department's activities include such things as inventory valuation, depreciation of capital items, contractor payments, intragovernmental transfers and reimbursements, and the valuation of retirement, environmental and other liabilities. Most business functions and many operational functions of the department have financial implications. Some are straightforward and are completed immediately, such as a small purchase of consumable items. Others are more complex, such the purchase of inventory for stock, a capital investment, a sale of equipment to a foreign country that pays in a foreign currency, or hiring a permanent employee who accrues retirement benefits. The systems that support the primary acts of purchasing, inventory, real property management, foreign military sales, and human resources must also interface with financial management systems to capture the financial effects of those actions.

The Legislative Requirement

The Constitution of the United States gives the Congress the power of the purse and most interpret that to mean appropriating funds for government programs. Section 9 of Article 1 of the Constitution states, "No Money shall be drawn from the Treasury, but in Consequence of Appropriations made by Law; and a regular Statement and Account of the Receipts and Expenditures of all public Money shall be published from time to time." Many forget about the second clause. Given that this falls under the powers of Congress, it is appropriate that they legislate the manner and form of such statements and accounts. Since the Budget and Accounting Act of 1921, Congress has been a driving force for change in federal accounting and financial management. The 1990s were a period of significant change with the passage of several key pieces of legislation that formed today's financial management landscape. Some of the more significant financial management laws are described in Figure 9.1.

Budget and Accounting Procedures Act of 1950 (31 U.S.C. 3511 (a)) directed the Comptroller General, in consultation with OMB and Treasury, to prescribe accounting principles, standards, and related requirements for executive agencies to follow. It directed the Treasury to render overall Government financial reports to the President, the Congress and the public.

The Federal Managers Financial Integrity Act of 1983 (P.L. 97–25) (FMFIA) amended the 1950 Budget Act and guides the management control program in the department. The act requires each agency implement a system of internal controls—and report management's assurance—that

- obligations and costs are in compliance with applicable law
- funds, property, and other assets are safeguarded against waste, loss, unauthorized use, or
- misappropriation; and
- revenues and expenditures applicable to agency operations are properly recorded and accounted for to permit the preparation of accounts and reliable financial and statistical reports and to maintain accountability over the assets.

The Chief Financial Officer and Federal Financial Reform Act of 1990 (P.L. 101-576), or CFO Act, reformed government financial management through a set of policy directives with the goal of improving government decision-making through the integration of budgeting, cost analysis, accounting, and performance management. Specifically,

- the Office of Management and Budget (OMB) was given greater authority over federal financial management through the creation of the position of Deputy Director for Management and the Office of Federal Financial Management
- The position of Chief Financial Officer was created for each of 24 federal departments and agencies
- A CFO Council was created, consisting of the CFOs and Deputy CFOs of the largest federal agencies and senior officials of OMB and Treasury, to provide

---

Figure 9.1   Financial Management Legislation.   *(continues)*

leadership and coordinate activities across the federal government related to modernization of financial systems, quality of financial information, financial data and information standards, internal controls, and legislation affecting financial operations.
- Agencies were directed to develop consolidated financial systems that met generally accepted accounting standards and were capable of producing corporate-style financial statements.
- The Federal Accounting Standards Advisory Board (FASAB) was created to develop a comprehensive set of accounting concepts and standards for the federal government.

The Government Performance Results Act of 1994 (P.L. 103-62) (GPRA) required executive agencies to produce 5 year strategic plans, annual performance plans with objective measurable goals to implement the strategy, and to report program performance against those goals. Performance reports are normally the responsibility of the agency CFO and often accompany either the annual president's budget request or the annual financial report. Because the DOD prepares a Quadrennial Defense Review, it was exempted from the requirement for a five-year plan.

The Government Management Reform Act of 1994 (P.L. 103-356) (GMRA), in addition to some human resource policy changes, mandated the use of direct deposit for federal pay, streamlined OMB financial reporting requirements, and required that the annual financial statements of executive agencies be audited before submission to OMB.

The Federal Financial Management Improvement Act of 1996 (P.L. 104-208) (FFMIA) built on the CFOA and GMRA by requiring each agency to "implement and maintain financial management systems that comply substantially with Federal financial management systems requirements, applicable Federal accounting standards, and the United States Government

Standard General Ledger (USSGL) at the transaction level". Until such time as the USSGL became the accounting standard, agencies had difficulty becoming auditable and it was extremely difficult to consolidate accounting data from multiple agencies.

NDAA for FY2008 (Section 904)—created the positions of Chief Management Officer (CMO) for DOD, Deputy Chief Management Officer (DCMO) for DOD and CMOs for the military departments who are responsible for enterprise-wide business transformation efforts. It also required a Strategic Management Plan (SMP) and annual progress reports that encompassed:
- Performance goals and measures for improving and evaluating the overall efficiency and effectiveness of business operations
- Key initiatives to be undertaken in meeting performance goals and measures
- Procedures to monitor the progress toward meeting performance goals and measures
- Procedures to approve plans and budget for changes in business operations
- Procedures to oversee all budget requests for defense business systems

Figure 9.1 (continued) Financial Management Legislation.

Figure 9.2, from a National Academy of Public Administration report on financial management, displays the four key phases of the budget process and the principal legislation and executive branch directives that prescribe them for the federal government.

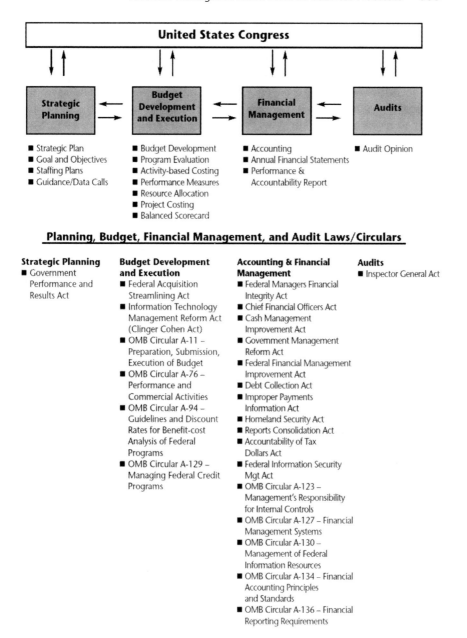

Figure 9.2   Governing Directives for Budget and Financial Management Processes.

ACCOUNTING IN THE DEPARTMENT OF DEFENSE

The requirement to maintain an accounting of government finances is not only embedded in the laws above, but hallmarks of public service are compliance with laws, the transparency of government operations and accountability for taxpayer funds. Defense managers should assume the responsibility to meet all three objectives and doing so requires a consistent flow of reliable financial information. Financial management processes seek to perform all three functions—compliance, transparency, and accountability—while supporting management decision-making. In this section, we examine the role of accounting in government and the DOD, especially. We consider the various bases for accounting stressing the information content of each base and will discuss the types of decisions and stakeholders that need such information. Portions of the DOD and federal government financial statements will be presented and described. Finally, we will discuss various views on the requirement to produce auditable statements—is the benefit worth the effort?

The bases for accounting

Ask a recent MBA graduate how many different types of accounting he or she studied and the likely response will be "two," financial accounting and managerial (or cost) accounting. Within the DOD, accounting is performed three ways: financial, managerial, and budgetary. *Financial accounting* is what one sees in a publicly-traded, for-profit corporation's annual report: balance sheets, income statements, and statements of cash flow and owner's equity. Terms associated with financial accounting include profit, earnings before interest and taxes, depreciation, and revenue recognition. Financial accounting is performed on an accrual basis in accordance with generally accepted accounting principles. That is, transactions are recorded when revenues are earned and costs incurred, which may not coincide with the actual receipt or payment. The rules are rigid and strict; there is governmental oversight. The focus is on the accurate and reliable capturing, recording, categorizing, and presenting of historical events. The intended audience is outsiders. For a corporation, the outsiders are those in the capital markets (potential lenders and investors); for a government agency, the outsiders are those with a financial interest (legislators, taxpayers, and institutional investors in government debt).

Managerial accounting is what one doesn't see in a corporation's annual report. It is the internal analysis conducted by the corporation to weigh one decision against the next: Should we expand this plant? Should we drop that product line? Should we buy or lease a specific capability? What does it cost to provide this service versus an alternative? Terms associated with managerial accounting include weighted average cost of capital, cost drivers, internal rates of return, profit center, analysis of variances, and cost-volume-profit analysis. Managerial accounting is performed on a cost basis, rather than an accrual basis. There are common tools and practices for managerial accounting, but no generally accepted management accounting principles; there is no government oversight.[1] The focus is on internal management decisions about the organization's mission and scope of operations; the goal is to increase the value of the firm through an understanding of costs and profits. The audience is internal; in fact, the data is so proprietary that corporations typically wouldn't dream of sharing it within their industries, with the government, or the public. Within the DOD, this type of accounting enables working capital fund activities to set their rates based on unit cost; it is also important information for determining cost estimates for investment decisions.

Budgetary accounting is what government employees are most familiar with. It is the process of budgeting, justifying, and accounting for appropriations. Terms associated with budgetary accounting include commitment, expenditure, apportionment, obligation, authority, and disbursement. Budgetary accounting is performed on a cash basis in which events are recorded as cash (or budget authority) is received, budget authority is obligated, and payments are made (obligations are liquidated). There are strict rules embedded in federal appropriation law and principles articulated by the Comptroller General; there is significant oversight. The focus is on compliance with the law—that funds have been spent in accordance with the purpose, time, and amount restrictions attached to the appropriation. The audience is both internal and external. There is no corporate analogy, however, e.g., the concept of "obligation" has no meaning to a commercially focused Certified Public Accountant (CPA) or most business school accounting faculty. Table 9.1 summarizes the three bases of accounting.

For example, we may consider the process of buying and using a piece of capital equipment to illustrate differences between public and private sector accounting methods and definitions. In financial, or accrual-based, accounting transactions are recorded when economic events occur. When the item is shipped (or received) the organization has an asset and a corresponding accounts payable transaction. When the item is paid for, liquid assets are reduced and the accounts payable is satisfied. When the item is used in the process of generating products or services for sale, the asset is depreciated and the depreciation expense matched to the revenue generated. Managerial accounting analysis may have supported a make-buy-lease decision or a financing decision when the item was procured. Managerial accounting is also concerned with recording the cost of the item in relation to the lines of businesses or cost centers it supports. Budgetary accounting

TABLE 9.1   Three Bases for Accounting

	Financial Accounting	Managerial Accounting	Budgetary Accounting
Associated Terms	Income Statement, Balance Sheet, Cash Flows, Revenue Recognition, EBITDA, EPS, P/E ratio, Quick ratio	Allocation, cost and profit centers, product line, ABC, NPV, IRR, ROI, WACC, non-financial indicators	Commitment, obligation, outlay, appropriation, apportionment, budget, phasing plan, color of money
Time Horizon	History – a record of what's already happened	Future – decisions about alternative courses of action	Past, present, and future
Audience	Capital Markets – shareholders and potential lenders	Internal Management*	Stakeholders: managers, taxpayer, congress, everyone
Rules	GAAP	Tools, no rules*	Law
Oversight	SEC, AICPA, FASB	None*	Congress, tax payers, OMB, chain of command
Objective	Transparency, accountability, and comparability (over time & across entities)	Maximizing profitability, adding economic value, decision-making by managers	Compliance

* Companies under contract with the federal government are required to follow certain cost accounting standards.

Source: Authors, 2011.

would have recorded a commitment at the time a contracting approach was identified, an obligation when the contract was signed, and an outlay (expenditure) when the bill for the equipment was paid.

Notice that budgetary accounting merely records the promise to pay (obligation) and the payment, but does not capture the ownership of the asset or the impact of that asset on any line of business. Budgetary accounting does not tell a manager what is owned or how it is used, thus it is of limited utility for management decision-making. As the former DOD comptroller, Tina Jonas, once commented before a House committee, "It is important to note that our financial systems and processes were developed to support the budget and appropriations process. Therefore, they unfortunately do not generate the type of financial information that meets the needs of the Department's decision makers." (Jonas, 2002: 2) Financial and managerial accounting practices have long been viewed as adjuncts to budgetary accounting.

Financial management practices should ensure that all three forms of accounting occur in an integrated fashion. The Budget and Accounting Procedures Act of 1950 required heads of agencies to "establish and maintain systems of accounting designed to provide reliable accounting results to serve as the basis for agency budget requests and budget control and execu-

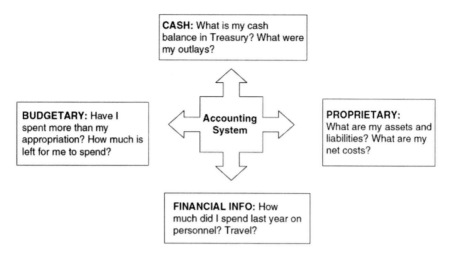

Figure 9.3 Functions of Accounting Systems.

tion" (Hoge and Martin, 2006). The example above—buying and using a capital asset—is a single series of actions that can be described as consisting of certain financial and economic transactions which would be captured in the various accounting systems. Since all three forms of accounting are describing the same series of actions, one should be able to cross-walk the information in the various accounting reports to get a comprehensive view of the organization. Hoge and Martin (2006: 123) graphically represent the different purposes for which accounting systems exist in Figure 9.3.

Let us look at the various financial statements produced by the federal government and the DOD and how they link together.

The Financial Report of the United States Government

Each year the Department of Treasury with the Office of Management and Budget produce the Financial Report of the United States Government, commonly referred to as the Consolidated Financial Report (CFR). The CFR and the budget are the two primary financial documents produced by the federal government. The CFR is retrospective—a record of what happened last accounting period and how has that affected the financial condition of the nation. The budget is prospective—a record of the aspirations, policy objectives, and specific program proposals, how much budget authority they require and what the expectation for revenue generation to pay for it. The CFR includes Table 9.2 to explain this further:

TABLE 9.2 Key Financial Documents Produced by U.S. Federal Government

President's Budget	Financial Report of the U.S. Government
Prepared primarily on a '*cash basis*'	Prepared on an '*accrual and modified cash basis*'
• Initiative-based and prospective: focus on current and future initiatives planned and how resources will be used to fund them.	• Agency-based and retrospective – prior and present resources used to implement initiatives.
• Receipts ('cash in'), taxes and other collections recorded when received.	• Revenue: Tax revenue (more than 90 percent of total revenue) recognized on modified cash basis (see Financial Statement Note 1.B). Remainder recognized when earned, but not necessarily received.
• Outlays ('cash out'), recorded when payment is made.	• Costs: recognized when owed, but not necessarily paid.

The CFR is analogous to a corporation's annual report. It provides a view of the financial results of operation and the financial position of the government; it provides an accounting for the money the government raised and spent; and it reports on operating performance and the effectiveness of accounting systems and internal controls. (GAO, 2009d) As its goal mirrors that of a corporate annual report, so does its structure: a management discussion, financial statements, notes to the statements, supplemental information, and an auditor's opinion. Additionally, the federal government provides stewardship information for intangible investments (e.g., basic research or education) and special categories of assets (e.g., national parks). Most of the financial statements in the CFR are supposed to be compiled using accrual-based accounting: the Statement of Net Cost, Statement of Operations and Changes in Net Position, and the Balance Sheet. They are also reconciled to the budget and budgetary accounting practices in the statement, Reconciliation of Net Operating Revenue (or Cost) and Unified Budget Surplus (or Deficit).

The *balance sheet* shows the nature and amount of the federal government's assets and liabilities, the difference between them being the net position. An abbreviated balance sheet is displayed in Table 9.3.

The *statement of net cost* displays the cost to operate the federal government by agency and department (Table 9.4). Costs are displayed on an accrual basis, and not based on budget authority or actual cash outlays. Thus, items such as long-term liabilities are recorded as incurred, not when paid.

The *statement of operations and changes in net position* displays the annual financial result of the government's operations in a manner similar to the way a corporate income statement displays revenue, cost and income Table 9.5). Since the federal government does not generate an "income," the

TABLE 9.3   Balance Sheet of U.S. Government

(In billions of dollars)	As of Sept. 30 2008	As of Sept. 30 2007 (Restated)
■ **Assets**		
Cash and other monetary assets	424.5	128.0
Loans receivable, net	263.4	231.9
Inventories and related property, net	289.6	277.1
Property, plant, and equipment, net	737.7	691.1
Other assets	259.5	253.0
Total assets	1,974.7	1,581.1
✳ **Stewardship Land and Heritage Assets**		
▲ **Liabilities**		
Federal debt securities held by the public and accrued interest	5,836.2	5,077.7
Federal employee and veteran benefits payable	5,318.9	4,769.1
Environmental and disposal liabilities	342.8	342.0
Other liabilities	680.3	598.1
Total liabilities	12,178.2	10,786.9
● **Commitments and Contingencies**		
◆ **Net Position**		
Earmarked funds	704.6	620.2
Nonearmarked funds	(10,908.1)	(9,826.0)
Total net position	(10,203.5)	(9,205.8)
Total liabilities and net position	1,974.7	1,581.1

■ **Assets** are the operational resources the federal government has available as of the end of the fiscal year. The largest category—property, plant, and equipment, net—includes land and buildings and the federal government's military equipment, such as ships, aircraft, and tanks, after subtracting accumulated depreciation.

✳ **Stewardship Land and Heritage Assets,** which include national parks and natural resources, are reported in nonfinancial units in the notes to the financial statements.

▲ **Liabilities** are the financial responsibilities of the federal government as of the end of the fiscal year. In addition to federal debt securities, liabilities include federal employee and veteran benefits payable—the amount the federal government estimates that it owes or will owe its military and civilian employees and veterans under its life and health insurance and pension plans.

● Not all **Commitments and Contingencies** of the federal government require recognition as liabilities on the Balance Sheet. Commitments that require the future use of resources, such as long-term leases, as well as loss contingencies that are assessed to be at least reasonably possible, are disclosed in the notes to the financial statements.

◆ **Net Position** is the difference between the federal government's assets and liabilities.

difference between revenue (mainly taxes) and cost is referred to as the change in net position, and is reflected in the government's balance sheet.

The *reconciliation of net operating revenue (or cost) and unified budget surplus (or deficit)* displays the relationship between the net cost of government and the annual budget deficit (Table 9.6). Most citizens are familiar with the annual budget deficit—the net difference between revenue and outlays—but are less familiar with the concept of net cost. This statement reconciles the budgetary account with the accrual-based account of the annual financial result.

TABLE 9.4   Statement of Net Cost of U.S. Government

(In billions of dollars)	Year ended Sept. 30 2008			Year ended Sept. 30 2007		
	●	■	▲			
	Gross Cost	Earned Revenue	Net Cost	Gross Cost	Earned Revenue	Net Cost
Department of Defense	767.6	26.8	740.8	689.6	25.1	664.5
Department of Health and Human Services	769.1	56.4	712.7	718.6	51.8	666.8
Social Security Administration	663.9	.3	663.6	626.4	.3	626.1
Department of Veterans Affairs	434.6	4.2	430.4	63.1	3.7	59.4
Interest on Treasury securities held by the public	241.6	—	241.6	238.9	—	238.9
Department of Energy	35.8	4.2	31.6	67.5	4.3	63.2
All other entities	979.0	159.0	820.0	753.2	162.6	590.6
Total	3,891.6	250.9	3,640.7	3,157.3	247.8	2,909.5

● **Gross Cost** is the accrual-based total cost of the federal government's operations for the year. The statement presents gross cost, earned revenue, and net cost by federal agency and department. Also, it includes the cost of interest on Treasury securities held by the public, which includes foreign investors.

■ **Earned Revenue** comes from fees charged for goods and services. The fees charged for postal services such as stamps are a well-known example of earned revenue. Earned revenue shows how much the federal agencies and departments earn from their operations to cover their gross costs, as opposed to relying on taxes and borrowing to cover the costs.

▲ **Net Cost** is the portion of the gross cost left after subtracting earned revenue. The federal government funds the net cost of government from tax revenue and, as needed, by borrowing.

Source: GAO, 2009d

One of the strengths of accrual based accounting is a display of the long-term liabilities of an organization compared to its current assets and income-producing capital assets. Unfortunately, the federal government's financial statements are limited in that respect. Unlike a corporation such as an automobile manufacturer, the US government does not use a factory to produce goods that are sold to generate revenue. The government generates revenue through its sovereign power to tax. For various reasons, not the least of which is the difficulty in valuing them, the government's balance sheet does not include assets such as national parks. It is hard to say what the Grand Canyon is worth and whatever the value, it is not a meaningful figure if the government does not use its value to cover liabilities or generate revenue. Also, the balance sheet does not include all long-term liabilities, including the long-term liabilities of Social Security and Medicare.

TABLE 9.5 Statement of Operations and Changes in Net Position of U.S. Government

(In billions of dollars)		Year ended Sept. 30, 2008	
(Operations)	• Nonearmarked funds	✤ Earmarked funds	Consolidated totals
Revenue:			
Individual income tax and tax withholdings	1,210.0	868.4	2,078.4
Corporate income and other taxes and miscellaneous revenues	451.7	131.3	583.0
✤ Intragovernmental interest	—	201.0	201.0
Total Revenue	1,661.7	1,200.7	2,862.4
Eliminations			(201.0)
▲ Consolidated Revenue			2,661.4
Net Cost:			
Net Cost	2,186.4	1,454.3	3,640.7
✤ Intragovernmental interest	201.0	—	201.0
Total net cost	2,387.4	1,454.3	3,841.7
Eliminations			(201.0)
▶ Consolidated net cost			3,640.7
✦ Intragovernmental transfers	(338.0)	338.0	—
● Unmatched transactions and balances	(29.8)	—	(29.8)
✳ Net Operating (Cost) Revenue	(1,093.5)	84.4	(1,009.1)
(Changes in Net Position)			
⊐ Net Position, Beginning of Period	(9,826.0)	620.2	(9,205.8)
Adjustments to beginning balances	11.4	—	11.4
Net Operating (Cost) Revenue	(1,093.5)	84.4	(1,009.1)
⊐ Net Position, End of Period	(10,908.1)	704.6	(10,203.5)

- **Nonearmarked Funds** are financed by nonearmarked receipts and proceeds of general borrowing. These funds finance federal government activities that are not financed by earmarked funds.

✤ **Earmarked Funds** are financed by specific revenue that remains available over time for designated purposes. Major earmarked funds relate to Social Security and Medicare programs, and military and civil service retirements and disability benefits.

✤ **Intragovernmental Interest** is the interest earned by earmarked funds and the interest paid by nonearmarked funds for money borrowed by the nonearmarked funds from the earmarked funds.

▲ **Consolidated Revenue** primarily comes from federal income tax collections, which includes taxes earmarked for Social Security and Medicare.

▶ **Consolidated Net Cost** is the net cost from the Statement of Net Cost.

✦ **Intragovernmental Transfers** supplement earmarked-fund activities, generally through appropriations from the Treasury General Fund.

● **Unmatched Transactions and Balances** are the unexplained differences between the Net Operating Cost and the Changes in Net Position.

✳ **Consolidated Net Operating (Cost) Revenue** is the financial results of operations—the difference between the revenue and the net cost of government operations for the fiscal year (plus or minus) the net amount of any unmatched transactions and balances.

⊐ **Net Position** is the difference between the assets and liabilities reported on the Balance Sheet.

Source: GAO, 2009d.

TABLE 9.6   Reconciliation of Net Operating Cost and Unified Budget Deficit of U.S. Government

(In billions of dollars)	Years ended Sept. 30	
	2008	2007
● Net Operating Cost (Results of Operations)	(1,009.1)	(275.5)
■ Components of Net Operating Cost Not Part of the Budget Deficit		
Increase in liabilities for employee and veteran benefits	549.8	90.1
Increase in environmental and disposal liabilities	0.8	36.8
Depreciation expense	54.8	45.3
Other	70.4	25.0
▲ Components of the Budget Deficit Not Part of the Net Operating Cost		
Cash outlays for capitalized fixed assets	(106.4)	(58.8)
Other	(15.1)	(25.7)
✳ The Unified Budget Deficit	(454.8)	(162.8)

● **The Net Operating Cost** comes from the Statement of Operations and Changes in Net Position. It primarily represents the difference between the federal government's tax revenue and expenses.

■ **Components of Net Operating Cost Not Part of the Budget Deficit** are mostly current-year expenses under accrual accounting that do not involve current-year cash outlays. Increases in liabilities such as employee and veteran benefits and depreciation expense are recognized as current-year operating expenses under accrual accounting, but are not in the unified budget total.

▲ **Components of the Budget Deficit Not Part of Net Operating Cost** consist mostly of current-year cash outlays for transactions that do not involve current-year expenses, such as outlays to purchase buildings and equipment that the federal government capitalizes (records on its Balance Sheet as assets) and depreciates (expenses) as they are used in operations. The outlays to purchase these assets increase the unified budget deficit but not the current net operating cost.

✳ **The Unified Budget Deficit** represents the difference between cash receipts (primarily from taxes) and cash outlays for the year for all programs, on- and off-budget.

Source: GAO, 2009d

The *Statement of Social Insurance* fills that void (Table 9.7). Using actuarial projections based on current laws regarding eligibility and payments, the statement provides a present value estimate of government's liabilities under these programs. We see, on a present value basis, that Social Security payments are expected to exceed revenue by about $6.55 trillion over the next 75 years. That difference pales in comparison to the $36.3 trillion net liability for Medicare programs.

The CFR for the entire federal government is aggregated data from the statements of the various department and agencies that comprise the government. The Department of Defense compiles its own financial statements. Those statements, in turn, are the aggregation of 33 reporting entities ranging from the three military departments to the defense

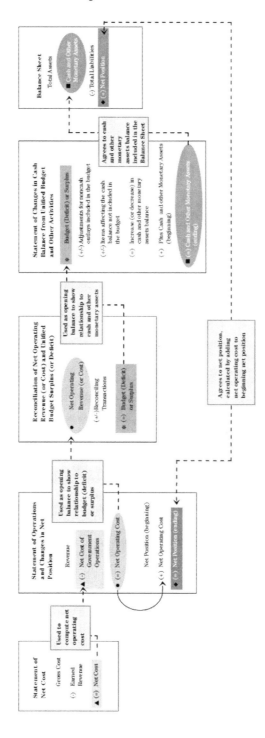

Figure 9.4 Relationships Among Financial Statements of U.S. Government. *Source:* GAO, 2009d.

TABLE 9.7   Statement of Social Insurance of U.S. Government

Present Value of Long-Range (75 Years) Actuarial Projections	As of January 1				
				Unaudited	
(In billions of dollars)	2008	2007	2006	2005	2004
● Federal Old-Age, Survivors and Disability Insurance (Social Security)					
Revenue (Contributions and Earmarked Taxes)	36,357	34,113	32,107	29,450	27,699
Expenditures for Scheduled Future Benefits	(42,911)	(40,876)	(38,557)	(35,154)	(32,928)
Present Value of Future Expenditures					
in Excess of Future Revenue	(6,555)	(6,763)	(6,449)	(5,704)	(5,229)
■ Federal Hospital Insurance (Medicare Part A)					
Revenue (Contributions and Earmarked Taxes)	11,883	11,023	10,644	9,435	8,976
Expenditures for Scheduled Future Benefits	(24,619)	(23,315)	(21,934)	(18,264)	(17,468)
Present Value of Future Expenditures					
in Excess of Future Revenue	(12,736)	(12,292)	(11,290)	(8,829)	(8,492)
▲ Federal Supplementary Medical Insurance (Medicare Part B)					
Revenue (Premiums)	5,478	4,789	4,481	4,187	3,889
Expenditures for Scheduled Future Benefits	(21,197)	(18,221)	(17,613)	(16,571)	(15,329)
Present Value of Future Expenditures					
in Excess of Future Revenue	(15,719)	(13,432)	(13,131)	(12,384)	(11,440)
▲ Federal Supplementary Medical Insurance (Medicare Part D)					
Revenue (Premiums and State Transfers)	2,107	2,405	2,366	2,547	2,651
Expenditures for Scheduled Future Benefits	(9,964)	(10,766)	(10,250)	(11,233)	(10,770)
Present Value of Future Expenditures					
in Excess of Future Revenue	(7,857)	(8,361)	(7,884)	(8,686)	(8,119)
Other - Present Value of Future Expenditures					
in Excess of Future Revenue	(104)	(100)	(97)	(86)	(83)
Total	(42,970)	(40,948)	(38,851)	(35,689)	(33,363)

- ● **Social Security** provides **Federal Old-Age and Survivors Insurance and Disability Insurance.** Both programs are financed by taxes on employees and employers, including the self-employed, and are administered by the Social Security Administration.

- ■ **Federal Hospital Insurance (Medicare Part A)** covers inpatient hospital and related care. It is financed primarily by a payroll tax on employers and employees, including the self-employed, and a portion of the income taxes paid on Social Security benefits. Federal Hospital Insurance is administered by the Department of Health and Human Services.

- ▲ **Federal Supplementary Medical Insurance,** which also is administered by the Department of Health and Human Services, consists of two parts, called Medicare Part B and Medicare Part D. Medicare Part B covers hospital outpatient services, physicians' services, and other assorted products and services. Part D covers the federal government's prescription drug program. Both parts are financed primarily by transfers from the general fund of the Treasury and premiums from participants.

Source: GAO, 2009d.

agencies to the Military Retirement Trust Fund. The DOD prepares four principal statements as described in Table 9.8.

The CFO Act requirement to produce auditable, corporate-style, accrual-based financial statements is a controversial one. There has been controversy over the utility of the statements, the appropriateness of the statements, and the value of the audit opinion. First, some have argued that the

TABLE 9.8　Key Financial Statements for the U.S. Department of Defense

Statement	What Information It Provides
Balance Sheet	Reflects the Department's financial position as of the statement date (September 30, 2009). The assets are the amount of future economic benefits owned or managed by the Department. The liabilities are amounts owed by the Department. The net position is the difference between the assets and liabilities.
Statement of Net Cost	Shows separately the components of the net cost of the Department's operations for the period. Net cost is equal to the gross cost incurred by the Department less any exchange revenue earned from its activities.
Statement of Changes in Net Position	Presents the sum of the cumulative results of operations since inception and unexpended appropriations provided to the Department that remain unused at the end of the fiscal year. The statement focuses on how the net cost of operations is financed. The resulting financial position represents the difference between assets and liabilities as shown on the consolidated balance sheet.
Statement of Budgetary Resources	Provides information about how budgetary resources were made available as well as their status at the end of the period. It is the only financial statement exclusively derived from the Department's budgetary general ledger in accordance with budgetary accounting rules.

Source: Department of Defense, 2009: 44.

statements are of modest or very little value since few actors actually use the information. (Jones and McCaffery, 1997; Brook, 2010; GAO, 2007b) Much of that criticism stems from the failure of agencies or legislatures to use the information in the financial statements for budgetary decision-making. Some studies show that institutional investors and municipal debt credit agencies are among the very few who use the statements (Smith and Chen, 2006). This is not surprising when one reflects back on Table 9.1 which summarized the various bases of accounting, their information content, and key audiences. The criticism asserts that the value of statements is less significant because it fails to meet a certain expectation, an expectation, however, that is unfounded. Unfortunately, both OMB (1993) and GAO (2005) created the expectation that such statements will lead to improved management and budgetary outcomes. One cannot fault the statements for not doing something they were never equipped to do. Improved management should come primarily from managerial accounting and improved budgeting through budgetary accounting.

Others question the appropriateness of the requirement given fundamental differences between governments and businesses. Chan (2003) and GASB (2006) outline several critical differences between the sectors that suggest financial reporting should be fundamentally different. First, governments do not have owners rendering the basic accounting equation (Assets = Liabilities + Owners Equity) problematic. Government assets are frequently acquired through non-economic transactions (e.g., conquered territories) and historical costs are often meaningless. Government's contractual and legal liabilities may be straightforward, but its political and social liabilities for the general welfare of the populace are not. Business accounting requires a link between expenses and the revenue they generate; that is, no revenue can be recognized until goods or services are provided. In government there is rarely a link between payment (taxes paid) and

TABLE 9.9   Balance Sheet of the U.S. Department of Defense

Department of Defense Consolidated Balance Sheet Agency Wide	2009 Consolidated	*Dollars in Millions* Restated 2008 Consolidated
ASSETS (Note 2)		
Intragovernmental:		
Fund Balance with Treasury (Note 3)	$ 502,754.3	$ 468,396.9
Investments (Note 4)	434,884.0	394,508.5
Accounts Receivable (Note 5)	1,219.6	1,326.2
Other Assets (Note 6)	2,594.7	1,282.9
Total Intragovernmental Assets	$ 941,452.6	$ 865,514.5
Cash and Other Monetary Assets (Note 7)	2,316.8	2,804.8
Accounts Receivable, Net (Note 5)	8,234.1	7,825.3
Loans Receivable (Note 8)	398.5	236.0
Inventory and Related Property, Net (Note 9)	228,796.6	233,586.6
General Property, Plant and Equipment, Net (Note 10)	559,614.3	506,953.6
Investments (Note 4)	2,017.1	1,861.5
Other Assets (Note 6)	56,637.5	55,829.4
Stewardship Property, Plant & Equipment (Note 10)		
TOTAL ASSETS	$ 1,799,467.5	$ 1,674,611.7
LIABILITIES (Note 11)		
Intragovernmental:		
Accounts Payable (Note 12)	$ 2,207.8	$ 1,687.4
Debt (Note 13)	391.7	262.6
Other Liabilities (Note 15)	11,485.7	12,047.0
Total Intragovernmental Liabilities	$ 14,085.2	$ 13,997.0
Accounts Payable (Note 12)	35,519.8	32,622.8
Military Retirement and Other Federal Employment Benefits (Note 17)	2,012,166.1	1,984,605.4
Environmental and Disposal Liabilities (Note 14)	66,230.0	70,505.9
Loan Guarantee Liability (Note 8)	21.1	24.5
Other Liabilities (Note 15)	33,839.9	34,107.6
Commitments & Contingencies (Note 16)		
TOTAL LIABILITIES	$ 2,161,862.1	$ 2,135,863.2
NET POSITION		
Unexpended Appropriations - Earmarked Funds (Note 23)	$ 5.6	$ 4.9
Unexpended Appropriations - Other Funds	504,339.3	472,983.4
Cumulative Results of Operations - Earmarked Funds	(1,342,858.4)	(1,345,925.0)
Cumulative Results of Operations - Other Funds	476,118.9	411,685.2
TOTAL NET POSITION	$ (362,394.6)	$ (461,251.5)
TOTAL LIABILITIES AND NET POSITION	$ 1,799,467.5	$ 1,674,611.7

Source: Department of Defense, 2009: 45.

benefits dispensed. Further, appropriation law often prohibits the incurrence of costs until budget authority (revenue) is provided; reversing the conditionality of the accounting logic. Further, many government transactions are involuntary and do not conform to the assumptions of market transactions that underlie corporate accounting standards.

Of the 33 reporting entities within the DOD, only six[2] were expected to receive a qualified or unqualified audit opinion in 2009 (Department of

**TABLE 9.10   Consolidated Statement of Net Cost of the U.S.
Department of Defense**

Department of Defense Consolidated Statement of Net Cost Agency Wide	2009 Consolidated	Restated 2008 Consolidated
Program Costs		
Gross Costs	**$ 697,813.3**	**$ 726,427.3**
Military Retirement Benefits	87,880.4	160,282.8
Civil Works	11,712.0	8,849.4
Military Pay & Benefits	144,750.6	137,701.7
Family Housing & Facilities	11,639.6	9,326.0
Operations, Readiness & Support	268,007.0	251,313.0
Strategic Modernization	173,823.7	158,954.4
(Less: Earned Revenue)	**(46,167.9)**	**(56,272.5)**
Military Retirement Benefits	(4,313.9)	(24,099.5)
Civil Works	(1,657.1)	(1,339.0)
Military Pay & Benefits	(1,019.4)	(976.3)
Family Housing & Facilities	(7,465.5)	(5,919.4)
Operations, Readiness & Support	(18,349.5)	(11,503.1)
Strategic Modernization	(13,362.5)	(12,435.2)
Net Cost of Operations	**$ 651,645.4**	**$ 670,154.8**

Dollars in Millions

Source: Department of Defense, 2009: 6.

Defense, 2009). Of the 24 CFO Act agencies in the federal government, the DOD is one of only five that still cannot pass the test of an independent audit (Treasury, 2010). The DOD has spent hundreds of millions of dollars in pursuit of a clean opinion, and some question the value of the effort. Achieving auditability takes considerable managerial and institutional effort (Brook, 2001) and given the size and complexity of the DOD relative to other federal agencies, it is not surprising the task has been challenging. Given the debate over the value of the statements themselves, senior leadership attention to the requirement has been minimal. There are some who argue that despite the low utilitarian value of the financial statements, there is value in achieving the clean audit opinion. Brook and Candreva (2008) argue that there is democratic and managerial value. There is democratic value in the public perception that the DOD has made progress putting its financial house in order. Ever since the GAO began publishing a High Risk list in 1990, DOD financial management has been highlighted as a problem (GAO, 2009). Achieving a clean audit opinion is public evidence of a threshold standard of competence and accountability. The pursuit of that level of financial acumen should have secondary managerial benefits. Achieving auditability requires improvements to financial management processes and accounting systems that contribute to better financial man-

TABLE 9.11 Consolidated Statement of Changes in Net Position of the U.S. Department of Defense

Department of Defense Consolidated Statement of Changes in Net Position
Agency Wide

	2009 Earmarked Funds	2009 All Other Funds	2009 Eliminations	2009 Consolidated
Cumulative Results Of Operations				
Beginning Balances	$ (1,259,693.0)	$ 335,935.7	$ 0.0	$ (923,757.3)
Prior Period Adjustments:				
Changes in accounting principles	0.0	1,689.6	0.0	1,689.6
Corrections of errors	0.0	(10,482.5)	0.0	(10,482.5)
Beginning balances, as adjusted	$ (1,259,693.0)	$ 327,142.8	$	$ (932,550.2)
Budgetary Financing Sources:				
Appropriations used	3.8	708,653.2	0.0	708,657.0
Nonexchange revenue	2,767.7	(13.7)	0.0	2,754.0
Donations and forfeitures of cash and cash equivalents	38.0	0.2	0.0	38.2
Transfers(in/out without reimbursement)	(836.7)	930.9	0.0	94.2
Other	0.0	(4.2)	0.0	(4.2)
Other Financing Sources (Non-Exchange)				
Donations and forfeitures of property	0.7	1.5	0.0	2.2
Transfers(in/out without reimbursement)	(88.2)	45.4	0.0	(42.8)
Imputed financing	0.0	17,017.3	12,313.5	4,703.8
Other	(19.4)	1,273.1	0.0	1,253.7
Total Financing Sources	$ 1,865.9	$ 727,903.7	$ 12,313.5	$ 717,456.1
Net Cost of Operations	(5,561.6)	669,520.5	12,313.5	651,645.4
Net Change	$ 7,427.5	$ 58,383.2	$ 0.0	$ 65,810.7
Cumulative Results of Operations	$ (1,252,265.5)	$ 385,526.0	$ 0.0	$ (866,739.5)
Unexpended Appropriations				
Beginning Balances	4.9	446,864.3	0.0	446,869.2
Prior Period Adjustments:				
Correction of Error	0.0	26,119.1	0.0	26,119.1
Beginning balances, as adjusted	$ 4.9	$ 472,983.4	$ 0.0	$ 472,988.3
Budgetary Financing Sources:				
Appropriations received	4.5	754,440.5	0.0	754,445.0
Appropriations transferred (in/out)	0.0	(188.4)	0.0	(188.4)
Other adjustments	0.0	(14,243.0)	0.0	(14,243.0)
Appropriations used	(3.8)	(708,653.2)	0.0	(708,657.0)
Total Budgetary Financing Sources	$ 0.7	$ 31,355.9	$ 0.0	$ 31,356.6
Unexpended Appropriations	$ 5.6	$ 504,339.3	$ 0.0	$ 504,344.9
Net Position	$ (1,252,259.9)	$ 889,865.3	$ 0.0	$ (362,394.6)

Source: Department of Defense, 2009: 33.

agement performance—more timely and accurate financial information—that, in turn, meets requirements for compliance with laws and policies and contributes evidence of responsibility.

DOD FINANCIAL AND BUSINESS MANAGEMENT REFORMS

What has the DOD done about the fact that more than 20 years after the passage of the CFO Act it remains incapable of producing auditable financial reports? The answer is "plenty." There has been a series of reforms over the past two decades, some aimed directly at financial management weaknesses, others seeking to improve related business processes. Some were initiated by the military leadership, others by the civilian leadership;

TABLE 9.12   Statement of Budgetary Resources for the U.S. Department of Defense

Department of Defense Combined Statement Of Budgetary Resources Agency Wide *Dollars in Millions*	Budgetary Financing Accounts		Nonbudgetary Financing Accounts	
	2009 Combined	Restated 2008 Combined	2009 Combined	Restated 2008 Combined
Budgetary Resources				
Unobligated balance, brought forward, October 1	$ 135,669.8	$ 111,980.6	$ 26.3	$ 25.5
Recoveries of prior year unpaid obligations	63,272.9	49,744.1	47.5	0.0
Budget authority				
Appropriation	855,564.3	859,403.8	0.0	0.0
Borrowing authority	0.0	0.0	58.4	130.0
Contract authority	67,626.3	78,927.8	0.0	0.0
Spending authority from offsetting collections				
Earned:				
Collected	178,143.5	174,493.0	45.1	53.9
Change in receivables from federal sources	1,188.4	791.8	0.0	0.0
Change in unfilled customer orders:				
Advance received	1,031.0	753.2	0.0	0.0
Without advance from federal sources	(88.3)	5,679.1	(10.8)	12.6
Expenditure transfers from trust funds	862.5	766.0	0.0	0.0
Subtotal	$ 1,104,327.7	$ 1,120,814.7	$ 92.7	$ 196.5
Nonexpenditure transfers, net, anticipated and actual	(15.7)	(264.1)	0.0	0.0
Temporarily not available pursuant to Public Law	(39,190.4)	(59,949.4)	0.0	0.0
Permanently not available	(85,969.3)	(85,156.8)	(0.1)	(27.6)
Total Budgetary Resources	$ 1,178,095.0	$ 1,137,169.1	$ 166.4	$ 194.4
Status of Budgetary Resources				
Obligations incurred:				
Direct	848,896.2	811,662.0	142.7	168.1
Reimbursable	183,271.6	189,837.4	0.0	0.0
Subtotal	$ 1,032,167.8	$ 1,001,499.4	$ 142.7	$ 168.1
Unobligated balance:				
Apportioned	128,048.6	120,047.8	0.3	0.3
Exempt from apportionment	3,373.5	1,060.7	0.0	0.0
Subtotal	$ 131,422.1	$ 121,108.5	$ 0.3	$ 0.3
Unobligated balance not available	14,505.1	14,561.2	23.4	26.0
Total status of budgetary resources	$ 1,178,095.0	$ 1,137,169.1	$ 166.4	$ 194.4

Source: Department of Defense, 2009: 24.

some were inspired by OMB and others in response to pressure from Congress. The breadth and amount of reform should not be interpreted to mean those reforms were all effective. Significant progress has been made in some areas and much work remains in others.

The early 1990s were a pivotal time for defense management reforms. The CFO Act of 1990 has already been discussed. It was followed by the Defense Acquisition Workforce Improvement Act (DAWIA, 1991) designed to improve the outcomes of the defense acquisition process and to improve the professionalism of the acquisition workforce, including those involved in acquisition cost estimating, budgeting and financial management.

In 1991, Secretary of Defense Cheney created the Defense Finance and Accounting Service "to standardize, consolidate, and improve accounting and financial functions throughout the DOD. The intent was to reduce the cost of the Department's finance and accounting operations while strengthening its financial management" (DFAS, 2010: 15). DFAS pays all DOD uniformed and civilian personnel, retirees, and major contractors, and does the majority of DOD accounting. At its inception, DFAS took responsibility for more than 300 installation-level finance and accounting offices, 330 finance and accounting systems and 27,000 personnel. Over the next 20 years, those operations would be drastically consolidated and reduced.

In the mid-1990s, the DOD created a Senior Financial Management Oversight Council and a CFO Master Plan to reengineer financial processes, ensure DFAS continued to consolidate and improve financial operations and to fulfill requirements of the CFO Act. The financial management and acquisition communities established electronic data interchange (EDI) tools for contracts and contractor payments. DOD was also reengineering business processes (such as travel processing) as part of the broader federal National Performance Review initiative. By the late 1990s, EDI had expanded to a broader e-business program complete with the language of "paperless" contracting and finance. Pay had moved to electronic funds transfer (ETF). (Francis and Walther, 2006).

By the year 2000, DOD adopted the OMB Circular A-76 competitive sourcing guidance to review over 270,000 positions for possible outsourcing. Financial operations took further advantage of automation and industry partnerships. Government purchase cards and travel cards were used more extensively. Cash was moved off ships and installations and replaced with electronic payment tools. By 2000, DFAS had reduced the number of finance and accounting systems to 76 with a goal of 37 by 2005 through the use of corporate level IT architectures. Payroll technician productivity increased more than five-fold. The Program Budget Accounting System (PBAS) reduced the number of problem disbursements and contractor overpayments. (Francis and Walther, 2006)

The use of commercial practices continued through the 2000s with an emphasis on measurement, outsourcing, quality improvements, best practices, and information sharing. President G.W. Bush's President's Management Agenda sought to link resources to performance goals. The National Security Personnel System (now disestablished) sought to do the same on an individual level. Secretary Rumsfeld had a particularly active business reform agenda for the department. He created a Senior Executive Council, a Business Initiatives Council and the Defense Business Practices Implementation Board. A Financial Management Modernization Program was estab-

lished in the office of the USD(Comptroller) to continue the trend to consolidate and improve finance and accounting systems and information flow
and to move the department closer to the objective of auditable financial
statements. The FMMP evolved into the Business Management Modernization Program (BMMP) under the leadership of the USD(Acquisition, Technology, and Logistics) when it was determined that the root cause of internal control problems in accounting was in the non-financial systems that
feed the accounting systems. The BMMP was a broad program designed
to improve the interfaces between accounting systems and the feeder systems that manage logistics, human resources, installation management,
and medical care. By the middle of the decade, the department created
the Financial Improvement and Audit Readiness (FIAR) Plan within the
comptroller's office that explicitly described the steps to get a clean audit
opinion. (Francis and Walther, 2006)

In February 2005, the BMMP was replaced by the Defense Business Systems Management Committee (DBSMC). The DBSMC was established "to
advance the development of world-class business operations in support of
the warfighter [...]The DBSMC will recommend policies and procedures
required to integrate DOD business transformation and to review and approve the defense business enterprise architecture and cross-Department,
end-to-end interoperability of business systems and processes" (Wolfowitz,
2005: 12). Later that year, the Business Transformation Agency was created to lead and coordinate enterprise-wide business transformation efforts
for the department. They do that through the Enterprise Transition Plan
(ETP). BTA took over program management responsibilities for overarching systems such as the Defense Travel System. They also set standards for
defense-wide business systems through the Business Enterprise Architecture (BEA). (Francis and Walther, 2006)

While DOD had been busy improving operations, progress was slow and
not well organized. GAO continued to highlight weaknesses in DOD's business practices and in 2005 added DOD's *approach* to business transformation
to the High Risk list (GAO, 2009c). By the fall of 2007, GAO had been pressing for DOD to establish a new Chief Management Officer (CMO) position
at Level II of the Executive Schedule (equivalent to the Deputy Secretary
of Defense) to provide consistent and dedicated leadership to the department's business operations (GAO, 2007). DOD resisted and the Fiscal Year
2008 National Defense Authorization Act compromised by giving the Deputy
Secretary of Defense himself the role of Chief Management Officer and established the office of the Deputy Chief Management Officer (DCMO). In
2002, more than a decade after the passage of the CFO Act, the Pentagon
estimated that gaining a clean audit opinion might still be eight to ten years
away (Kucinich, 2002a: 3); in 2010, the estimate is now 2017 (Department of
Defense, 2009). While reported internal control weaknesses are down and

nearly one-third of the department's balance sheet receives a clean audit opinion, progress has been much slower than predicted.

At the time this is written, DFAS had consolidated operations into only 13 sites with a goal of reducing that further. There are only a few dozen finance and accounting systems remaining. Whereas the department was faulted for paying excess interest charges for late vendor payments, recent audits fault the department for paying too quickly. Most vendor payments are done in a paperless fashion through tools such as Wide-Area Workflow and electronic invoicing. The associated work force has been reduced from about 27,000 to just over 12,000 personnel. As yet, there appears to be overlap of functions among the DCMO, BTA and DBSMC and their responsibilities for improvement of enterprise-wide business systems and financial management. Complicating the landscape, each military department has its own business transformation agenda, in some cases there are two: one run by the secretariat and one by the uniformed service chief. For instance, in the Navy, the Financial Improvement Program is largely run by the ASN(FM&C), Financial Management Operations branch. Meanwhile, the Navy Enterprise program is run by the uniformed leadership on the OPNAV staff and Fleet Forces Command. Most official departmental pronouncements of reform policy come from the secretariat. As such, the DOD DCMO prepared the departmental consolidated 2009 Strategic Management Plan which expressed the following desired outcomes, goals, measures, and key initiatives (See Table 9.13).

Effective management of business functions has been a concern for the Defense (and earlier, War and Navy) departments since they were first created. In the early 19th century, the construction of the first six ships of the fledgling navy were replete with management problems ranging from the provision of raw materials, shipyard capacity, qualified sailors, cost overruns, requirements changes, and controversial designs. In addition, the policies affecting management of the programs were confounded by international events and congressional pork barreling (Toll, 2006) (Navy, 1997). A century ago, Navy Secretary George von L. Meyer, a Massachusetts businessman, sought to reorganize the navy and to apply business principles to its operations, including modern cost accounting at shipyards and state-of-the-art scientific management techniques (Anonymous, 1911). As noted above, in the modern era, every defense secretary has had a business management reform agenda and the department has been subjected to nearly every contemporary management idea, many from the business sector. A detailed inventory and analysis of those reform agendas show that there are some common themes among the reforms: maximizing the utility or efficiency of assets (capital, human or financial), applying assets towards policy goals, measuring progress, and removing non-value-added organizational elements or processes (Francis and Walther, 2006: 125). If the department

TABLE 9.13   U.S. Department of Defense 2009 Strategic Management Plan Outline

Outcome	Goal	Measure	Key Initiatives
Acquire needed resources and make the best of them USD(C)	Spend American Reinvestment and Recovery Act (ARRA) funds quickly and effectively	Percent obligation rates vs. project ground-breaking	• Optimize ARRA transparency through full automated reporting of high-visibility costs • Obligate to annual spending targets
		Percent obligations rates vs. planned obligations	
	Maintain an effective budget execution function	Achieve monthly planned obligations	• Optimize transparency through full automated reporting of obligation rates and cancelling balances
		Manage cancelling balances	
Pay people and vendors on time and accurately USD(C)	Sustain timely and accurate pay	Sustain 2009 accuracy rate for Defense Joint Military Pay System (DJMS)	• Renew DJMS change-control board process
		Sustain military and civilian pay timeliness at 2009 rates or better	• Work to enhance integrated personnel and pay services
		Reduce pay improper payments	• Address pay issues unique to the Reserve component • Conduct full statistical analyses of pay improper payments
		Reduce vendor invoice backlog	• Promote the use of e-Commerce
Demonstrate good stewardship of public funds USD(C)	Maximize Anti-Deficiency Act (ADA) compliance	Reduce late investigations	• Launch new Financial Improvement and Audit Readiness (FIAR) strategy • Re-synchronize Enterprise Transition Plan milestones to support a revised FIAR strategy
		Reduce actual violations	
	Increase the audit readiness of individual DoD components	Achieve auditability	
		Implement effective financial systems	
Improve Real Property Installation Management USD(AT&L)	Make visible real property assets and link with direct and indirect costs, consistent with revised FIAR strategy emphasis on existence and completeness	Percent of real property management systems compliant with the DoD enterprise-wide standard called Real Property Inventory Requirements (RPIR)	Implement real property inventory data standards, data elements and sustainable DoD business processes
		Percent of Defense Agencies' real property assets that are reconciled and captured in an enterprise-wide system	Develop a standard DoD real property inventory reconciliation process and tool for use across all DoD
		Percent of DoD financial management systems reporting expenditures by site-unique identifier and/or real property – unique identifier	Track actual expenditures at the asset level for real property and installations support

Source: Department of Defense, 2009: 24.

is in a state of perpetual reform, why are the business outcomes not better improved? Why have high-risk areas not been resolved?

PERSISTENT CHALLENGES TO DOD FINANCIAL AND BUSINESS MANAGEMENT

Examiners have developed different lists of the more significant business management challenges facing the department, depending on their perspective and tasking. The Government Accountability Office High Risk list includes (GAO, 2009c):

- DOD Approach to Business Transformation
- Business Systems Modernization
- Personnel Security Clearance Program
- Support Infrastructure Management
- Financial Management
- Supply Chain Management
- Weapon Systems Acquisition
- Contract Management

The DOD Inspector General also identified eight management and performance challenges facing the department. Among them are financial management, acquisition processes and contract management, and information assurance (Department of Defense, 2009). A recent survey of defense financial managers showed that the PPBE system is not as responsive as it should be, there are knowledge gaps in the financial management workforce, and the financial information generated by the department is not as valuable as it should be. (American Society for Military Comptrollers, 2010)

As we have seen, these are not new problems. And we have seen that the department is continually reforming its practices. So why have the problems not been resolved? What are the persistent challenges that have prevented DOD from successfully overcoming them? The problems fall into a few categories: failures in strategic direction and practice, system complexity and modernization challenges, business process limitations, and oversight and incentives.

Failures in Strategic Direction and Practice

The sheer size of the task of fixing DOD financial management requires long term commitment, but structural factors make that commitment hard to sustain. While leadership may be committed, its tenure in the job is too short and frequent changeover slows down the process of reform. New administrations bring on their own team and the size of DOD means that it takes a relatively long time to get the team in place. Filling the top several spots may be done quickly, but much of the implementation depends on the second and third tier political appointments in the military departments and in DOD itself. A year after being inaugurated, the Obama Administration had only filled 315 of 526 government-wide positions that required Senate approval (60%); another 91 were still awaiting Senate confirmation, leaving 120 positions (23%) without someone identified to fill them (Anonymous, 2010). These are the people who would actually develop and implement reforms and it takes awhile for them to get onboard, learn what is needed, and pull together as a team. Then, hardly is a team in place before turnover begins to occur.

Top DOD political appointees have averaged only 1.7 years of on the job tenure and this hinders long term planning and follow through (Kutz, 2002a: 25). Some suggest that the lack of a champion for long-term reform has contributed to the failure of previous initiatives, and even when there is a champion, he must remain in place for years to make a difference. This thinking was the foundation for the GAO's recommendation for a Chief Management Officer with an extended term. The existing policy of having the Deputy Secretary serve as CMO does not provide the recommended continuity (Candreva and Brook, 2008).

It is ironic that a new appointee may not understand enough about DOD and its systems at the start to be able to move forward expeditiously with reform, and by the time he does his time left in that position is too short to do much good. Another irony is that the really good people who could make major changes are often the first to leave, moving up in the hierarchy to better positions, because they have demonstrated competence and reliability. This phenomenon not only affects political appointments, but also critical military officer assignments, where tours of duty are as short as two years and the up-or-out personnel system encourages "quick fixes" over persistent and systematic change. As Franklin Spinney, a tactical air analyst for DOD, put it, "We don't have the kind of corporate memory, so a lot of people come in and they go along with this stuff in the short term, they don't really get the big picture until they leave." (Spinney, 2002: 29)

While the executive level DOD civilians do lend a greater degree of continuity, leadership, and technical expertise, the turnover among the political and uniformed military classes have a direct affect on the attitude of the long-term civil service employees. Speaking to the causes of the lack of continuity in reform efforts, Stephen Friedman said: "If people there in the Defense Department believe this is a flavor of the month and that their bosses are going to be leaving in—whatever the actuarially measurable time is—a year and a half for senior people and then this will not be a continuing priority, you will not have the sustained effort." (Friedman, 2002: 11)

One of the reasons that leadership is so important is that a cultural resistance to change is evident in DOD where many feel that not passing an audit or not making progress to fix an ineffective financial system will result in few consequences for them. Thus, followers feel they do not necessarily have to go where leaders want to take them on matters of reform. Some say there is a culture throughout DOD that supports the belief that DOD can virtually ignore changes with which it disagrees. DOD Inspector General Robert Lieberman spoke of the pessimism regarding implementing a new financial architecture for the department, "The DOD might lack the discipline to stick to its blueprint. The DOD does not have a good track record for deploying large information systems that fully meet user expectations, conform to applicable standards, stay within budget estimate, and meet planned schedules." (Lieberman, 2002: 24). While DOD has documented and published a Business Enterprise Architecture in the ensuing years, it remains incomplete and there is scant evidence that it has actually driven significant improvement in business operations (GAO, 2009b).

The belief in the superiority of one's own system has contributed to the proliferation of stovepipe systems and service- and even command-unique feeder systems that have allowed the services and DOD agencies to develop redundant and conflicting solutions to business needs. Despite basic similarities in military personnel management, for example, DOD has been incapable of building a single integrated human resources management system for the uniformed force. The military departments fought against foregoing their individual systems in favor of a joint system. As a result, in 2010, DOD abandoned a billion dollar investment in the Defense Integrated Military Human Resources System (DIMHRS) because after a decade of development it had "little to show and limited prospects" (Peters, 2010: 1).

Senior defense management attention also tends to focus on programs that have direct military utility: tactical aircraft, shipbuilding, mine-resistant ambush-protected vehicles, command-and-control systems, and the like. Systems that merely support such functions—the contracting, human resources management, and accounting systems—are considered less prestigious and receive comparatively less management attention. In many cases, the requirements placed on business systems are designed to optimize a military goal such as support to wounded soldiers or reducing wait time for parts. Functionality that supports a business goal, such as reducing the cost of health care or more streamlined inventory management is compromised in the process.

System Complexity and Modernization Challenges

The GAO recently reported, "DOD is a massive and complex organization. In support of its military operations, the department performs an assortment of interrelated and interdependent business functions—using thousands of business systems—related to major business areas such as weapon systems management, supply chain management, procurement, health care management, and financial management. [. . .]the DOD systems environment that supports these business functions is overly complex; error-prone; and characterized by little standardization across the department, multiple systems performing the same tasks, the same data stored in multiple systems, and the need for data to be entered manually into multiple systems. Moreover, DOD recently reported that this systems environment is comprised of approximately 3,000 separate business systems" (GAO, 2008b).

Extant information systems not only were created by organizational stovepipes, they evolved over decades of organizational realignments, changing reform initiatives, and frequent leadership turnover. The Fiscal Year 2010 version of the Enterprise Transition Plan (BTA, 2009) includes a master list of over 2,500 business systems grouped in 28 categories. The Business Enterprise Architecture "defines the Department's business transformation priorities; the business capabilities required to support those priorities; and the combinations of enterprise systems and initiatives that enable those capabilities" (BTA, 2010). The BEA documents critical business processes and maps the relationships among these business systems. To illustrate the complexity of the DOD's systems, Figure 9.5—which the authors know is illegible due to the small font necessary to display it—shows the interfaces among the financial management systems in the department.

This level of complexity makes reform particularly problematic. Whereas a for-profit corporation may implement an enterprise-wide system to capture the bulk of the business functionality of disparate systems, among DOD's thousands of systems are *multiple* enterprise resource planning systems. DOD is not just an airline manufacturer, they are an operator. They not only operate tactical aircraft, they operate airlift and trainers, fixed and rotary wing. They also build and operate the airports. They maintain a huge fleet of ships and ground vehicles. The DOD runs several major universities, a health care system, tens of thousands of housing units, and more. The DOD is involved in hundreds of types of businesses.

Business Process Limitations

Large businesses like Wal-Mart and Sears have excellent inventory management practices that include standardization of data, little or no manual processing and systems that provide complete asset visibility. Wal-Mart requires all components and subsidiaries to operate within its framework and does not foster stovepipe system development. GAO found Wal-Mart and Sears had visibility over inventory at the corporate distribution center and retail store level. GAO noted that in contrast, DOD does not have visibility at the department, military service, or unit levels.

DOD business operations are huge: the Defense Logistics Agency manages about 5 million line items for 1600 weapon systems, integrates eight supply chains and manages 25 distribution depots with a capacity surpassed by only FedEx and UPS. DLA operates in 28 countries and process over 100,000 requisitions a day (DLA, 2010). By comparison, Walmart operates

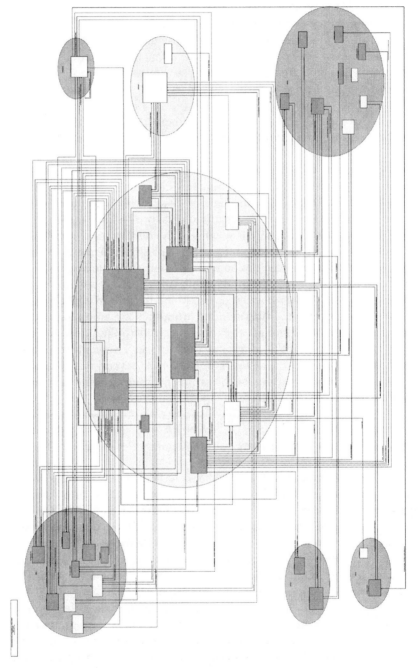

Figure 9.5 Financial Management System Interface Diagram. Source: (BTA, 2010).

8448 retail stores in 15 countries and processes 28 million sales per day. They have 40 distribution centers (Walmart, 2010). Wal-Mart requires all components and subsidiaries to operate within its framework and does not foster stovepipe system development. GAO found Wal-Mart and Sears had visibility over inventory at the corporate distribution center and retail store level. GAO noted that in contrast, DOD does not have visibility at the department, military service, or unit levels.

Integrated or interfaced systems and standardized data allowed both Sears and Wal-Mart to specifically identify inventory items. Wal-Mart headquarters staff was readily able to identify the number of 6.4-ounce tubes of brand name toothpaste that were available at a Fairfax, Virginia store. Other information was also available, such as daily sales volume (Kutz, 2002b:8). A similar system does not exist at DOD, yet it suffers persistent inventory management weaknesses. A recent GAO audit identified seven factors they grouped into three categories that reflect "a deficiency in DLA's current ability to determine the extent to which it is fulfilling DOD guidance directing the military components to size secondary item inventories to minimize DOD's investment while providing the inventory needed" (GAO, 2010: 21)

- processes for determining how many parts to buy
- initiatives that have not been fully implemented
- assessing and tracking the overall cost efficiency of its inventory management

Another way to interpret GAO's findings is that DOD is unable to achieve Walmart's level of efficiency of operations because DLA, unlike Walmart, must respond to the demands of a disparate set of customers. It does not have the free will to invest in new systems modernization and must compete with warfighting programs for funding on an annual basis. It does not have an incentive to improve the efficiency of operations, rather there are strong incentives to be inefficient if those inefficiencies result in gains in warfighting effectiveness.

Beyond inventory management at other business processes, contracting effectiveness has also plagued the department. Contracting management has been on the GAO's High Risk list since 1992 and the GAO cites five key weaknesses: sustained senior leadership, capable acquisition workforce, adequate pricing, appropriate contracting approaches and techniques, and sufficient contract surveillance (GAO, 2006). In fiscal year 2008, DOD obligated over $200 billion on contracts for services which was more than half of all the funds on contracts, including those that buy major end items like ships, tanks, aircraft and buildings. Service contracting is particularly vulnerable to waste due to the intangible nature of deliverables. Weaknesses in DOD contracting practices for services include:

- widespread use of "undefinitized" [not yet completed] contracts
  (those contracts authorizing contractors to work before the contract
  terms, including price, are final) for urgent work
- over $10 billion for professional services under time-and-materials
  contracts that have no incentive for contractor efficiency or basis for
  effective oversight
- incomplete records and files
- inadequate contract surveillance

While the number of contracting actions and the corresponding dollar
value doubled from 2001 to 2008, the number of personnel in the contracting field remained relatively constant. In a quest for efficiency (reducing
contracting overhead costs), the department increased risk for waste. There
is simply less oversight and fewer qualified managers overseeing a process
that is gaining in importance (GAO, 2009a). Contracting in the theater of
operations is even more error-prone as contracts are written hurriedly, paperwork is even less complete, yet critical operations cannot be successfully
accomplished without the contracted assistance. In an 18 month period
from the fall of 2006 to the spring of 2008, over 55,600 contracts obligating
over $30 billion were let in Iraq and Afghanistan. Less than 60% were com-
peted; few are audited. As of April 2008, there were nearly 200,000 contractors working on those projects (GAO, 2008a). That exceeds the number of
uniformed personnel. In Afghanistan in 2010, there are about 2 contractors
for every soldier in the theater of operations. Despite the critical nature of
contractor support, contingency contracting practices and oversight have
not been elevated to the level they receive sufficient attention in planning
and supporting military operations.

The turbulence of DOD contracts has caused inaccurate payments and
led to potentially thousands of man-hours of financial reconciliation to correct errors. DOD data for fiscal year 1999 showed that almost $1 of every $3
in contract payment transactions was for adjustments to previously recorded payments- $51 billion of adjustments out of $157 billion in transactions
(GAO, 2002:3). GAO found that DOD contracts containing multiple fund
citations and complex payment allocation terms were more likely to have
payment errors because of the amount of manually entered data and the
consequent opportunity for error in the manual processing.

Oversight and Incentives

Not only are DOD business processes complex, the structure of oversight and governance of business management reform in the DOD is also
complex. Previously, we noted that the DOD Deputy Chief Management

Officer publishes the Strategic Management Plan (SMP) that contains priorities, desired outcomes and goals to meet those outcomes, measures and specific initiatives. This document directs the department to align the Business Enterprise Architecture (BEA) to the strategy. It is important to note that the BEA is not just IT system interfaces and data structures, but it also defines core business processes. Should the BEA not be aligned, or if there are other business improvements that are desired (e.g., continuous process improvement efforts), then the desired changes are described in the Enterprise Transition Plan. The ETP is a roadmap of systems modernization and process improvement designed to implement the SMP. Earlier, we noted the existence of the Defense Business Systems Management Council; they are a governing body that approves major changes to defense business systems. Also, major system changes are also subject to the same defense acquisition regulations as other major procurement programs; and are subject to the constraints of the Clinger-Cohen Act of 1996 which regulates information technology (IT) investments for the federal government. All the while, Congress has its own ideas about what is good for defense. DOD has been criticized for the fluidity and complexity of its governance structures, re-

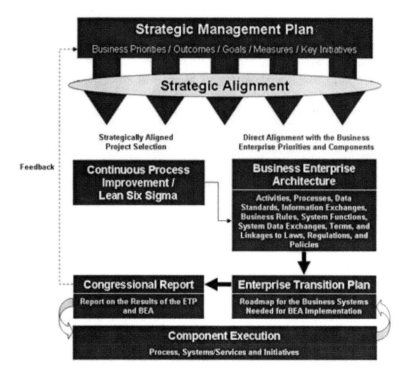

Figure 9.6 Governance of Business Management Reform. *Source*: BTA, 2009.

sulting in more planning than progress. Figure 9.6 shows the relationships among several of these oversight components.

Given the high turnover of key personnel, political and military, keeping these strategies, programs and plans aligned takes considerable effort. A change in emphasis between presidential administrations can be disruptive, such as when the preference for out-sourcing under the G.W. Bush administration was replaced with a preference for in-sourcing under the Obama administration. DOD is hampered in its reform efforts by rules and regulations that are time-consuming and restrictive, including cumbersome appropriation accounting requirements, detailed record keeping and reporting mandates, and obstacles to private sector partnering in areas that are inherently commercial. To enact significant reform, DOD must overcome budget language and strict procurement regulations, but it must do so with processes that still leave Congress with appropriate oversight controls and a procurement process that guarantees freedom from fraud and abuse and an efficient procurement cycle.

Traditionally, spending for information technology has been decentralized throughout the department and funds were provided to functional managers who would procure IT to better manage their functions. Enforcing IT architecture standards across the department was problematic as there was not good visibility into the IT spending. In the 1990s, IT procurement became a focus of acquisition management reform and CIOs were installed throughout the department to set and enforce standards. Congress even mandated a new budget exhibit that details spending on the procurement and operation of the department's IT. Interestingly, that budget exhibit is not found on the USD (Comptroller) website with the other budget exhibits, but is instead available through the CIO. Section VII of the IT budget outlines business modernization and contains about 1500 line items totaling nearly $6.5 billion. Of that, only 129 initiatives totaling $631 million are under the purview of DFAS or the BTA; about 10% of the IT spending for business modernization (Department of Defense, 2010). Given the decentralized nature of defense business systems and processes, oversight is particularly challenging.

The way in which oversight is conducted creates or modifies incentives on the actors in the system. Part of the reform challenge is related to the perversity of the incentive system in DOD. The stereotype of the civil servant is as someone who does just enough to get by, is hard do fire, and can indulge in mute insubordination if irritated. While overdrawn, the path to change leads through a better incentive system. DOD financial management employees must be motivated to pursue change. Comparing managers in the Defense Department with those in the private sector, Friedman stated:

If you looked at a manager's incentives, he hasn't gotten any material bonus for doing better work. It's hard to measure whether he's in fact, done better work. It's hard for him to discharge an employee that he considers to be incompetent, and then at the end when he looks at it, there aren't the incentives to really stick your neck out and do anything other than manage the budget. (Friedman, 2002: 11)

Some believe that just as rewards are hard to come by, so are penalties. They say that no penalty is imposed on DOD for failure to comply with modern financial management practices because the defense function is simply too important, and that imposing budgetary penalties on DOD would harm national security. For example, Rep. Dennis Kucinich has said:

I do not believe the Department of Defense will fix this broken, unsustainable system on its own. What motivation does it have? Despite its routinely dreadful performance, Congress almost never rejects a Pentagon request for more money. The time has come for Congress to treat the Department of Defense as the market treats any commercial enterprise. Just as investors withhold their supply of capital to a company that fails to meet its expectations, Congress must refuse to supply additional funds to the Pentagon until its books are in order. If Congress keeps appropriating more and more money, despite these horrendous practices, what's the incentive for the Pentagon to reform? (Kucinich, 2002a: 3).

Representative Shays added, "We need defense so we keep operating. But if we knew that we couldn't function unless we got our act together, I think it (reform) would happen more quickly." (Shays, 2002: 6)

Perverse incentives also operate to the disadvantage of government. DOD has to pay interest if it does not pay promptly, but a contractor who is over-paid faces no such penalty. In effect, DOD has made him an interest free loan. Comptroller General David Walker noted,

We have some perverse incentives. We have a situation where the government ends up having to pay penalties if it doesn't promptly pay. On the other hand, if the contractor has been overpaid, they do not have an affirmative responsibility to tell the government that they were overpaid. And, in addition to that, if they don't end up telling the government, and if they don't refund the money within a reasonable period of time, they're not charged any interest. And so, therefore, we have an [uneven] playing field... (Walker, 2001)

Walker noted that $900 million dollars was refunded to the government by contractors who were overpaid in the 2000. He said that this amount was lower than usual and that the situation was getting better.

There is little doubt that DOD financial managers believe that they are in a culture that demands they use up their money. Here this cultural imperative surfaces in the practice of using up unspent funds in closed accounts, legal or not. This was discussed at length in the chapter on budget execution. Another instance of perverse incentives involves procurement of weapons systems. Both within and without, DOD planners and contractors know that less expensive systems stand a better chance of selection than more expensive systems, all things roughly equal. This gives rise to a low ball gaming strategy. This has been referred to by DOD insiders as the "political engineering process." One part of it includes bidding the lowest price with the intent to make it back later in contract changes as requirements are added back. Another part of it involves spreading the contract around geographically to create a wide constituent base so that many members of Congress will have a vested interest in supporting the project. A third part involves using optimistic cost estimates in budgeting to garner support for the program during POM deliberations. Then once a program has been started, it attains a momentum that makes it hard to stop, even as the costs increase. Representative Kucinich observed, "The contention is that once the out years are reached and the true costs of production become evident, there is no longer the political will to cancel the program" (Kucinich, 2002a: 29). The Defense Acquisition Program Assessment called this combination of incentives, "the conspiracy of hope" (Kadish, Abbott, Cappuccio, Hawley, Kern, and Kozlowski, 2006).

CONCLUSIONS

The introduction to this chapter noted that budgeting is more important than financial management. Budgeting is prospective and confronts important questions about priorities and which programs will succeed or fail. Financial management is retrospective and provides an accounting for what previously transpired. Financial management is also integrative and sees that processes that support warfighting are aligned, effective, efficient and accounted for. For an organization focused on planning for the future and being ready for contingencies, budgeting is far more interesting and rewarding. But when it comes to giving an accounting, of demonstrating performance, and of ensuring "a well-oiled machine", one must also attend to financial management processes. This is particularly true when contractors outnumber soldiers on the battlefield and when more money is spent on service contracts for management support than on new ships, tanks and aircraft.

This chapter defined financial management as those functions and supporting processes that encompass the department's ability to give a public accounting for performance of programs and the associated budget authority entrusted to it, including the financial impacts of the department's activities. It outlined a stream of legislative requirements for financial management in the DOD and it is clear that the Congress has become frustrated with DOD progress and thus is more active in directing specific actions toward improvement. Whether the legislated actions, or DOD's own, are more effective is debatable.

This chapter also related obligational accounting to financial and managerial accounting and presented the financial statements for the DOD and federal government as a whole. The three bases of accounting should be considered complementary and not substitutes as each describes the activities of the department in different ways for different audiences and to support different types of decisions. It is important to understand the quality of those accounting practices and, as they improve, policy-makers should learn to use more sophisticated financial information. There are times when cost data or accrual data will result in better decisions than obligational data. Whether corporate-style practices, statements, and audit standards are appropriate for government entities is open to debate and opposing views were presented.

Finally, this chapter addressed the phenomenon of perpetual management reform in the department. There is a long history of successive attempts to resolve the persistent management problems that plague DOD. Progress has been very slow, sometimes spotty, but measurable. It took the U.S. only nine years from President Kennedy's speech in 1961 to send a man to the moon and return him safely to earth. Current estimates are that it will take three times as long for DOD to produce auditable financial statements. Accounting is not rocket science, but the requirements of the CFO Act have been more difficult to meet.

We would suggest that part or much of the inability or failure to fully implement the CFO Act this is attributable to four key factors: (a) flaws in the Act itself that need to be corrected before further implementation efforts are made (Anthony, 2002: 297–312), (b) the reports produced under CFO requirements relate in no way to the appropriation structure used for tax and spending decision making by Congress, the format of OMB's President's Budget, the financial structure of PPBES or any DOD financial system, (c) there is no evidence that the reports required by the CFO Act once completed and audited are used by anyone in or out of government for any decision making purpose–not OMB, not the US Treasury, not Congress–or that they serve any discernable purpose at all, i.e., nobody cares what they report so the incentive to implement the Act fully is highly diminished. Only GAO continues to think these reports are worthwhile, (d) the costs of implementation are very high, in the billions of dollars. The key question that results when these factors are considered is whether implementation of the CFO Act passes the benefit-cost test?

NOTES

1. There is an important exception to this general statement. In those instances where a corporation does work for the federal government under a cost reimbursement contract, the government will need to examine cost accounting information to assure only those costs that are allowable and allocable to the contract are invoiced and paid.
2. Those DOD entities that received favorable audit opinions are: US Army Corps of Engineers, Military Retirement Fund, Medicare-Eligible Retiree Health Care Fund, Defense Commissary Agency, Defense Finance and Accounting Service, Defense Contract Audit Agency, and the Office of the Inspector General.

CHAPTER 10

REFORM OF PPBES

Where to Next?

INTRODUCTION

Under former Defense Secretaries Rumsfeld and Gates a number of changes were been made and implemented to varying degrees with the intention of improving the manner in which the PPBES serves as a planning and budgeting decision system for DOD, in part to better integrate financial decisions with Presidential preferences and acquisition decision making. The linkage between PPBES and the acquisition system has been strengthened somewhat, although not enough, through program review by the JCS (J8) where all DOD acquisition programs now are assessed for jointness, capability and feasibility. Whether and if so what types of reform might be put in place by current Defense Secretary Leon Panetta is unknown.

THE FUTURE OF PPBES

With respect to budget formulation as opposed to execution we might wonder what would happen to DOD resource decision making if the POM were eliminated and replaced by a process of longer-term budgeting. In traditional budgeting, budget submitting offices (BSOs) have to answer several

Financing National Defense: Policy and Process, pages 367–372
Copyright © 2012 by Information Age Publishing

important questions as they ascertain what they need in the budget and as they justify their requests to funding sources. These questions include "what," "why," "when," "where," and "how." The answer to "how much" flows from the answers to the prior questions. All of these questions are important, but possibly the two most important questions in this set are the "what" and "why" questions. They set the stage for the fact-finding that causes answers to the how, where, and when questions to surface.

For example, if there is no need for a ship or a tank, then there is no need to define when you might need it, where you might need it, or how it might be configured or delivered. This interrogative pattern is the whole cloth upon which budget decisions are based. Much academic research has focused on the concept of incrementalism, i.e., that budgets change only by small amounts on the margin and not much as a percentage of the total from one year to the next. This is a tested analytic finding, but not one that is useful for the PPBES decision makers because they do not build budgets by focusing on percent of change. Rather, they first determine what it is they need (capability and requirement). They do this by analyzing the world around them and its impact on the organization and its systems. They then establish what is needed to improve or operate more efficiently or effectively than in the previous planning period or fiscal year. Finally, they evaluate in detail what this will cost and what can be executed in the annual budget.

With the implementation of the PPBS in 1964 under Robert McNamara, the defense budget system split the focus of these questions into three parts. The planning and programming functions (in which the SPG and POM are built) deal with the "what" and "why" questions, and to some extent "where" and "when." Most of what is left for the budget process is the task of answering the question, "how much this year?" Still, budget formulators do have to present their fully justified budget to reviewers in the DOD, the OMB, and Congress. This means that they have to convey the part of the POM that answers the "how" and "what" questions along with the request for "how much." To do this, budget offices have to put back together the pieces of the program that is built in different places for different purposes by different sponsors. Asking what the best profile for the ingredients for an aircraft carrier battle group over the next ten years (a planning and programming question) is different from asking how much is needed to operate the battle group for the next year. However, in PPBES, to decide "how much," the budgeters have to know what the total program will look like in practice.

As long as there is clear articulation and separation of these processes and one feeds carefully into the other, this system can work—as long as the POM feeds information into the budget process. For the most part, budget-ers may have been happy to have many of the big resource questions decided for them, leaving them to focus on pricing out next year's needs. For their part, programmers have developed rules that allowed them to develop a good POM for each cycle. Usually, this means everyone gets something, but nobody gets everything they want.

With the passage of time, dysfunctions appeared in this scenario. First, the military departments created POMs that were more conducive to their needs than to joint warfighting needs. The Goldwater-Nichols Act reforms (1986) were intended to rectify this situation. Then, with the drawdown after the fall of the Soviet Union, budget offices were placed in the awkward position of having to make decisions because the calendar said it was time to do so—even when the POM had not been completed—because those who built the POM could not decide which was the best way to downsize while maintaining the capacity to deter or fight future wars. Military department and DOD budget offices were, by and large, unhappy at having to make programmatic drawdown decisions under this circumstance. However, now in the past few years, the program decision-making process has not been completed in time to meet the needs of the budget.

Most recently, this is allegedly due to the combined program- and budget-review process under the PPBES. Also, various changes have been made to the processes of planning and programming for weapons acquisition, but none has been fully successful. Part of the problem is the overly complicated programming and budgeting process. Former Secretary of Defense Rumsfeld and others have characterized the PPBES process as too slow and too complicated. As part of his transformation effort, Rumsfeld and DOD staff changed PPBES so that the programming and budgeting analysis and decision phases could be roughly concurrent. The POM process begins first, but both the budget and the POM process are supposed to end at the same time. In effect, the failure of the programming system to reach decisions may be viewed as having broken the budget process.

In reality, the budget process can only reach the "how much" question by answering the "what" and "why" questions. If the answers to these questions all appear at the same time, or when they are not answered at all, then the budget process has to, in effect, duplicate what is supposed to be done in the POM process to produce a budget on time. Indeed, under the new PPBES process, some parts of the budget process have had to operate as if there was no POM process.

This leads to the question: is there a genuine need to prepare a POM, especially if budgeting were done on a longer-term basis of two to five years? Perhaps it would be useful to take the transformation PPBES reform one step further and discard the separate POM process by simply incorporating the POM questions and POM process outputs into the budget process? This is a more sizeable task than it appears due to the existence of a bureaucracy which produces the POM. Conversely, it is perhaps a less sizeable task than it seems because the military staff involved in the POM process have other career lines and can perform functions as war fighters, and/or players in the defense-acquisition process or the warfare-requirements-setting system. There would be some civilian positions, mainly those in the Pentagon, that would disappear in this new integrated POM/budget cycle—a cycle that could perhaps be called the planning, budgeting and execution system (PBES). However, replacement of the entire PPBE system with longer range budgeting is the option we prefer.

RELATED BUDGET REFORMS TO IMPROVE EFFICIENCY

While creating a two-phase planning and budgeting system would rationalize the operation of PBE within the DOD, a useful further step would be to create a longer-term appropriation period. DOD fiscal execution patterns are needlessly complicated by the rush to spend one-year appropriations before the close of the fiscal year. And the mixing of different appropriation periods for different appropriations needlessly complicates administration for those who execute budgets.

Most of the DOD budget functions on a multi-year pattern—longer for military construction and procurement of long-lived assets such as ships and aircraft, and shorter for personnel and supporting expenses (O&M). However, even if personnel is legally an annual appropriation, in reality the force size and composition is relatively fixed and will remain so until some external crisis event forces review and change. Personnel could as well be a two, three, of even five-year appropriation. We suggest that the DOD budget is, in effect, a multiple-year budget now. It would make sense to recognize it as such and to appropriate for multiple-year periods for all accounts, and to extend the obligation period for short-term accounts beyond one year at minimum.

A two-year appropriation (or obligation period) for personnel and O&M accounts would be a useful starting point for Congress, as we have noted. Critics of such an approach often point to Congress's need to exercise oversight through the budget. However, Congress can exercise whatever oversight it cares to in various ways, for example by focusing on execution review in off-budget years in a two-year cycle. A two-year budget also would reduce the opportunity for Congress and the President to insert what all recognize as "pork" into defense appropriations. The suggestions we make here would reduce opportunities for pork, but would also allow for meaningful oversight by Congress, and would reduce the size of the Pentagon bureaucracy while releasing additional military officers from administrative jobs for return to duty in their warfare specialties.

It must be noted that the task of defense resource planning and budgeting is part managerial and part political. Thus, from our perspective, no amount of budget process, PPBES or business process reform will reconcile the different value systems and funding priorities for national defense and security represented by opposing political parties, nor will it eliminate the budgetary influence of special-interest politics. Value conflict was evident in the early 1980s when public support, combined with strong Presidential will and successful budget strategy, produced unprecedented peacetime growth in the defense budget, in particular in the investment accounts. Constituent and special-interest pressures make it difficult for Congress and the DOD to realign the defense budget. While we applaud the spirit of many of the changes made in DOD during the period 2001–2005, reform of defense budgeting process does not mean that producing a budget for national defense politically will be much easier in the future than it has been in the past. Threat perception, capabilities assessment and politics drive the defense budget, not the budget process itself (McCaffery & Jones, 2004). Additionally, the size of the deficit and rate of increase in mandatory expenditures make top-line financial relief for the DOD unlikely.

We also may observe that a sequence of annual budget increases for national defense in the early and mid-2000s have not brought relief to many accounts within the DOD budget. At the same time, requirements of fighting the War on Terrorism and other military operations have intensified the use of DOD assets and the costs of military operations. For example, funding to support US operations in Libya in 2011 were drawn from the DOD base budget and not from supplemental appropriations or out of country operations (OCO) accounts. Because the need for major asset renewal has been postponed for too long, new appropriations have gone and will go in the future largely to pay for new weapons system acquisition, and for war fighting against terrorism. What this means is that accounts such as those for Operations and Maintenance for all branches of the armed services will continue to be under pressure and budget instability; restraint will remain a way of life for DOD for the foreseeable future. This places a heavy burden on DOD leadership, budget comptrollers and analysts and resource-process participants to achieve balance in all phases of defense budgeting and resource management.

Ending what we know as programming and the POM would be a major change to PPBES. In our view, programming is only effective, if at all, at the end-game anyway but preparing and processing the POM wastes huge amounts of valuable DOD staff time and energy that can be put to better use. Also, ideally, the period for obligation of *all* accounts in the new DOD budget process would permit obligation over a period of two or three years for all accounts—including fast spend accounts including O&M, MIL-PERS, etc. The reason for multiple-year obligation for all accounts is to enable more effective budget execution and end the highly wasteful and inefficient end-of-year "spend it or lose it" incentive syndrome. This change would, of course, require the approval of Congress. However, the DOD could implement long-range budgeting (including capital budgeting) as a part of the overall reform—while Congress continues to operate on the annual budget cycle it prefers (for a number of reasons related to serving constituent and member interests). No change in the federal budget process can be made unless it permits Congress to continue to do its business according to the incentives faced by members. To think otherwise is naive. Still, as noted above, the only part of the reform advocated here that would require explicit congressional action is lengthening the obligation period for all accounts to two or three years, as has been done internationally, in the UK and other countries, for example. And, in fact, before the elimination of what was termed the "M" account in the early 1990s due to illegal use of this account by the Air Force in financing the B-1 bomber and other programs. DOD had substantially greater flexibility in managing money for which the obligation period had expired. Under the M account process expired funding was allowed to be retained and reallocated by DOD for a period of three years.

The change to extend the obligation period for one-year appropriations to two years would require Congress only to modify certain provisions of appropriation law. Otherwise, the DOD could implement a long-range accrual based budgeting system on its own, subject to gaining approval of and support for it from Congress—but this would not require change in law. In essence, it is incumbent on the DOD to persuade Congress to support such change, and this will only occur if the DOD is able to show members how they, the DOD and the American taxpayer will be better off as a result of this reform.

CHAPTER 11

RESTRUCTURING, REDUCING BUDGETS AND DEBT WHILE MANAGING FISCAL STRESS

INTRODUCTION

After approval of a national debt limit increase negotiated between the two houses of Congress and President Obama in early August 2011 and subsequent delayed passage of the defense department FY12 appropriation it became very clear that future budgets for defense would be reduced; what was unclear was by how much and when? The agreement struck by Congress at this time included creation of a select "super" committee to recommend budget and financing changes to reduce the size of annual government budget deficits until they were eliminated so the nation had a balanced budget, and reduction in the amount of the total national debt. If Congress and the President could not agree and approve these changes an automatic sequestration (budget cut across the board in discretionary accounts would take effect in January, 2013. These events also demonstrated the failure of Congress to respond to fiscal stress in a timely and expeditious manner through the normal budget process. This reduced public confidence in the ability of the US federal government to manage its finances responsibly, which had an immediate effect on the US economy as reflected in stock prices and general public frustration at a time when citizen criti-

Financing National Defense: Policy and Process, pages 373–395
Copyright © 2012 by Information Age Publishing
All rights of reproduction in any form reserved.

cism of government already was rampant due to the impact of a protracted economic recession.

This chapter is intended to provide information that elected and appointed Defense Department officials, defense comptrollers and budget office staff, public managers in general, and others need to understand and to better assess alternative methods for improving the management of economic and fiscal stress through restructuring and budget reduction. Analysis of restructuring may be segmented into a variety of sub-topic areas including the causes of economic and fiscal stress, methods for improved management of restructuring, issues and dilemmas faced by public officials and managers attempting to manage restructuring, The range of non-mutually exclusive approaches to management of restructuring includes (a) doing nothing, likely to work only for a short time if economic and fiscal stress persists, (b) increasing revenues through a variety of means including borrowing while reducing expenditures, (c) increasing employee and organizational productivity and employing a set of more innovative responses that are productivity-related. This chapter addresses all of the issue areas indicated above in the context of the ongoing US and global condition of economic stress. In terms of the immediacy of the topics addressed in this chapter, at the time of this writing the problem of reducing federal government debt to preserve the fiscal solvency and credit rating of the US, a fiscal crisis in Greece and several other financially weakened European nations, and the continued weak recession recovery condition of the US economy were stressful issues faced by elected officials and public managers. The motivation to address fiscal stress seriously in the US and the European Union was driven by the potential impact of government bankruptcy.

Dialogue on how to manage fiscal stress, defined here as the condition when a nation or other entity can still get loans to support its debt and operations, and crisis when either loans are not available or are needed to prevent absolute bankruptcy and cessation of government operations, should be informed by an understanding of the various approaches to coping with long-term revenue shortfalls leading to sizable long-term debt. Further, it is essential to understand the complex relationships between government spending and the direction of overall fiscal policy in combination with monetary and other policies and the productivity of the public and private sectors that is needed to drive the economy towards recovery.

RESTRUCTURING GOVERNMENT TO REDUCE GOVERNMENT BUDGETS AND DEBT: CONTEXTUAL FACTORS

Why should the Department of Defense and governments in general restructure financially and reduce budgets as a means of coping with fiscal and financial stress? The most evident answer is the need to cope with the serious consequences of a US and worldwide recession that began with the collapse of the housing market in the US in August, 2008 and continued through 2012. The basic problem facing the US and many governments in most regions of the world and at all levels is that expenditures needed to sustain existing programs exceed current revenues, including revenues generated through borrowing. The problem is so significant that once relatively wealthy or at least stable governments now carry large and increasing debt loads. This includes the US federal government, the State of California and many US state and local governments, Greece, Ireland, Portugal, Italy, Spain and other nations in Europe, the worst Iceland and Romania that actually went bankrupt. The problem in the Euro zone has required financial restructuring as part of bail out plans negotiated and required by the International Monetary Fund (IMF), ECB and perhaps workout European Union intervention.

In the US and elsewhere from a longer range perspective, part of the answer to the question why restructuring and cutting costs lies in the relative performance and cost of the services sector of the economy. In the US the services sector now constitutes more than 75% of gross domestic product (GDP) and the trend projections are that the relative size of the service sector will continue to grow. This means that the United States must increase service productivity to grow wealthier. However, in recent decades productivity increases have been low and slow. Government productivity poses an especially severe problem. The roughly 20% of the American workforce employed by government generates less than 15% of GDP. Value-added per worker is only 5% lower in private services than in manufacturing, but government productivity lags manufacturing productivity by a third. In contrast, in 1958 the value-added per government worker was 40% higher than in the goods sector.

A second part of the answer goes to the organization of service delivery. Arguably, bureaucracies are relatively effective for mass service production and delivery. Bureaucratic organizational arrangements have successfully provided security, jobs and economic stability, ensured fairness and equity, and delivered the "one size fits all" services needed during the era of government infrastructure development that lasted from the turn of the last century to the 1960s (Osborne and Gaebler, 1992: 14) and even to the present, especially in developing nations. During crises such as the Great Depression and two world wars, bureaucratically organized government has helped solve many problems. In the meantime, however, the hierarchical, centralized, rule-driven organizational arrangements invented during the industrial era seemed to some observers to have become increasingly anachronistic as a result of failure to deal with unpredictable challenges (Jones and Thompson, 2007).

Global economic competition, international capital mobility, and breath-taking improvements in telecommunications and information storage, processing, and retrieval have produced a knowledge-based economy—one where workers demand autonomy and citizens/customers demand high quality products, superior service, and extensive choice. Hieratically organized bureaucratic organizations in many cases cannot meet these demands effectively, e.g., US federal government response in hurricane relief for New Orleans in 2005–2006. Meeting these demands requires flexible, adaptable, innovative, and citizen/customer-focused organizations or networks of organizations that offer an array of high quality services tailored to individual wants and needs, e.g., as demonstrated in the extreme in responses to the earthquakes in Haiti in 2010 and Chile in 2011.

Absence of Experience with Restructuring

A basic problem in meeting such challenges with appropriate responses is that few public officials or managers have had much experience restructuring. Consequently, better methods for management and management reform are needed. Additionally, better evaluative frameworks against which organizational reduction options and actions may be assessed also are required. Even after economic recovery has occurred it is necessary to deal with its lingering effects. In the US economic growth was 2.2% in 2010 but dropped in 2011 while the level of unemployment was not expected to fall below 9% in the near future. In Europe, the Organization for Economic Cooperation and Development (OECD) advised that while the worst of the extremes of economic crisis probably had passed by 2011, while some real problems remain in some nations as noted previously, the focus must remain on structural reform of economies in the region and reducing unemployment (OECD, 2010). To quote one OECD report:

> OECD countries have taken a wide range of measures in response to the crisis, notably in the areas of infrastructure investment, taxes, the labour market, regulatory reforms and trade policy ... OECD countries have so far avoided major mistakes—in particular concerning trade and labour market policies—but some risks remain. The crisis has in general reinforced the need for structural reforms. These reforms could help to speed up the ongoing recovery, strengthen public finances while protecting long-term growth and, in some cases, contribute to the resolution of global current account imbalances (OECD, 2010).

In the US the need for restructuring was underscored by the public dialogue that accompanied the debt ceiling debate in July 2011 when it became evident that government debt could not be reduced merely by cutting the discretionary part of the federal budget, including defense, without doing something to reduce the long-term costs of funding federal health care and Social Security entitlement programs.

PATTERNS OF RESTRUCTURING GOVERNMENT AND REDUCING BUDGETS AND DEBT

This chapter does not explore all approaches to restructuring governments. However, we know from past experience that managing public sector fiscal stress and crises requires more than cutting government spending. No nation can resolve fiscal stress or crisis solely through budget reduction. It is pure folly to think this is the case. History shows that workouts have to involve loan assistance, debt restructuring, and sometimes forgiveness or reduction, (i.e., defeasance as it is referred to in the private sector) along with means to increase revenue, increase government productivity and of the greatest importance stimulation of the economy so as over time to increase revenue flows to government. For example, given the way the both the US and the United Kingdom attempted to solve their debt and fiscal stress condition through drastic budget cuts it is no wonder that recent forecasts for GDP in 2011 were reduced in August 2011 in the US and were estimated to be only slightly positive or possibly negative in the UK. Quick and deep government budget cuts have a direct negative impact on economic growth that tend to worsen rather than help resolve longer-term fiscal problems that are to some extent structural in nature in the economy.

Fiscal stress management approaches may be divided into non-mutually exclusive categories according to variances in the measures and the timing of their application in attempt to restore fiscal and economic stability. Let us explore these approaches to coping with fiscal stress more closely while placing them into perspective relative to the other measures governments use to stabilize their financial condition.

1. Budget reductions typically appear in three non-mutually exclusive and often overlapping forms, (a) politically acceptable across the board reductions (e.g., 10%) that punish effective agencies the same as ineffective ones, (b) specific program reductions where some programs are cut by an additional percentage based on evidence that they are not efficient or their services are not needed as much as more important programs, (c) program termination and merger where some programs are terminated completely while others are merged functionally into other continuing programs (Jones, 2010). Some efforts to raise general and specific taxes may be attempted accompanying this approach but they are almost always politically unpopular.

Budget reductions in government operations accounts such as the cuts made in by Congress and the President in negotiating federal spending plans and priorities for Fiscal Years 2011, 2012 and beyond. The problem with the approaches used in these instances is that budget cuts have tended not address areas of high rates of growth in spending, and the motives for reduction typically have been political in nature, i.e., posturing to persuade potential voters in the next presidential and congressional elections to elect more members of one party or the other. Significant and often inaccurately reported media attention to such reductions also has a real impact on citizens who are presented an inaccurate and misguided perception that something meaningful has been done to help resolve fiscal stress conditions. The critical question with these types of budget cuts is which programs and beneficiaries are hit hardest and what is the consequent effect on their welfare and the condition of the economy? Predictable spillover consequences also include shifts in the preferences of many private investors to perceived safe havens such as gold, silver and precious metals, contributing to drops in currency values and reduced consumer confidence.

2. Budget reductions with initial cuts in operational accounts accompanied by efforts at spending control and in some cases revenue increases in government trust fund accounts of the type that provide long-term income, health and other types of social security, particularly to older and low income portions of the population. In the US federal government such measures eventually will lead to restructuring of Social Security through increasing the retirement age, limiting the annual inflation indexed cost of living increases paid to program recipients (something that has already been done for several years), and increasing social security taxes paid by both employees and employers. Similar measures would have to be made to medical care trust fund programs, e.g., Medicare and Medicaid programs, to ration benefits in some way and to increase user fees shifting more costs to consumers and away from government. These programs are far more resistant politically to rationing and cost shifting due to the size of the populations affected and the specific impact on the poor. In the US and in most developed nations much more is spent on these entitlement programs that in support of defense and social service government operations, e.g., roughly 70% of annual spending in the US is on these programs where around 30% is spent from accounts that pay for daily government operations, including national defense.

3. Debt bailouts by the US Congress and President of the type made in 2008 and 2009 under the Troubled Assets Relief Program (TARP) in 2008 and the Economic Recovery Act of 2009, when government invested directly in numerous banks and in some firms to protect jobs (the US auto industry) and made loans to other entities deemed "too large to fail." Outside the US such bailouts were orchestrated by international organizations that provided

relatively low interest loans over long terms and, in some cases, direct infusion of capital into the receiving nations and their banks, e.g., loans to Ireland and other financially troubled European nations. In the case of Greece lending from the European Union (EU) and the International Monetary Fund (IMF) were predicated on the receiving nation's willingness to engage in longer term budget reductions and measures to increase tax revenues. While some debt restructuring was part of these types of relief packages this approach was used less extensively than under the next option.

4. Debt workouts orchestrated by Congress and the President in the US, and by international organizations (e.g., the EU, IMF and others) that provided relatively low interest loans over long terms and also required and enabled significant debt restructuring, spreading and extending pay back periods and incorporating some form of defeasance, i.e., writing off some debt (forgiveness) to domestic and international lending institutions. The so called Brady Plan orchestrated by the US in the period 1999–2001 in cooperation with international lenders that pulled Argentina out of bankruptcy involved forgiveness and collateralizing (backing up) some of the long term debt using US Treasury bonds. This approach also assumes government budget reductions and some schedule for future tax increases when the economies of debtor nations begin to recover. In the case of Argentina these policies succeeded; the Argentine recovery continued until the global fiscal collapse in late 2008. This option reveals the advantage of incorporating some degree of debt forgiveness into restructuring programs where this is an option. This model could have been employed to a greater extent by US banks and other lending institutions to resolve the long lingering problems of excess debt and inventory in the housing industry. However, banks in general did not perceive this option to be to their advantage.

Although monetary policy is not the focus of this chapter, some attention to this area of policy warrants attention in that the history of fiscal stress should not be ignored. When the Japanese housing bubble burst in the 1980s, Japan's monetary policy decision makers held interest rates relatively high for a long period. The result was a decade of virtually zero GDP growth for the nation. Efforts by the US Federal Reserve Bank (FRB or "The Fed") to reassure markets that a low interest policy would be extended for the near future, with a constant vigil for signs of inflation, spurred the US stock market and international stock prices to rise in 2010 and through July 2011, and sent signals to industry that investment required to contribute to sustained economic recovery would be supported. The FRB also successfully stimulated economic recovery in the US in 2010 and until early August 2011 by buying US Treasury debt on two occasions (these actions were termed QE1 and QE2; QE stands for Quantitative Easing).

Unfortunately, this confidence was seriously weakened in early August by the failure of the US House of Representatives to agree with the Senate and the President to raise the US debt limit until the nation was pushed to the edge of bankruptcy. Although a temporary agreement on raising the debt limit finally was approved by Congress, this "debt hostage taking" behavior by House Republican conservatives caused the stock market to move sharply downward. In addition, and it resulted in one credit rating agency (Standard & Poor's) to cut the US federal government credit rating from AAA to AA- in early August 2011. Given that the debt limit could have been raised easily as early as March 2011 as requested by President Obama, the Secretary of the Treasury, the Chairman of the FRB, most House and Senate Democrats and many others without causing any financial crisis, this intentional and completely unnecessary brinksmanship by House Republicans pushed the US economy back into moderate recession. Investor and public confidence that Congress could manage its' fiscal affairs responsibly was sacrificed by the Republican party in the House of Representatives whose objectives were motivated by their partisan zeal to regain control of the Senate and the Presidency in the elections of 2012. This behavior showed the willingness of House Republicans to sacrifice the financial welfare and stability of US citizens and the US economy to meet their own narrow political agenda.

In contrast, the European Central Bank (ECB) raised interest rates in 2011, in part due to fear of rising inflation, and by October 2011 had not completed the process required to approve and implement an agreement similar to that applied by the FRB in the US to buy substantial amounts of debt from economically failing European Union nations (e.g., Greece) to stabilize the Euro and European debt and equity markets. While such caution is understandable and predictable given that the EU does not have a central bank with sole debt management and other financial powers equivalent to the FRB in the US, EU economic recovery continued to suffer as a consequence. Continuing uncertainty about whether the EU could manage its fiscal affairs responsibly also destabilized the US and world economy. In addition, it is notable that during the period 2009 to 2012 the US Treasury and Federal Reserve Bank took only moderate steps to support the falling US dollar. One consequence of this policy made it easier to sell cheaper US products and services abroad. At the same time, the high valued of the Euro and UK Pound made it more difficult to sell products abroad from these markets.

RESTRUCTURING IMPLEMENTATION: DEFINING MISSIONS, CORE RESPONSIBILITIES, METHODS AND CONSEQUENCES

A guiding principle in the beginning of restructuring is that each component of the organization should be evaluated in terms of its contribution of value to the delivery of services that meet citizen demand. According to the principles of mission or responsibility accounting, budgeting, and management control, all components of the organization may be classified according to their primary mission or responsibility, for example, mission centers, revenue centers, expense or cost centers, internal service centers, investment centers (Anthony and Young, 1988). This framework can be useful in designing restructuring initiatives at the organizational level.

Restructuring may result in significant "delayering" or flattening of organizational structure, considerable delegation of authority, responsibility and decision making on day to day operations to levels of the organization closer to its constituency. And restructuring in fact usually means shedding jobs. Indeed, large-scale productivity improvement in government (and the private sector) may depend to a great extent upon elimination of many, perhaps even thousands of jobs. However, accompanying such reductions we should expect that restructuring also to include increased employee education and training, especially in the use of new technologies and work processes.

How has the public sector tended to approach restructuring in the past? A reasonably similar pattern or sequence of events appears to characterize the restructuring initiatives of many governments. The first response to restructuring typically involves denial. This is usually followed by short-term measures to reduce spending, accompanied by efforts to assign blame. Organizations at this point must chose between reducing services and cutting positions. Many public officials balk at this choice, arguing that budget cuts should be made gradually, that organizations are better off relying on employee attrition, withdrawal of vacant positions, cuts in support budgets, and even deferral of maintenance than on across-the-board personnel reductions or cuts targeted at specific programs or services. When push comes to shove, they cut "soft" services first. Then they cut things that are relatively invisible to the public, for example, support operations and maintenance such as O&M and in some cases acquisition funding in the Department of Defense, employee training, and capital asset replacement. These cuts eat up an organization's accumulated capital and often demoralize its managers and employees who understand that such reductions often result in job losses, and that ignoring maintenance and investment will lead to higher costs later. Another factor that budget cutters often fail to realize up front is that massive layoffs are costly. When severance pay, loss of morale and valuable skills, not to mention the dislocation experienced by the employees who lose their jobs, are considered, staff reductions through attrition may actually be more cost-effective than termination—especially poorly targeted terminations.

The next phase of restructuring usually involves deeper, across-the-board budget cuts, often accompanied by hiring and salary freezes, increased use of part-time and nonpermanent employees, and other initiatives to reduce total salary costs. However, the across-the-board strategy tends to weaken organizations throughout. It is especially damaging in that high-demand programs and high quality personnel are cut the same amount as programs with lower demand and quality (Jones and McCaffery, 1989). Unfortunately, elected officials and many public managers and employees appear to initially prefer the across-the-board and attrition approach, with nonessential services cut first and the last-hired employees the first to lose their jobs.

Application of length of service rather than merit criteria often eliminates less experienced and younger staff. Unfortunately, these employees may be more adaptable to change than those with longer records of service. Employees cut in this phase in some cases may include a higher proportion of new entrant women and minorities than is represented in the total organizational workforce. There also may be an accompanying loss of highly skilled employees who find better employment opportunities elsewhere in a less stressful working environment.

In the military employee reductions often are done by rank or by type within operational communities. For example, at certain times during the past five years or so the Navy has at different times reduced through forced retirement the number of flag officers, and officers at the 06, 05 and 04 levels. Also, where specific warfare platforms have been phased out the corresponding operator communities have been reduced, e.g., in aviation at the end of A6 and F-14 Tomcat use life. It is a challenge to the Navy and Marine Corps to manage the transitions of operators, maintenance and repair and other employees when one platform is stood down while standing up its replacement, e.g., transitioning from the F/A-18 Legacy and Super Hornets to the Joint Strike Fighter (F-35C) or the P-3 to the P-8 aircraft. More than once Navy operator communities have been cut too deeply during such transitions resulting in insufficient numbers of operators for newer platforms and subsequent efforts to change retirement schedules and attempt to persuade operators and maintenance personnel to remain on active duty. This push and jerk approach to employee reduction has some very negative consequences in terms of military personnel morale and stress, it confuses career planning for the employees affected and causes continuity of operations problems for the organization.

How Should Restructuring and Budget Reduction Be Better Managed?

A number of variables in addition to those mentioned at the beginning of this chapter must be addressed in restructuring management. Once the organization has reviewed and assessed its mission, determined its core competence areas, performed value chain analyses, decided on what work processes should be retained in-house, contracted out or eliminated, then it must address a number of issues related to managing reduction. Typically, the first issue is how to deal with personnel reduction. As noted, personnel may be reduced by attrition or by layoff or termination. It is essential for the organization to develop a deliberate strategy for personnel reduction and then to stick to it. Governments often attempt to reduce through attrition, a slow yet effective method given that the cost of this approach is more affordable. Attrition takes longer and is costly in terms of skills and competencies lost while termination may appear to be more cost-effective in the long-term but costs more up-front because of the necessity for paying employees for accrued benefits, and in some instances bonuses. Legal constraints, union contracts and other strictures make termination more difficult in the public versus the private sector. Under many if not most circumstances seniority rules over performance, which is not always the best approach to reduction for employees or the organization. Using seniority alone can result in loss of more productive employees at the cost of retaining those who are less productive.

Numerous political and other constraints may force governments toward the attrition approach. However, it should be kept in mind that termination is used more often in the private sector because of its apparent long-term cost advantage and for other reasons, including sustaining employee morale for those not terminated and the advantages gained from shifting the organization rapidly toward the achievement of new goals and market opportunities. With respect to strategy, use of both attrition and termination may be the best course of action depending on the circumstance and degree of political and market pressure to reduce costs quickly.

To manage employee reduction, the criteria for cutting jobs and rules on how cuts will be made must be developed and communicated clearly to managers and employees. Some governments have developed procedures to mix length of service (seniority) and merit criteria in defining layoff or employment termination plans. Employee performance evaluation systems may be designed so that employees generate service credits for high performance ratings. These credits are then added to other credits earned through length of service and total service credits are then used to define employee layoff order and rights to move into the positions (bumping) held by less senior employees in the same or similar job classification elsewhere in the organization.

The service credit system also may be used to set priorities for employee reassignment to new positions within the organization, either at the point where personnel cuts are made or after a period of layoff. While reduction-in-force rules often include the provision of replacement or bumping rights across organizational units, unlimited bumping is stressful, disruptive and may cause serious losses in employee morale and productivity. A "single bump within class" system that restricts movement rights to a single choice and compares the qualifications of those seeking to replace other employees with requirements for the contested positions appears to be a preferable option. Union contracts may constrain the close application of qualification and requirement definitions, as may civil service rules.

One of the most important dimensions of personnel management under restructuring is the extent to which the organization invests in education, training and placement of employees whose jobs have been cut. Education and retraining may be necessary to enable reassignment of employees to new positions within the organization. Investment in education, training and placement services, and job search assistance is costly but desirable in most instances. Responsible management of job loss can build rather than reduce morale, and may help sustain or even promote the productivity of employees whose jobs are not cut, particularly for those continue to face the threat of elimination. It should be reiterated that strategic planning and the establishment of program, personnel and budgetary priorities should guide restructuring. Maintenance of service quality and the retention of valuable employees must continue to be of paramount importance to the restructuring organization.

Under conditions of program reduction and termination, enforcing priorities requires strong political support, effective strategic planning, a sophisticated information base for decision making, and considerable attention to negotiation with employees, citizens receiving services and stakeholders. All of this can be assisted through definition of critical mass program and core resource operation levels, i.e., resource levels below which programs cannot operate and still achieve their mission and objectives satisfactorily.

Assessing Value Added and the Need for Participation in Restructuring

Assessment of value added to the services delivered to citizens and employee and stakeholder participants is critical to effective restructuring. The capable manager will recognize that cuts ought not to be based on organizational prestige, program longevity and employee seniority, budget size, or other convenient but non-value added criteria. Regrettably, the program information needed to do otherwise is often lacking. At this point, the implementation of a strategic planning process that generates accurate and reliable information about market and citizen demand patterns and shifts in demand and to enable comparisons between programs, service production costs and productivity is critical. The planning process and strategic plans also ought to fit with longer-term financial, debt management, and capital planning. In attempting to plan and execute restructuring effectively, managers are likely to be frustrated as they recognize the extent to which they have under invested in or simply squandered valuable accounting, planning, program and policy evaluation, information technology, and other analytical resources in the past.

The issue of participation in restructuring decision making is sensitive and may be dominated by management-labor contracts to a considerable extent. Arguments for broader participation of employees are often made on the grounds of fairness, contribution to employee morale, and adherence to democratic management values. A much stronger argument for participation is that employees and program constituents have information that needs to be assessed by program managers in deciding whether and how to restructure. Many of the best suggestions on how to save money and increase efficiency are likely to come from program managers, their staffs, and citizen consumers if they are asked.

Centralization of planning and the reassessment of priorities are typical in public sector restructuring. Prolonged dependence on one or a few individuals to make restructuring decisions can result in reintroduction of many of the bureaucratic weaknesses that contributed to the need to restructure in the first place. A "mandarin" system of management by personal influence is inimical to effective restructuring because of its effect on workplace moral and on the openness of the organization to restructuring, reengineering, reinvention, redesign, and rethinking. However, the degree of centralization of authority in restructuring may be less important than other variables in explaining successful restructuring.

Smoothing the impact of cuts, continuity of leadership, the extent to which restructuring is politicized, the ability to define and communicate organizational mission and goals, the extent to which service priorities are established and budgeted, form of government, and degree of cooperation between and within executive and legislative branches all appear to be more important variables than centralization *per se.*

The dilemma of centralization of decision authority versus broader participation in restructuring comes down to a fundamental trade-off—either centralize and limit representation for purposes of decision and execution efficiency, or allow decision participation to be more open and, consequently, open to greater fragmentation and delay of the type apparent in the US when Congress debates and votes on budgets and related fiscal measures. Open access and participation in restructuring more fully utilizes the knowledge extant in the minds of employees, those served by the organization and stakeholders. However, broad participation often limits the ability of governments to establish new priorities quickly and to target cuts. Broad participation may make significant restructuring impossible as all parties have the opportunity to articulate reasons why the organization should remain as it is rather than adapting to new market, social, and political conditions. Either way, something of value is sacrificed, which reinforces the view that the best approach may be a combination of the two wherein politicians and managers cooperate but employ a procedural mechanism to limit choice and constrain the time in which choices can be made and appealed. For example, in the closure of military facilities in the United States, the Congress and executive branches of government used a self-constraining procedural mechanism (Base Realignment and Closure or BRAC) to close hundreds of military bases that had been proposed for termination, in some cases for decades, but were not closed due to political opposition motivated by the need to preserve employment in the areas where targeted bases were located. This and other examples indicate that under "perfect storm" conditions elected officials can act to discipline themselves to do things they have not been willing to undertake when fiscal crisis is not present (Thompson & Jones, 1994: 210–215).

Restructuring Under Serious Fiscal Stress Conditions

Where governments face serious fiscal stress that if not managed appropriately will turn into crisis as has been the case for many governments since late 2008, prolonged and acute mismatches between jurisdictional means and policy commitments may severely inhibit the ability to continue the status quo approach to budgetary and debt management. Budget deficits accrue and interest payments increase as a percentage of total revenues and spending and Gross Domestic Product (GDP). Credit ratings may be threatened, and the ability of governments to finance operational and capital budgets and entitlement programs and to borrow to meet short-term cash shortages may be seriously impaired. In many instances, the need for restructuring may not be recognized until the government as a whole, or specific public agency (e.g., the Federal Aviation Administration in 2011), faces a financial crisis and cannot continue to obtain funding either through the political/budget process or from borrowing in the bond market, then the need for restructuring becomes readily apparent. Where governments, including municipal governments, rely in part on deficit financing and borrowing, the ability to get credit from external lenders may become impaired or lost. At minimum the cost of borrowing typically increases so as to push a higher fiscal burden of debt repayment responsibility onto future taxpayers.

When the confidence in elected officials to resolve fiscal stress through negotiations and the regular budget process is lost and willingness to fund programs including defense is jeopardized, and where loss of revenues and credit-worthiness has occurred, longer-term financial planning and action is needed to evaluate the effects of program and service demand shifts and, from a financial perspective, to improve cash flow, cash management, and long-term spending and investment practices. Under conditions of fiscal crisis, special attention must be given to insure long-term entitlement and pension fund solvency and affordability, to limit debt loads to fit revenues derived from the tax base and tax rates, to meet debt service requirements, to assess property and equipment leasing or liquidation options, to develop more accurate capital asset depreciation and replacement costs and schedules, to improve inventory management and, where possible, to establish fund reserves to support entitlement programs and even the operating budget in the event of future revenue short-falls.

Long-range program and financial planning may require the participation of a wide range of public officials and others including for the US federal government the Federal Reserve Bank and its Chairman, private sector bankers, bond market advisors and other fiscal policy experts to define and assess alternatives and requirements and to rebuild investor confidence in the accountability and credit-worthiness of the government. Governments often discover, as have many private organizations, that long-term productivity improvement requires risk capital for investment in new equipment, employee training, performance of additional program analysis and improved projections of future service demand. In addition, governments have to determine how to convince the public that tax increases and increased fees for health care and other services are needed. Because restructuring requires analysis of service value, additional costs for accounting system modification, data collection, analysis are inevitable.

In too many cases governments are penny wise and dollar foolish in attempting to manage fiscal stress—they choose to reduce costs by cutting deeply across-the-board without regard to the impact on their capacity to deliver quality services in the future, e.g., in national defense. This approach is not a model for successful restructuring despite the obvious advantages of expedience and ease of compromise it gives to elected and appointed officials. In addition, this approach is possible only where governments such as the US practice cash rather than accrual accounting, which means that budget cutters can convey a false impression of fiscal health by playing games (often in violation of the law) with the timing of recording income and expenditures, e.g., scorekeeping of the relationships between appropriations, obligations and outlays by congressional appropriation committees. One of the best case examples is New York City in the fiscal crisis of the mid-1970s. Fiscal smoke and mirrors gimmicks sometimes will suffice for a period of time to persuade the news media and the public that a crisis has been averted, apparently without serious long-term loss, when in fact it has only been temporarily delayed. Unfortunately, the costs of mismanaging the financial component of restructuring are high and are borne for a long time, for example, in the Orange County, California bankruptcy in the mid-2000s and more recently to varying extents in a wide range of other venues including the State of California and other US state and local governments, Iceland, Ireland, Latvia and the Baltic nations, Greece, Spain, Italy, Romania, Ukraine and other Central and Eastern European nations, in selected provinces of China and elsewhere in Asia and the pacific region (Wescott et al, 2009), particularly in developing nations hard hit by the economic stress of 2008 through 2010. Further, it should be noted that poor management of the debt ceiling crisis of August 2011 by federal government decision makers had significant impact on the national economy.

OPPORTUNITIES PRESENTED IN RESTRUCTURING

Perhaps the most important contribution restructuring can make to increasing public sector and government productivity comes from the replacement of out-of-date technology. In many governments shortsighted, across-the-board budget cuts made over a multiyear period create substantial technology and employee education and training gaps. Also, significant process reengineering may be needed to better utilize new technologies to gain full benefit from the productivity increases expected but sometimes not realized despite costly investment, for example in information technologies and the development of comprehensive enterprise system architecture, e.g., for accounting and funds management and reporting.

Under restructuring, some of the ideas proposed to resolve organizational and citizen problems may initially appear to be too radical but may later prove to be workable. For example, under the pressure of fiscal necessity the city of Oakland, California sold the public building that housed its museum to private investors. The city continued to provide museum services under lease agreement with the new owners. Similar sale and lease-back agreements have been successful for other cities. These arrangements enable local governments to reduce their operations and maintenance costs while private investment incentives help to insure proper maintenance and care for facilities. Public users of government services (e.g., health care) and facilities may be required to bear a larger proportion of costs through fees and increased co-payments. Better cost accounting can help in setting prices and appropriate fee levels relative to measures of ability to pay if political decision makers want to spread the burden of increased costs to the public more fairly relative to income and wealth.

Justifications for provision of services by government must be thoroughly reevaluated by elected and appointed public officials and managers in making decisions about fees increases, program reductions, or whether to continue provide selected services at all. The trend toward privatization of government service provision in the period 2000–2008 was driven by recognition that many of the services provided traditionally by government can and in some cases ought to be provided by the private or not-for-profit sectors of the economy. Restructuring that includes contracting-out, privatization and other options often associated with what has been termed New Public Management (NPM) (Hood, 1991) may be useful in some instances as a means of reducing the scale and scope of government while increasing productivity. However, out-sourcing, public-private partnerships (PPP), privatization and other alternative service delivery methods have revealed many weaknesses over the past decade or so and thus need to be thoroughly evaluated relative to the nature of perceived problems, their contexts and the range of alternative actions. Regrettably, we have learned that some alternative service provision methods that appeared to promise cost reduction and increased productivity failed to achieve either outcome.

Further, it is fair to observe that restructuring has to be managed taking into account the rigidities and constraints built into hierarchical public bureaucracies. Typical manifestations of such constraints include overspecialization of function, devotion of inordinate amounts of time to self-defense rather than to problem solving, problem avoidance through obfuscation, resistance to the implications of new information, and a fear of adaptation to new social and economic conditions. Inability to adapt reduces the probability of survival. Recognition that these rigidities and constraints exist should cause us to devote more resources to the study of restructuring and to the education and training of public decision makers, managers, and service providers in methods for diagnosing the need for and managing public organizational change.

THE CHALLENGING POLITICAL CONTEXT OF GOVERNMENT RESTRUCTURING

Beginning in the United States in the mid-1970s and continuing through 2012 much public discussion has focused on the need to reduce the size, scope and role of government in the economy. Ironically, this dialogue continues as many governments add to the size and scope of their responsibilities including buying into or assuming ownership of private and third sector entities, e.g., private sector banking institutions and the auto manufacturing industry in the US in 2009. To a considerable extent the debate in the United States has and continues to parallel that occurring elsewhere in the world; such dialogue has taken place and presently continues in fiscally threatened nations around the world. The stimulus for serious dialogue about the size, scope, and role of government and the need for government restructuring are essentially economic in origin as noted at the beginning of this chapter. However, politics always plays an important role in influencing decisions and typically reflects the dilemma posed to elected officials when they ponder how to respond to both economic stress and public opinion. For example, in the US most citizens understand and dislike the high costs of health case but opposition among the public to reform intended in part to reduce costs is widespread. Consequently, in managing fiscal stress elected officials have to consider two important variables: how costs and services are to be affected by reform, and public fear of the negative consequences of any and all changes proposed.

Views on the pervasiveness and relative disadvantage resulting from the broad social and economic role played by government in general, and perceptions of inadequacy of public service performance specifically are subject to interpretation based on different political and ideological perspectives. Further, the condition of the economy may be, in some cases, somewhat independent of views on the need for restructuring government. In the United States, political demand for restructuring continued during a period of unprecedented economic growth that produced a balanced federal government budget in the late 1990s and then into the period of federal budget deficits in the 2000s, with the budget deficit dialogue presently a high priority in 2012 given the extensive neo-Keynesian stimulus spending undertaken by the federal government to rescue the economy and accompanying large annual budget deficits and total debt incurred. Added to stoke this fire were the general elections for the Presidency and Congress in 2012. The US Congressional Budget Office has projected that total US debt will rise from 68% in 2010 to 69% or more of GDP by 2020 if some means of reducing annual deficits and total debt are not developed and implemented by Congress and the President (Congressional Budget Office. 2011). While this level of debt is not high compared to that of Italy (125% in 2011) for example, it is high for the US, and the level of debt in Italy, Portugal, Spain, Ireland and Greece is of continuing concern to Western European nations and the European Union. Based on a number of economic assumptions a US debt load of 69% in 2020 is estimated to more than triple the amount of the annual interest payment on the federal government debt in that year compared to that projected for 2010 (from $207 billion to $723 billion). Increased debt costs have an opportunity cost in that spending more for paying the cost of the debt means spending less for provision of services of all types, including national defense, unless taxes revenue rises to compensate for debt service costs. Revenues could be increase in two ways: through tax increases or through increased tax revenues from a thriving economy. There is no magic fix to solve debt loads and costs in the US or in other nations that have sizable national debts.

There is another worrying aspect of government indebtedness. In mid to late 2011 consumers and financial markets were stuck in what may be termed "debt panic." Debt panic is the result of political mismanagement of debt and to a significant extent the news media coverage of this failure. Scare tactics in congressional budgeting are not new. (Meyers, 1997: 23–42) Debate in Congress over raising the debt ceiling of the US was covered in detail by the media and drew great public attention to the debt. However, the facts about the US federal government debt raised during this period of dialogue were not new to anyone familiar with federal budgeting and budgets. Throughout its history the US has financed wars and economic stimulus financing using debt. This is normal practice. Further, the need to reduce spending in entitlement programs, especially for health care, has been evident since the mid-1990s. And the future costs of Social Security should not have been introduced into the debate because this program is self-financing. Fixing the projected Social Security financial solvency is simple compared to controlling or reducing the government costs of financing health care for and aging population. Social Security financing is restructured routinely by Congress and the President when needed. The last such fix was done under the Reagan administration in the mid-1980. It involved the same measures that will be employed to put this program on sound financial footing previously: the Social Security tax will be increases along with the retirement age for beneficiaries. Cost of living increases will be adjusted downward. The debate over raising the debt ceiling raised public fears unnecessarily about the credit-worthiness of the federal government. Politics and posturing in advance of the elections of 2012 contributed to debt panic. Debt panic resulted in loss of investor confidence in the economy which, in turn, reduced investment of the type needed to stimulate economic growth that would produce the increase in tax revenues needed to reduce annual budget deficits and total federal government debt. In this respect debt panic stimulated by politics and media frenzy was highly damaging to an economy in the process of recovering from a deep recession. Whatever was gained politically as a result of this mismanagement of the debt ceiling issue was offset by losses in the economy and especially for middle income class taxpayers who bear the burden of paying the costs of government services and debt.

This event in the mid-2011 demonstrates that where public dissatisfaction with government increases, some politicians are able to orchestrate public frustration resulting from reduced economic opportunity and employment uncertainty to their own benefit, as was the case with the U.S. economy in the early 1990s which resulted in a change in the US presidency, and in the period 2008 to 2012 where some members of Congress opposed restructuring because they knew their constituents feared the consequences of any major changes in federal government financing and fiscal policy. Typically the party out of power attempts to exploit public fear into voter support against government programmatic and fiscal restructuring in hopes of increasing their opportunity to win the Presidency and a majority of seats in the two houses of Congress. However, in fairness it must be noted that the debt panic of August 2011 was the result of political exploitation by both political parties and not just the party out of power.

In the meantime, what was happening with the productivity of the federal government? Earlier in this chapter we have explained the importance of increasing government productivity. A variety of responses to the budget cutting required by the debt ceiling workout of August 2011 and FY2012 appropriation decisions once they were reached were two trends that were easily identifiable: (a) federal departments and agencies financed in the discretionary part of the federal budget began to examine ways to cut their budgets as required by Congress. This had an immediate impact of the Department of Defense. Secondly, throughout the federal government attention of program managers and their employees turned to justifying the worthiness and value of their programs. When this happens, often employee productivity suffers. Managers and employees perceive the need to fight to retain their programs and jobs. Most notably, the early August debt ceiling agreement tasked Congress to establish a joint select committee on the deficit to report recommendations on deficit and debt reduction by December, 2011. Still, even if this committee proposes specific actions, Congress still will need to approve these measures into law for them to come into effect. Whether Congress would be able to do this during an election year was a question many asked.

Additionally, a number of approaches to increasing productivity began to be examined more carefully. In this regard President, Barack Obama articulated the importance of performance evaluation of government agencies in numerous public pronouncements. As of 2012, OMB continued to emphasize use of performance information in the budget process. It was clear that the expressed interest of the Obama administration in assessing the performance of government agencies had taken place under the leadership of OMB (Joyce, 2011). However, there is no evidence to suggest that the Obama administration would attempt on its own to implement some type of performance budgeting or other budgetary innovation as a means to drive spending reduction in efforts to produce a balanced budget. Rather, performance was and continues to be reviewed by OMB as one input to budget decision making, which is perhaps the best that advocates of restructuring and budgeting can hope for in the medium term without action by Congress to force such measures along with straight-forward budget cutting. Budget cutting occurred despite the fact that cutting, for example, 20% of the 30% of federal spending in the discretionary part of the budget would have little impact on the size of the total US federal debt.

CONCLUSIONS

The political as well as economic dilemma complicating the need for restructuring the public sector results in large part from the fact that for more than 50 years developed nations and their citizens have become accustomed to and to a greater extent dependant upon continued growth of their governments and economies. Growth has fit well with the motives of political decision-makers seeking electoral and financial support on the basis of providing jobs, public works projects and welfare assistance, and the preferences of citizens desiring the benefits of increased political representation of their demands, needs, and preferences. Labor union power, although diminished in some nations including the United States, typically resists restructuring (e.g., Greece in 2011 and 2012), although the inevitability of budget and salary and staff reductions in government and in the private sector eventually has to be faced and negotiated carefully by union leadership.

Restructuring and budget cutting are not particularly attractive to politicians who, because of it, are no longer able to reward constituents in the traditional manner through increased funding to the benefit of local interests. Neither is it attractive to public managers desiring to preserve their programs and staffs, nor to citizens having benefited from the provision of payments and services by government. It is little wonder that, collectively, we tend to want to avoid thinking about public sector restructuring given that its outcomes are likely to displease great numbers of citizens, managers, employees and politicians.

From the early 1990s through 2012 in the United States and also globally, during certain periods public sector spending has increased and revenues have declined for number of reasons including (a) increased global economic competition, (b) recessions that have reduced economic growth, tax revenues and employment, (c) major political events including the end of the Cold War and corresponding economic and social reformulation in Europe and nations emerging from the former Soviet Union, the terrorist attacks on the US on September 11, 2001 and the subsequent wars and other actions to fight the threat of terrorism, (d) tax reductions and spending approved by the US Congress and President and other legislative bodies and executive officials internationally, some of which have been intended to stimulate economic growth during periods of recession, (e) reductions in national government transfer payments to subordinate state and local governments as a result of revenue shortfalls at the national level, (f) changes in government funding priorities ranging from national defense spending to support wars against terrorism and nation building in the middle east to spending for a wide variety of domestic services including education, social assistance programs, health care and social security and other social service programmatic areas. Expenditure cuts by government decision makers have been made necessary in response to sizable reductions in tax and non-tax revenues. Thus it is important to note that political motivation to increase, redistribute or reduce the rates of growth of government expenditures, or to actually reduce spending levels in nominal or constant dollars in some parts of the budget, emanate from a variety of factors independent of public dissatisfaction with government or public sector performance.

Issues that must be addressed when public sector officials and managers face restructuring in response to economic challenges and changes in patterns of political and social demand include the following: should the scope of public policies, programs and organizations be reduced? Why do government policies become immune to review, modification, and termination? How can we determine which policies and programs should survive and which should be modified, reduced, or terminated? How should decision-makers attempt to reduce or terminate public programs or organizations where this is necessary?

The challenges confronted presently probably are the most serious for the US and many other nations since those faced during the depression of the 1930s and by Europe and Japan the end of the Second World War in 1945. Still, over the past six decades or so elected officials, public managers, fiscal and policy analysts, and the public have had to respond to the effects of changes in government policies, shifting political and spending priorities and restructuring of government organizations.

Two factors are highly evident in observing responses to current fiscal conditions. The first is that public memory of past periods of fiscal stress appears to be almost nonexistent. Secondly, many public officials or managers now in office have had little or no experience with serious restructuring of government. Thus we are left to wonder not only how well present challenges will be managed but whether the lessons of success and failure resulting from attempts to cope with current circumstances will be learned and internalized so that we may manage such conditions better in the future.

This chapter is intended to provide information that US government and Department of Defense officials, DOD budget and program managers, and elected officials and public managers need to understand to better assess alternative methods for improving approaches to government and organizational restructuring during times of economic and fiscal stress. Evaluation of restructuring may be segmented into the determination of (a) the causes of economic and fiscal stress, (b) methods for improved management of restructuring, and (c) the issues and dilemmas faced by public officials and managers attempting to manage restructuring.

The range of non-mutually exclusive approaches to management of restructuring described and analyzed in this chapter includes:

- Doing nothing, likely to work only for a short time if economic and fiscal stress persists,
- Reducing spending and controlling growth rates in expenditures in selected areas of government,
- Increasing revenues through raising taxes and various types of service charges,
- Maintaining credit-worthiness and continued capacity for borrowing,
- Increasing employee and organizational productivity through application of innovative responses to fiscal stress.

The final category of productivity-related responses may be viewed to include a number of approaches including increased cooperation between and networking between organizations, programs and constituents, increased citizen volunteerism in provision of services to the public, mission and program rethinking and modification, joint service agreements within and between governments, and in some cases contracting out and privatization, given the caveats about such arrangements identified in this chapter.

Before developing plans to manage restructuring, prudent elected officials and public managers need and likely will recognize and better define the characteristics of demand for restructuring. In doing so, they may wish to consider responding to this demand in a holistic manner rather than in the piecemeal fashion that has characterized attempts to restructure the public sector and public sector organizations in the past.

Finally, it is important to note that restructuring initiatives must be tailored to the contextual demands of each national, regional, local, and organizational circumstance and cultural environment in which they are applied. There is no one best way to restructure. Therefore, in managing economic and fiscal stress we are better off thinking about better rather than best practice. What works well in one circumstance many not work well or at all in others.

CHAPTER 12

UNITED STATES DEFENSE AND BUDGETARY POLICY IN THE GLOBAL CONTEXT

INTRODUCTION

Although by domestic laws, rules and norms shape U.S. defense budgeting and financial management, the United States' decisions on how to spend defense resources have a powerful impact on the international security environment and other nations' strategic choices. Understanding the interaction effects between U.S. defense budgeting and other nations' reactions to it is vital if the defense-management process is to succeed in its objective of preparing the United States' armed forces to fulfill the government's foreign policy aims. This final chapter examines U.S. defense budgeting from an international perspective.[1] Within this context, answers will be sought to the following questions.

1. What is the effect of U.S. defense budgets on the international development and diffusion of new military technologies?
2. What factors drive the United States' arms export decisions?
3. How do U.S. policies shape the international market for armaments?
4. Do the United States' decisions about how many resources to consecrate to defense drive those of other nations through arms racing or burden-sharing?

Financing National Defense: Policy and Process, pages 397–454
Copyright © 2012 by Information Age Publishing
All rights of reproduction in any form reserved.

At base, the United States' role as the world's largest investor in defense has led to it assuming a preponderant role in the development of new military technologies. This relegates most other nations to the position of selective imitators insofar as they observe the United States Department of Defense (DOD) decisions attentively and emulate practices and technologies that appear successful. In principle, while this dynamic confers *military* advantages on the United States because it generally fields innovative weaponry before other nations, it confers *economic* benefits on nations that accept the status of technological second-movers because they can dispense with many of the risks (and inevitable economic losses) inherent in striving for innovation. In their efforts to appropriate U.S. military technologies in an economic and timely manner, allies actively seek technology transfers from the United States. When permitted, such transfers can be unilateral, reciprocal or commercial in nature. Denied these opportunities, the United States' potential rivals have turned to espionage or attempted to acquire U.S. technology via third parties.

Besides granting the United States a preponderant role in the development and diffusion of new military technology, the United States' large defense expenditures also shape the international arms trade. Because the United States procures sizeable quantities of weaponry for its armed forces, its defense industries benefit from the scale and learning economies generated by the world's largest internal defense market. When combined with the effects of high defense research and development (R&D) budgets, this enables U.S. defense industries to, as a rule, offer weapons that are more innovative and cost-effective than the competition. For this reason, U.S. arms manufacturers win a large proportion of the contracts for which they compete. However, despite the economic advantages of exporting weaponry, there are strong countervailing reasons for not selling specific weapons systems to certain nations. Injudicious defense exports can compromise sensitive technologies, strengthen potential adversaries and fuel regional arms races.

To weigh the merits of a given arms sales, the United States has developed procedures that incorporate a plurality of interest groups and government actors into a (comparatively) transparent process. Within this context, the White House, Congress, the State Department and Department of Defense all contribute to decisions about what weapons to export and to whom. Overall, while the U.S. arms export decision making process may appear balanced from the point of view of the United States' domestic politics and foreign relations, would-be importers perceive it to be restrictive, unpredictable and unreliable. As a consequence, foreign nations frequently face a dilemma as to whether they should adopt an *efficient* course of importing cost-effective U.S. weaponry or pay considerable premiums in terms of more expensive weapons and foregone military capabilities to achieve a greater degree of defense-industrial *autonomy.*

In their efforts to compromise between the competing goals of efficiency and autonomy, many nations purchase weapons from multiple suppliers, manufacture U.S.-designed weapons under license or develop indigenous weapons based on U.S. technology. Because arms export statistics ignore licensed production and many sub-systems exports, the fact that U.S. companies account for approximately 35 percent of international arms export contracts (by value) understates the United States critical role in the international transfer of military capabilities (Stockholm International Peace Research Institute [SIPRI], 2011). For nations that seek the highest level of defense-industrial autonomy-attainable through the indigenous development and production of weaponry, a combination of unrestrained arms sales and generous export subsidies are a sine qua non for achieving the volumes of export sales needed to sustain an independent defense-industrial base.

While United States defense budgeting plays a crucial role in shaping how weaponry is developed, produced and diffused across the globe, the impact of U.S. defense budgets on other nations' decisions about how many resources to dedicate to defense appears comparatively modest. Despite the historic importance of arms racing amongst rival great powers and burden-sharing amongst allied nations, there exists no convincing evidence that either rivals or allies are basing their decisions about how much to spend on defense on U.S. defense budgeting trends. Ironically, the reasons for the disconnectedness of the United States' defense expenditures with those of other nations, whether allied or not, with the United States at a high level of U.S. expenditures. Thus, while potential rivals accept as economically counter-productive any attempt to imitate U.S. trends in defense expenditure, its allies are most willing to free ride on the United States' provision of security than share the economic burden of providing for mutual defense.

DEFENSE BUDGETING AND MILITARY INNOVATION

Any analysis of the United States defense budget's impact on other nations must begin with an examination of the sheer predominance of the United States' investment in defense. No state has attempted to match the United States' defense spending since the substantial increases to the United States' defense budget implemented in the early 1980s (SIPRI, 1979: 1987). At that time, Soviet policymakers accepted that their already over-taxed economy could not afford to dedicate more resources than the 15 to 40 percent of its GDP that was already dedicated to defense (Strayer, 1998; Odom, 1998). Since the dissolution of the Soviet Union in 1991, few nations' defense budgets have even approached the same order of magnitude as the United States'. Figure 12.1, below, illustrates the gap between the United States' defense budget and those of the world's other principal military powers (e.g., Brazil, China, India, Japan, Russia and Western Europe's four largest nations) (SIPRI, 2010).

As may be observed, in no year since 1991 has any state spent even a fifth as much on defense as the United States. Even accepting that certain budgetary estimates (i.e., China or Russia) may be conservative and that purchasing power parity may enable other nations to achieve better value-for-money (a problematic assumption), the size of the United States' defense budget is unrivaled.

In domains crucial to a nations' future military power, such as defense R&D and procurement expenditures, the gap between the United States and other great powers is even more significant. On an annual basis, the United States spends six times as much on defense R&D as *all* 27 member nations of the European Union combined ($79 billion versus $12 billion for Europe). However, Europe is not a unified nation state and U.S. defense R&D expenditures exceed those of the largest European nations (France and the United Kingdom) by a factor of 15 (EDA, 2009). Even, China, whose defense R&D budget has grown rapidly to a figure of $4–6

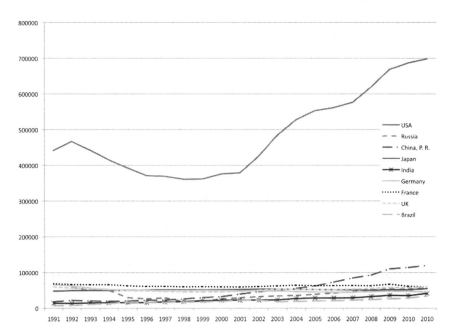

Figure 12.1 US and other Great Power Expenditures Compared (Figures in Billions of Constant 2010 Dollars).

billion per annum, spends less than a tenth as much as the United States (Henrotin, 2010: 56; Bitzinger, 2011: 447). Meanwhile, Asia's other major arms producers, India, Japan and South Korea, each spend approximately $1.5 billion annually on defense R&D, which is barely one-fiftieth the United States' investment ,(Bitzinger, 2011: 445).[2] In short, the United States' defense R&D effort dwarfs those of any other state and exceeds that of the rest of the world combined. Moreover, the dimensions of the United States' force structure and the scope of its infrastructure for conducting large-scale operational experiments (i.e., the combined land areas of the Air Force's Nevada Test and Training Range and the Army's National Training Center are larger than Belgium) render it possible to realistically test potential innovations in unique ways. For example, few nations can convert brigade-sized units to an unproved table of organization, as the United States did to develop the airmobile concept in the 1960s or the network-centric light armored concept in the 1990s- for the purpose of examining a potentially innovative idea (Tolson, 1973; Stanton, 1987; Jones and Thompson, 2007).

As a result of its comparatively large expenditures, the United States has consistently led other powers in introducing new technologies. From stealth aircraft to satellite navigation, electronic flight control systems and network centric warfare, the United States has been the first to field many of the technologies shaping contemporary warfare. The development of new technologies is an inherently risky process and one reason the high U.S. research and procurement budgets has produced innovation is that they are large enough to absorb failures. Alongside those U.S. projects that have produced genuinely helpful new products, many others failed to live up to expectations. For example, in addition to pioneering the successful stealth technologies used in the F-22 and F-35, the Air Force first invested considerable resources in the severely limited "faceted" stealth design of the F-117, while the Navy spent approximately $5 billion on the ill-fated A-12 project, which never reached even the prototype stage (Stevenson, 2000). Besides the risk of pursuing false paths towards genuine innovations, many of the innovative concepts pursued by the United States eventually proved unworkable (DeVore, 2010). Such has been the case for exotic satellite-based weaponry, pentomic divisions and nuclear powered bombers. To a degree, cancelled projects and discarded ideas are an inevitable by-product of the U.S. system of military innovation.

Most nations lack the resources needed for the trial-and-error process of innovation that the United States pursues. Although these nations understand that the first nations to field innovative weapons can reap *military* advantages, they also recognize that there are substantial *economic* advantages to accepting the status of a technological second-mover. As a result, most nations tend only to adopt a technology *once* the United States has already demonstrated its viability and cost-effectiveness. Ingemar Dörfer applied

402 ▪ Financing National Defense

the term "sub-optimization" to describe this tendency for nations to deliberately rely on only proven technologies and content themselves with performance characteristics falling short of what may be theoretically possible (Dörfer, 1973: 18). Norman Augustine, Lockheed's onetime chairman, supports Dörfer's contention that substantial economies can be achieved by pursuing lower performance goals. As Augustine argues, "the last 10 percent of performance generates one-third of the cost and two-thirds of the problems (Augustine, 1997: 103)."

For great powers other than the United States, developing an acceptable weapon system at an affordable price is more important than fielding a product more advanced than those possessed by other nations. For example, even though three European fighter aircraft (the Eurofighter, Rafale and Gripen) were developed roughly in parallel to the United States' F-22, and one Chinese project (the J-10) was even launched somewhat later, none of the non-U.S. projects sought to incorporate radically new technologies, such as stealth technologies, Active Electronically Scanned Array (AESA) radars or super-cruise engines. Only recently, two decades after the F-22 prototypes' first flights have Russia (2010) and China (2011) unveiled prototypes of equivalent aircraft. While conceding technological leadership, the economic advantages of such an approach are clear. For example, the R&D costs of recent French and Swedish combat aircraft ($13 and $3 billion respectively) were substantially lower than those for either the U.S. F-22 or F-35 ($37 and $49 billion respectively), (Hartley 2001; GAO, 2006: 56–57; GAO 2010).[3] Although the Chinese and Russian stealth aircraft costs are unknown, both designs adopted established U.S. principles for reducing radar signatures rather than exploring alternative configurations. Thus, while the United States pays a premium for innovation, other nations achieve economies by following its technological lead.

Because nations wait in many instances for the United States to prove the value of an innovation before pursuing it themselves, they naturally also dedicate substantial efforts to understanding the nature and results of U.S. investments in new military capabilities. Much information is transferred more-or-less voluntarily to U.S. allies through joint exercises, alliance institutions and bilateral agreements. However, the mechanisms whereby individual allies receive U.S. technology differ. Israel has principally received unilateral technology transfers (Clarke, 1995), while a range of Anglo-American technological exchange agreements facilitate reciprocal transfers between the United States and United Kingdom. Finally, Japan, South Korea and Taiwan have been permitted to acquire certain technologies commercially from U.S. defense contractors (Lorell, 1995; Bitzinger and Kim, 2005). As a consequence of these different transfer mechanisms, Israeli Python missiles, British Astute-class submarines, Japanese F-2 fighters, South Korea's T-50 training/strike aircraft, and Taiwan's F-CK-1 fighters all bear a notable U.S. technological paternity.

Aside from formal technology transfers, allies emulate many of the promising U.S. projects and organizational innovations they are exposed to. Dedicating far fewer resources than the United States to developing and experimenting with new military capabilities, most allies wait for the United States to prove the value of an approach before investing their own resources in it. For example, the U.S. drive for military transformation-based on the thorough exploitation of digital networks-inspired similar, albeit smaller programs in France, Germany and the United Kingdom (Lungu, 2004). In addition, the British Army has followed the United States' lead (Blakeman, et al., 2010) in launching a project to equip its expeditionary forces with Mine Resistant Ambush Protected (MRAP) vehicles, while France emulated the United States' Special Operations Command with its own Commandement des Opérations Speciales in 1992 (Micheletti, 1999; NAO, 2009). However, the fact that allies selectively adopt U.S. innovations should not be misconstrued as blind emulation. In general, they adapt U.S. concepts to suit their own budgets, needs and doctrine. One example of this is the British and Canadian embrace of Network-Enabled Operations as a more conservative and less costly version of Network Centric Warfare, as is advocated by the United States.

Nations that are not allied with the United States have historically scrutinized the United States' defense budget and military operations for what lessons they may convey, while also seeking to appropriate U.S. technologies by whatever means possible. For example, many of China's dramatic reforms of both its armed forces and defense-industrial base can be traced to an exhaustive Chinese study of U.S. military operations during the 1991 Gulf War (Pollpeter, 2010). Since then, China has attempted to acquire additional insights into U.S. weaponry by purchasing Israeli weapons that incorporate U.S. technology, buying debris from U.S. weapons recovered over South Asia and technologically exploiting what they obtained from the 2001 Hainan Island Incident, when a SIGINT EP-3 aircraft was briefly interned on the island (Clarke, 1995; Fisher, 2007: 145; Hewson, 2008; Migdalovitz, 2008: 29-33). Sometimes, other great powers' efforts to understand the implications of U.S. military developments have resulted in analyses superior to those conducted in the United States itself. Indeed, the roots of contemporary U.S. debates on the revolution in military affairs (RMA) and in military transformation can be traced to Soviet studies from the 1980s on trends in U.S. military power (Salmonov, 1988; Krepinevich, 1992; Gareev, 1998; DeVore, 2010).

One indirect, yet consequential effect of the United States' preponderant investment in innovation is its role in setting international standards for military inter-operability. Through organizations such as the NATO Standardization Organization (NSO) and Partnership for Peace (PfP), a large number of nations are exposed on a regular basis to U.S. technological standards. More often than not, U.S. standards become either the de facto or *de jure* international standards because these organizations strive to improve inter-operability amongst partner nations and because U.S. standards are frequently the first on the table (Ferrari, 1995: 33–35; Hartley, 1997: 23; Měrtl, 1998: 113–115). The difficulty of engaging in international peace-keeping or exporting weapons to a world market has obliged even nations not formally aligned with the United States to adopt U.S. standards. For example, Swedish Gripen fighters are now built to U.S. digital communications standards (Link 16), and even Russia and China have been obliged to develop variants of their major weapons systems to NATO standards for export (Keijsper, 2003).

As a cautionary note, the United States' advantage in developing and fielding new military technologies will not always necessarily translate into commensurate battlefield successes. Because victory or defeat in warfare hinges on factors such as doctrine, force structure, training and strategy, it is frequently not the first state to introduce an innovation that reaps the fruit of its capabilities. Thus, although the United Kingdom introduced both the tank and aircraft carrier, Germany became the principal strategic beneficiary of the former invention during the Second World War, while Japan and the United States realized the potential of the latter. In both cases, the key to the successful exploitation of new technologies lay in new force structures (e.g., the combined arms panzer division and the integrated aircraft carrier battle group) and doctrines (e.g., deep armored exploitation into an enemy's rear and the launching of air strikes beyond visual range), rather than the production of new weapons per se.

Moreover, past experience also demonstrates that parochial considerations can lead military organizations to neglect existing low-tech challenges. For example, even though the United States pursued the objective of building a 600-ship Navy during the Reagan administration, virtually no resources were dedicated to the mundane task of sweeping naval mines. As a result, the United States Navy possessed only three Korean War-era minesweepers in service during the late-1980s and would have therefore been incapable of escorting Kuwaiti oil tankers in 1987–88 or conducting operations in the northern Persian Gulf in 1991 had European allies not assisted with their more comprehensive minesweeping capabilities (Craig, 1995: 168–254; DeVore, 2009). In short, despite the United States' advantages in developing and fielding new military technologies, it is in the organizational and conceptual domains of defense budgeting that the United States' armed forces are liable to be strategically surprised.

THE UNITED STATES' ARMS EXPORT PROCESS

The Sources of U.S. Comparative Advantages

While the United States' investments in defense R&D favor the precocious fielding of new military technologies, the scale of its procurement spending generates cost advantages for U.S. defense contractors competing in international markets. As with R&D, U.S. spending on the procurement of weapons systems dwarfs that of other nations. The United States invests $140 billion per year on defense procurement. By way of comparison, other great powers spend between one-fifth (China) and one-twentieth (Germany) as much as the United States. For example, China spends $26 billion, Russia $16 billion, the United Kingdom $11 billion, France $10 billion, Japan $9 billion and Germany $7 billion on defense procurement (EDA, 2009; Bitzinger, 2011).[4] All 27 members of the European Union collectively spend only $43 billion, which amounts to less than a third of U.S. procurement expenditures. (EDA, 2009) Moreover, a combination of genuine comparative advantages and protectionist laws (the "Buy American Act" and Congressional politics) ensures that a larger proportion of U.S. procurement spending goes to domestic defense industries than is the case in many other nations. (Neuman, 2009: 72)

In addition to providing the United States' armed forces with the wherewithal to accomplish their missions, this level of procurement spending provides U.S. arms manufacturers with substantial competitive advantages over foreign firms. Two distinct economic phenomena, learning economies and scale economies, explain why high domestic spending sustains international competitiveness. Since the 1950s, research has demonstrated that the ability of a labor force to build complex weapons systems increases with experience (Asher, 1956). This phenomenon of "learning by doing" means that the average cost of a product decreases as the cumulative number of units produced increases. Current research suggests that the man-hours needed to produce major weapons systems can decline by 20 to 25 percent for each doubling of output. Overall, learning economies have been demonstrated to result in 10 percent decreases in the production price of weapons over long orders (Hartley and Martin, 1993: 178–179).

Whereas learning economies are a product of the cumulative production, economies of scale are a function of production rates. When larger volumes of a weapon are produced, it becomes possible to organize the manufacturing process more efficiently and amortize the fixed overhead of production facilities over more units. Although data on the scale-economies of major weapons systems is limited, a British government study argues that a 10 percent decrease in the unitary cost of a product may be achieved with each doubling of output (NAO, 2001: 17). Although there is theoretically a point where increased output ceases to generate economies of scale and may even produce increased per-unit production prices (i.e., diseconomies of scale), the production runs of defense goods are rarely, if ever large enough to produce this effect (Hale, 1987). As a consequence, it is a general rule that the more units produced, the lower will be the unitary production prices of a defense product.

Together, learning- and scale-economies promise substantial savings on the unitary costs of weapons. If two nations manufactured identical weapons systems during a certain number of years, yet one state produced twice as many units as the other, then that state could theoretically achieve a 20 percent reduction in unitary production costs. In practice, U.S. production runs are frequently more than twice as large as those of other great powers. For example, while U.S. combat aircraft may be produced at a rate of 12 to 15 units per month, national British or French programs can at best achieve a monthly cadence of two to five (Hartley and Martin, 1993, 178–79; Hébert, 1995: 76–78). As a result of these larger production runs, U.S. defense corporations can generally sell weapons abroad at cheaper prices than foreign companies marketing equivalent products.

The United States' advantages in competing for export markets pose a major challenge to the viability of other nations' defense-industrial bases. In effect, U.S. defense manufacturers possess cost advantages in international markets because the United States' domestic market is so large. Even though the United States is the world's largest arms seller and annually exports nearly $15 billion (prices in current dollars) in weapons, exports constitute less than 10 percent of U.S. defense industries' output (SIPRI, 2011). In other words, approximately 90 percent of U.S. produced weapons end up in the hands of the United States' armed services. As a consequence, while arms exports are desirable for U.S. corporations and can yield certain benefits even for the state, they are not essential to the viability of the United States' defense industrial base. Insofar as the United States is highly capable of winning export orders, yet not dependent on doing so, it occupies a virtually unique position in the international market.

Compared with the United States, most nations depend on arms exports for the maintenance of a defense industrial base, yet have fewer competitive advantages for achieving them. To compensate for insufficient domestic production runs, many arms producing nations must export a substantial proportion of the arms they produce in order to achieve adequate economies of scale and avoid the necessity of closing production lines between national orders, which both generates unemployment and results in the loss of vital skills. To take an extreme example, the survival of Israel's defense industrial base structurally depends on exporting three-quarters of the arms produced in that state (Hughes, 2003). For other arms producers, the imperative to export is only slightly less onerous. Russia, for example, seeks to export roughly half its total output, and Europe's largest arms producers appear to be aiming to export one-third of their production (Bitzinger, 2003: 53–55; Kalinina and Kozyulin, 2010: 34–39; Interview, 2010). Given the apparent conundrum of many nations needing to export a large proportion of their defense output for domestic arms production to remain viable yet being unable to achieve the cost-effectiveness of United States contractors, certain scholars have argued that the United States could acquire a de facto monopoly over the international arms market (Kapstein, 1994; Caverley, 2007).

However, contrary to predictions that the size of the United States' protected domestic market would lead to an international monopoly on the sale of major weapons systems, the U.S. share of the international arms market remains more limited than one might expect. Figure 12.2, below, compares the sales of the world's eight largest arms exporters since the end of the Cold War.

Although the United States has been the world's largest arms exporter in every year excepting one (2002), its share of the international market has varied from a high of 58 percent (1992) and a low of 27 percent (2008) per annum. While substantial, such a market share is less than one might expect from a country that invests 10-fold more than any other state on military R&D and 5-fold more on procurement. Conversely, some nations

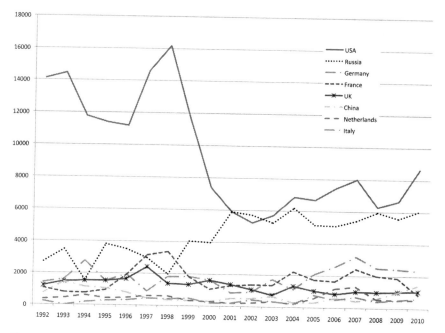

Figure 12.2 Arms Exports Since the End of the Cold War. (Figures in Billions of Constant 1990 Dollars)

export more weapons than their domestic defense-industrial investments would predict. To understand why the United State's exports the quantity of weapons it does and how its presence shapes international markets, it is necessary to examine both United States' and other nations' arms export policies and policymaking processes.

Advantages and Disadvantages of Exporting Armaments

U.S. policymakers have long been ambivalent as to the merits of exporting armaments. Within this context, certain economic, military and diplomatic arguments are regularly evoked both for and against arms exports. Table 12.1, below, illustrates the factors that usually weigh either for or against a given export agreement.

The case for arms exports has economic, military and foreign policy components. Because the United States' government has already born the substantial sunk costs needed to develop a weapons system, export orders are a cost-free (for U.S. taxpayers) means of securing a greater degree of profitability for U.S. firms and providing jobs for U.S. workers. Considering the sizeable proportion of the United States' national investment in high technology R&D dedicated to armaments, it would arguably be economically counter-productive to not vigorously export armaments. Besides being advantageous for the United States economy as a whole, arms exports can also improve the United States defense industrial bases' ability to cost-effectively support the armed services.

By expanding production runs, exports can result in lower acquisition costs for the United States armed services. Such a dynamic is apparent in the F-16 program, where 4,300 aircraft have been procured by 24 countries, and provides a primary rationale for current efforts to involve large numbers of nations in the F-35 program (Kapstein, 2004; Sorenson, 2009: 130–131). In certain cases, such as the development of the F-16 Block 60 for the United Arab Emirates, foreign clients have born many of the R&D

TABLE 12.1   Arguments For and Against Arms Exports

Arguments for Exports	Arguments Against Exports
1. creating jobs and corporate profits	1. compromising sensitive technologies
2. lowering domestic procurement costs	2. fueling arms races
3. keeping assembly lines open	3. strengthening potential adversaries
4. strengthening allies and friends	4. sanctioning nations' behavior/policies
5. promoting inter-operability amongst allies and friends	5. producing negative diplomatic consequences

costs of improving and/or upgrading U.S. weapons (Steuer, et al., 2011: 19–20). By lengthening production runs, exports can also sustain production lines during periods when a dearth of domestic orders would normally lead to their closure (Kemp, 1994, 155). For example, exports have at times kept assembly lines for F-15, F-16 and C-130 aircraft open when DOD contracts would not have. In the recent past, exports have been advanced as a means of keeping the C-17 and F-22 production lines open (Sorenson, 2009: 131–132).

In addition to the economic arguments for arms sales, several political arguments have been made for why selling arms can improve the United States' security and influence. One of the longest standing arguments in favor of arms sales has been a desire to strengthen the United States' friends and allies. In important respects, the arms given, sold or leased to U.S. allies facilitated the United States' victories during both world wars and the Cold War. Without U.S. military goods, it is uncertain whether the Entente could have repulsed the German spring offensives of 1918, whether the Soviet Red Army would have triumphed on the Eastern Front in 1943, or whether Greece, Turkey and Iran could have withstood communist pressures during the early Cold War (Soutou, 1989; Overy, 1995: 180–244).[5] More recently, a desire to provide U.S. allies with qualitative advantages over their opponents has underscored debates about arms exports to Israel, Saudi Arabia and Taiwan. Part and parcel to calls to strengthen America's friends and allies have been arguments that the United States needs to improve its ability to militarily operate alongside potential military partners. Because the United States' largest military interventions have all involved international coalitions, it is evident that inter-operability can, at times, be critical to military effectiveness. In principal, at least, a liberal arms export policy would enhance military inter-operability by ensuring a greater degree in equipment commonality between allies (Wolf and Leebaert, 1978).

While powerful arguments can be made in favor of arms exports, equally compelling considerations are frequently advanced for a restrictive arms export policy. For a country that invests so much in military innovation, any arms exports risk placing valuable military technologies in the hands of competitors capable of reverse engineering or otherwise imitating U.S. products. China has reportedly acquired much U.S. military technology through unauthorized re-transfers of U.S. weapons or designs. It allegedly acquired blueprints for Aegis air defense systems from a Japanese officer, an example of the F-16 fighter from Pakistan and a variety of U.S. military technologies from Israel (Clarke 1995; Fisher, 2007: 145; Hewson, 2008; Cheung, 2009: 137–42; Sorenson, 2009: 134).

Besides potentially compromising U.S. technology, the introduction of new or qualitatively superior weapons into sensitive regions can fuel arms races and create windows-of-opportunity for aggression. For example,

Egypt's unprecedented September 1955 arms deal with the Soviet Union (ostensibly Czechoslovakia) prompted Israel to attack in October 1956, before the Egyptian armed forces could assimilate the new weaponry (Kyle, 2011 [orig. 1991]: 62–85). To prevent situations such as this from arising, the United States has (imperfectly) followed a policy, enshrined in a Presidential directive from 1977, of not being the first state to export new categories of armaments to a region (Sampson, 177: 184–185; Kemp, 1994: 154; Le Roy, 2002).

Along with concerns that arms exports could destabilize regional balances-of-power, there is an equal concern that U.S. arms exports might strengthen nations that could become hostile to the United States. Because there is no guarantee as to how a recipient of U.S. arms will behave in the future, there is always the possibility that U.S. soldiers and seamen could find themselves confronted with U.S. weapons. Such indeed occurred after the Iranian Revolution of 1979, when a government hostile to the United States inherited the state of the art stockpile of U.S. weapons that had been sold to the previous pro-American regime. While there is a persistent fear that exported weapons could be used against the United States, another motivation for not exporting U.S. weapons lies in the desire to not sanction activities or policies that the United States disapproves of. For example, the United States used arms embargos to (unsuccessfully) deter India and Pakistan from acquiring nuclear weapons in the 1990s and to express displeasure with Turkey's 1974 invasion of Cyprus (Sampson, 1977: 311–312; Hackett, 1988).

A final reason for not exporting certain categories of weapons lies in the negative diplomatic consequences that could result from sales. China is sensitive to the nature of United States arms transfers to Taiwan, and Russia is concerned with U.S. exports to nations it considers to lie within its sphere of influence, such as Georgia. Exports of overly sophisticated or offensive armaments to either Taiwan or Georgia could, therefore, exacerbate relations with China and Russia respectively. Although neither sophisticated nor illegal, the export of weapons such as cluster bombs, depleted uranium munitions and napalm can tarnish the United States' public image because humanitarian organizations have made these weapons the object of lengthy negative publicity campaigns. For example, Israel's use of U.S.-made cluster bombs in populous areas during the 2006 Lebanon War generated much adverse publicity for the United States (Human Rights Watch, 2008; Migdalovitz [CRS], 2008: 31–32). To prevent arms exports that would negatively impact the United States' image in the world, Section 502B of the Foreign Assistance Act was enacted in 1974, banning arms sales, except in extraordinary circumstances, to governments that display a "consistent pattern of gross violations of internationally recognized human rights" (Schroeder, 2005: 34).

United States Export Policies and Procedures

Considering that powerful factors militate both for and against U.S. arms exports, determining the merits of any given arms transfer is subject to a process that is both more pluralistic and transparent than those of many foreign counterparts. The United States' arms export approval process is pluralistic in that it officially involves numerous actors within both the executive and legislative branches of government and unofficially embraces a wide range of interest groups. This process provides procedural mechanisms whereby many diverse perspectives-reflecting military, economic, diplomatic, humanitarian and parochial interests-weigh in on whether the United States will or will not transfer a weapon to another state. Because of the nature of this decision making process, the United States' arms export process is comparatively restrictive and reflects an ever-changing balance between economic, military and diplomatic factors (Schroeder, 2005). From the perspective of foreign nations desirous of importing U.S. armaments, the arms export process frequently appears unreliable, unpredictable and laden with conditions. To better appreciate how this process functions, we will examine first the role of the executive branch and then the legislative branch in the arms export process.

The executive branch of the United States government plays the crucial gate-keeping role of deciding whether the United States government should permit negotiations or reject out-of-hand a nations' request for U.S. weaponry. Within this context, two distinct procedures exist for negotiating an arms deal—the Foreign Military Sales (FMS) program and the Direct Commercial Sales (DCS) process. The FMS and DCS procedures differ substantially, with the former being a DOD administered program and the latter involving the State Department approving direct commercial negotiations between U.S. firms and foreign nations.

The FMS program was established as a consequence of the Foreign Assistance Act of 1961 and the Arms Export Control Act of 1976. What distinguishes the FMS program from the DCS process is that the former involves government-to-government contracts, administered by the United States DOD, rather than contracts between U.S. corporations and foreign nations, as does the DCS process. This means that the United States Government contracts for FMS weapons from U.S. firms, before transferring the weapons to the foreign client. In principle, FMS contracts are administered on a no profit, no loss basis by the DOD.[6] To this end, a 3.8 percent fee, levied on contracts, is used to fund a specialized DOD agency, the Defense Security Cooperation Agency (DSCA), which administers the contracts. Because FMS is managed by the DOD, both the armed services and the Office of the Secretary of Defense can exert a direct influence on what equipment is offered for sale and under which conditions (GAO, 1999; Sorenson, 2009: 132–133).

For importing nations, FMS has the advantage that the United States DOD undertakes the complex tasks of monitoring a contract guaranteeing the quality of the goods delivered and ensuring that the training and service provisions are adequately fulfilled. For nations with limited administrative capabilities, these FMS services can mean the difference between a state receiving a real military capability for its investment rather than being overwhelmed by the delivery of goods and services that a state's armed services are incapable of employing without additional set-up assistance. Despite these advantages, several factors have driven a long-term decline in the popularity of the FMS program relative to the DCS process. The FMS program's use of cost-based, rather than fixed-price, contracts and the lack of transparency about costs have convinced many nations that they can achieve better value-for-money through the DCS process (GAO, 1999: 6). Moreover, the fact that the FMS program only deals with equipment built to the same standards as those used by the United States armed services obliges nations desirous of acquiring customized equipment to do so via DCS (Sorenson, 2009: 133).

In contrast to FMS, the DCS process is administered by the State Department and involves a less proactive governmental role. In cases of DCS, a would-be purchaser of U.S. weapons must apply to the Office of Trade Controls at the State Department's Bureau of Political and Military Affairs for permission to begin direct negotiations with contractors. Once permission has been given, a contract's modalities will be negotiated directly between the U.S. firm and its potential foreign client. Contrary to FMS, the United States' government applies no surcharge for DCS sales and fixed-price contracts can be employed (GAO, 1999: 6). Moreover, if clients so wish, they can through the DCS process order customized or modified products that are not being used in the U.S. military services. However, the downside of the DCS process from a client's perspective is that it shifts the significant burdens of contract administration and oversight to the purchasing government, rather than that of the United States (Sorenson, 2009: 134–135).

Regardless as to whether FMS or DCS procedures are employed for a sale, both the State Department and DOD usually have an input into an arms sale. Although the DOD manages FMS, federal law mandates that the State Department must also approve all government-to-government (i.e., FMS) sales. While not mandated, the State Department frequently consults with the DOD on DCS sales. Formally at least, the State Department refers 30 percent of DCS requests to other agencies (including the DOD) for review (Schroeder, 2005). Table 12.2, below, illustrates the different instances in the State Department and DOD that are involved in arms export decisions.

TABLE 12.2 Arms Export Decisions and the DOD and State Department

DOD Actors/Agencies	State Department Actors/Agencies
1. the Defense Technology Security Administration 2. the Office of the Secretary of Defense 3. the Joint Chiefs of Staff 4. Combatant Commanders 5. the Defense Security Cooperation Agency	1. the Office of the Legal Advisor for Political-Military Affairs 2. the Under Secretary for Political Affairs 3. the Under Secretary for Arms Control and International Security Affairs 4. the Bureau of Democracy, Human Rights and Labor 5. regional bureaus 6. the Bureau of Legislative Affairs

As already noted, while the State Department must approve all arms exports (FMS and DCS), the DOD must approve FMS sales and is regularly consulted on DCS sales. During certain high-profile sales, other arms of the executive branch, such as the National Security Council, also intervene in the decision making process (Schroeder, 2005: 30).

In principle, it is after the executive branch has approved a sale that the legislative branch can exercise its right to either approve or invalidate a sale. According to the Arms Exports Control Act (AECA) of 1976, the executive branch must notify Congress 30 days before an agreement can be concluded for the provision of defense goods valued at $14 million or more (15 for NATO members, Australia, Japan and New Zealand). Once such notification is given, Congress has an allotted period of time to debate and, if appropriate, pass a joint resolution blocking the sale. However, even if Congress does not act during the mandatory notification period, it remains free to use its normal legislative tools to block or modify a sale *at any time* prior to the delivery of the equipment in question (Grimmett, 2010).

Concentrating overmuch on Congress' formal role, specified in the AECA, of legislating against arms sales approved by the executive branch could lead one to misconstrue the legislative branch's true influence over arms export decisions, which is both less direct and more pervasive than might first appear to be the case. In fact, Congress' officially mandated tools for shaping arms sales are rather unwieldy. Not once since the AECA's promulgation (1976) has a congressional joint resolution blocked a potential arms sale (Schroeder, 2005: 31). Hypothetically, even if passed, such a resolution could be subject to a presidential veto. In such a case, two-thirds of the members of both the House of Representatives and Senate would be needed to sustain a congressional joint resolution prohibiting an arms transfer. Thus, barring opposition from a strong majority in both the House of Representatives and the Senate, Congress cannot formally prevent an arms sale (Grimmett [CRS], 2010: 3–5).

Despite the limitations of its formal policy tools, Congress possesses substantial influence over arms export decisions. Although never actually implemented, the mere threat that Congress *could* block a sale can pressure the executive branch into either forgoing or modifying sales. Moreover, by opening formal hearings on arms export decisions and, thereby, focusing public scrutiny on the issue, Congress can politicize export decisions that would otherwise be treated administratively. In general, the executive branch is willing to go to considerable lengths to include influential members of Congress in the early stages of a defense export decision in order to avoid having to spend political capital in defense of potentially unpopular export decisions. One way Congress can be engaged prior to the formal notification of a sale is seen in a non-statutory commitment the executive branch undertook in 1976 to provide a classified briefing to the House Foreign Affairs Committee and Senate Foreign Relations Committee. This occurred 20 days prior to the formal notification provided Congress, as stipulated by the AECA (Grimmett [CRS], 2010). In potentially controversial cases, such as existed in 1976, influential members of Congress and their staff are brought into discussion far earlier (Schroeder, 2005: 32). Because the executive branch includes congressional representatives in the arms export decision making process earlier than federal law demands, divergences between the legislative and executive branches of government are generally resolved by compromise *before* the formal notification process begins.

Besides its ability to either directly influence or overturn the executive branch's arms export decisions, Congress can shape the overall political context with which decisions are made. Through two congressionally funded research centers, the Congressional Research Service (CRS) and the General Accounting Office (GAO), members of Congress commission studies that can both raise the visibility of arms transfers and provide valuable policy inputs. Within recent years, the CRS has examined arms sales to Pakistan, aircraft sales to South Asia and whether the F-22 fighter should be exported. Meanwhile, the GAO has investigated controls on cruise missiles and on unmanned aerial vehicle and F-35 aircraft technology. In certain instances, Congress can go further by passing legislation that establishes distinct conditions for sales to particular nations or of particular items. For example, the Pressler amendment to the 1985 Foreign Assistance Act rendered arms exports to Pakistan conditional on the president annually affirming that Pakistan did not possess a nuclear device (Grimmett [CRS], 2009). While the Pressler amendment was clearly designed as a restraint on exports, Congress has repeatedly used legislation to cajole the executive branch into greater sales to Taiwan (Kan, 2002). After surveying the cumulative impact of Congress' diverse tools for shaping arms transfers, Matt Schroeder concluded, "The levers of influence held by the legislative branch are remarkably effective.... The potential downside to this capacity is that well-placed lawmakers who know how to work the system can single-handedly derail major policy initiatives (Schroeder, 2005: 33).

The virtue of the U.S. arms export decision making process is that it includes a plurality of legitimate interests within a (comparatively) transparent process. The executive departments entrusted with the United States' military security and foreign relations can express their points of view. Meanwhile, members of Congress can voice concerns about issues as diverse as protecting defense-industrial jobs in their districts, preventing the transfer of sensitive technologies and punishing nations for policies of which they disapprove. Interest groups ranging from defense industries to humanitarian groups and lobbies dedicated to certain nations' interests (e.g., pro-Greek, pro-Israel and pro-Taiwan lobbies) can all weigh in on this process through the lobbying and public relations tools afforded them by United States laws. As a result of incorporating so many diverse groups into the arms export decision making process, the United States' decisions about whether to export a given weapon to a particular state reflect a complex balance of interests.

Foreign Views of the United States' Arms Export Process

While the U.S. arms export decision making process may appear balanced from the point of view of both the United States' domestic policies and its foreign relations, the process is all too frequently perceived to be unpredictable, unreliable and, at times, incompatible with importing nations' desires to be seen as fully sovereign powers. Because of the diverse interests represented and variegated channels of influence embedded in the United States' decision making process, there is an unpredictability in U.S. arms exports that confounds nations' ability to rely, in the long-term, on acquiring weapons from the United States. The diverse examples drawn from the experiences of Pakistan, Saudi Arabia, Japan and the United Kingdom will suffice to illustrate the dilemma that many nations face in deciding whether or not to buy U.S. weapons.

As far back as the 1950s, Pakistan was considered an important partner of the United States and could therefore import a wide range of U.S. arms. By 1979, however, concerns about Pakistan's nuclear weapons aspirations and the human rights record of its military regime prompted the United States State Department to ban new arms export contracts with Pakistan. Two years later, the new administration of President Ronald Reagan liberalized arms exports to Pakistan to an exceptional degree in order to strengthen that state as a regional counterweight to the Soviet Union. At this time (1981) the United States agreed to sell Pakistan F-16 fighters, which represented at the time the technological cutting-edge (Kemp, 1994: 151). From

1985 onwards, Congress steadily applied pressure on the executive branch, via the Pressler amendment, to link continued arms exports to Pakistan's abandonment of its nuclear program. In 1990, these pressures culminated, under the new administration of George H.W. Bush, in a fresh arms embargo on Pakistan. Most galling for Pakistan, this embargo extended even to products that had already been bought and paid for, including 28 F-16 fighters purchased the preceding year (Grimmett, 2009).

More than a decade later, President George W. Bush lifted the embargo on Pakistan after the 11 September 2001 terrorist attacks on the United States. As a powerful display of its willingness to sell Pakistan weapons, in 2006 the United States concluded $3.5 billion in arms export agreements with Pakistan, rendering that country the largest customer of U.S. arms at that time (Grimmett, 2009). Thus, in little more than two decades the United States twice embargoed Pakistan and twice lifted its embargos. Throughout this period, Pakistan's government has shown remarkable continuity in its commitment to a nuclear capability and in its connections with extremist groups and problems with democratic governance. From this point of view, changes in the United States' arms export policies towards Pakistan have been driven more by changing perceptions and politics in the United States than any actions on Pakistan's part.

While Pakistan may be a uniquely complex case, even long-standing allies of the United States have been subjected to the vagaries of its arms export process. Congress, for example, has repeatedly thwarted efforts by the executive branch to export arms to Saudi Arabia. In 1984–85, the executive branch sought to sell $2.8 billion worth of F-15 fighters to Saudi Arabia. However, Congressional opposition was such that the administration informed the Saudis it could not conclude the deal (Miller, 1990). The following year, in 1986, Congress threatened to block the sale of 2,400 Sidewinder, Harpoon and Stinger missiles to Saudi Arabia and, thereby, obliged the executive branch to withdraw all 600 Stinger missiles from the proposed sale. Later, in 1990, Congressional opposition to a proposed $20 billion arms deal with Saudi Arabia prompted the executive branch to settle for a more modest package of $7 billion worth of armaments (Grimmett, 2010: 6). Thus, although Saudi Arabia was able to import a steady flow of arms from the United States, it was never able to purchase all of the products it most desired. In each of these cases, the executive branch's goals of strengthening an American ally and winning lucrative contracts clashed with congressional fears that Saudi Arabia could use new high technology weapons against Israel.

Even America's closest allies are not immune from the unpredictability of the United States' arms export process. Japan's experience with the F-22 fighter is a case in point. Being one of the few clients that could afford the

F-22, Japan expressed its interest in acquiring the aircraft quite early. The United States Air Force's leaders initially voiced their opposition to *any* F-22 exports for fear of compromising the aircraft's technological edge.[7] As a result, Congress passed in 1998 a law prohibiting F-22 exports.[8] However, when it became apparent that the U.S. production run of 183 F-22's aircraft was nearing its end, the Air Force's leadership reversed itself and began *urging* F-22 exports to Japan as a means of keeping the production line open for a hypothetical future U.S. order. Members of Congress representing districts producing the F-22 (e.g., Marietta, GA; Fort Worth, TX; Palmdale, CA) joined in the effort to lift the export ban. However, since 2007 the House of Representatives has refused to authorize the F-22's export (Bolkcom and Chanlett-Avery, 2009). Even the United Kingdom, which enjoys the most privileged access to U.S. weaponry, has occasionally had cause to complain. Although the United States has sold the United Kingdom nuclear missiles since the bi-lateral Nassau agreement of 1962 and the two countries share submarine technology, the United States' hesitancy to communicate classified software codes nearly led the United Kingdom to withdraw from the F-35 project until a "painful" compromise was negotiated (Reinhard, 2006: 84–89; Sorenson, 2009: 134).

Frequently, even when the United States approves arms exports, it imposes strict conditions on what the purchasing country can do with the products they acquire. In certain cases, conditionality is specific to a product, while in others it applies to a state. An example of product-specific conditionality is the United States' regulations on Stinger missiles. When the United States sells Stinger missiles to any state, it requires that the purchasing state physically inventory the missiles on a monthly basis, regularly update the United States government on the whereabouts of all its' missile and accept the visits by U.S. inspectors on an annual basis. The United States also reserves the right to review the purchasing state's security procedures to make certain that they conform to U.S. standards (Schroeder, 2005: 31). An example of client-specific conditionality can be found in the United States' sale of AMRAAM air-to-air missiles to Taiwan. Although the United States agreed to sell Taiwan AMRAAM missiles in 2000, it decreed that the AMRAAMs would not actually be delivered *until* it was proven that China possessed an equivalent air-to-air missile (i.e., the Russian AA-12). Thus, Taiwan was essentially free to purchase missiles that would sit in U.S. warehouses until the United States government decided to export them (Kan, 2002: 10). No doubt, conditions such as those the United States imposed for the sale of Stingers and AMRAAMs constitute severe constraints on the ability of nations to freely use the arms they purchase in pursuit of their own foreign policy objectives.

AMERICA'S IMPACT ON THE INTERNATIONAL ARMS MARKET

Foreign Responses to U.S. Export Policies

Because of the restrictiveness, unpredictability and conditionalities inherent in the United States' arms export process, nations are forced to weigh the (generally) superior cost-effectiveness of U.S. weaponry against the risks of depending overmuch on the United States.[9] Put simply, the trade-off facing nations allied to or enjoying cordial relations with the United States is one between procurement *efficiency* and *autonomy* (Moravcsik, 1992). In general, the most efficient policy a state can adopt would be purchasing those weapons systems providing the greatest value-for-money that can be obtained on the world market. However, because a disproportionate number of these weapons would inevitably come from the United States, a foreign state would thereby become vulnerable to coercion, manipulation or punishment at the hands of any U.S. government willing to leverage its position in the arms market for foreign policy ends. Conversely, a state can obviate the risks of being blackmailed by arms suppliers (notably the United States) by pursuing a policy of complete defense-industrial autonomy, which would entail designing and building all of its weapons systems within its sovereign territory. However, such a course of action would be ruinously expensive for the vast majority of nations, which lack the budgetary, scientific and industrial resources to autonomously produce armaments with any degree of efficiency.

In actual fact, although complete defense-industrial autonomy is virtually unheard of amongst the United States' allies and friends, most of these nations are willing to pay considerable premiums in terms of more expensive weapons systems and forgone military capabilities in order to lessen their degree of defense-industrial dependence on the United States. Within this context, the autonomy-efficiency trade-off nations face is not one between two opposing policy alternatives, but rather one where a whole range of intermediary courses of action are available to nations. However, the underlying logic is such that each additional increment of defense-industrial autonomy a state wants to obtain can be bought only at the price of reducing its efficiency in arms procurement (and vice-versa) (Moravcsik, 1992: 23). Figure 12.3 below illustrates the different efficiency-autonomy trade-offs that exist between the maximum degree of efficiency provided by an economically liberal import policy, and the maximum degree of autonomy provided by the entirely indigenous development of armaments.

For many nations, the principal means of insuring against the unpredictability and restrictiveness inherent in U.S. arms exports lies in diversifying the sources from which they procure armaments. Even though U.S.

least	**Autonomy**		greatest
Importing weapons from the United States	Importing weapons from multiple foreign suppliers or Domestically producing U.S.-designed weapons under license	Domestically developing and producing weapons employing U.S. components or based on U.S. designs	Developing and producing weapons on an entirely indigenous basis
greatest	**Efficiency**		least

Figure 12.3 The Autonomy-Efficiency Trade-Off in Arms Procurement.

armaments frequently possess cost and performance advantages in comparison to competing products, other arms manufacturers are generally more predictable and less restrictive in their arms export policies.[10] As a consequence, a number of nations have adopted deliberate policies of splitting their major armaments purchases between the United States and other contractors. Oftentimes, this involves the purchase of similar products from two or more contractors. For example, both Greece and Taiwan have engaged in nearly simultaneous purchases of comparable U.S. and French fighter aircraft; Pakistan procures many categories of weaponry from both U.S. and Chinese sources; and Saudi Arabia redundantly purchases equipment from U.S., British and French sources (Huertas, 1996: 166–181; Phythian, 2000: 188–258; Carlier, 2002: 243–261).

Theoretically, procuring weapons from multiple foreign suppliers offers importing nations two major benefits when compared with the alternative of depending exclusively on the United States. Firstly, by purchasing a portion of its armaments from suppliers perceived as more reliable than the United States, a state can partially guarantee itself against the risk of the United States prohibiting future exports of either complete weapons systems or, even worse, spare parts for a state's existing U.S.-made weapons. Within this context, Pakistan's ability to weather the United States' arms embargo of the 1990s owes much to its having previously maintained China as a "second-source" for most categories of armaments (Medeiros and Gill, 2000). Secondly, by buying weapons from suppliers considered less restrictive than the United States, nations can both obtain types of weaponry that the United States is unwilling to export and exert pressure on the U.S. government to approve sales that would otherwise go to foreign suppliers.

Taiwan is a case in point. After the United States proved reluctant to sell it F-16 fighters and AMRAAM missiles in the early 1990s, Taiwan purchased comparable Mirage 2000 fighters and MICA missiles from France. Besides furnishing Taiwan with high technology weaponry the United States would not sell, this deal had the added benefit of prompting the U.S. government to revisit its arms export policy towards Taiwan. Because its unwillingness to sell Taiwan F-16s only resulted in French firms winning a $4 billion contract, the administration of President George H.W. Bush reversed course and offered Taiwan F-16s (Carlier, 2002: 244–251).

Although nations can lessen their vulnerability to a restriction in U.S. arms deliveries by diversifying their sources of supply, such a policy is economically inefficient. For one thing, nations that maintain a second-source for most categories of armaments must, by definition, buy large quantities of weapons that they consider suboptimal in terms of either cost or performance. Thus, by purchasing French Mirage 2000 fighters as a hedge against restrictions to their access to U.S. F-16s, Greece and Taiwan both ended up paying up to a third more per unit for aircraft whose radars and electronics lagged behind their U.S. counterparts (Simon, 1993: 63, 243–244; Carlier, 2002; List, 2005). Pakistan has, generally, been even less satisfied with the performance, reliability and financial conditions attached to its purchases of Chinese armaments (Medeiros and Gill, 2000: 9–10). Even when the second-source for weapons is nearly as cost-effective as the primary source, diversification is costly because of the greater administrative and logistical complexity it entails. For example, when nations acquire weapons systems from two suppliers rather than one, they must arrange for personnel to be trained to operate and maintain two dissimilar systems and also to interface with two distinct foreign supply systems for spare parts. Such duplication in administrative tasks inevitably results in higher overhead costs and/or lower levels of operational readiness (Huertas, 1996: 168)

Because of the short-comings of diversification, manufacturing American-designed weapons under license provides nations that possess adequate industrial bases with an appealing means of achieving a degree of defense-industrial autonomy while also retaining the advantages, provided by imports, of procuring high technology weapons developed and tested as part of the United States' unrivaled defense R&D effort. As demonstrated by the arms embargoes against apartheid South Africa and Serbia, nations that manufacture weapons under license can exhibit a much greater degree of resiliency when their former arms suppliers abruptly halt future shipments than nations that directly import weapons (Collet, 1993: 98–99; Huertas, 1996: 131–132). Moreover, for nations that already possess sizeable defense industries, licensed production can offer employment for factories that would otherwise be closed (Rich et al., 1984; Braddon, 1995). For all of these reasons, many nations have favored manufacturing U.S.

designs under license to be an optimal defense-industrial strategy. As a consequence, foreign nations have, over time, produced under license a wide array of U.S. weapons, including Aegis-equipped destroyers, M1 tanks and a veritable litany of iconic U.S. fighter aircraft, including the F-86 Saber, F-104 Starfighter, F-4 Phantom II, F-15 Eagle and F-16 Falcon.

However, despite its allure, manufacturing U.S. weapons under license has major drawbacks. For one thing, by establishing a separate production line rather than buying weapons directly off U.S. production lines, nations fail to benefit from the learning-and scale-economies that the United States generates as a result of its exceptionally large production runs. For example, the decision by six European nations to manufacture F-16 aircraft under license, rather than buy them directly from the U.S. manufacturer (General Dynamics), resulted in them paying 34 percent more per aircraft than would have otherwise been the case (Rich et al., 1984: 9–10). In other instances, the cost-penalties of licensed manufacturing is substantially greater. It is generally estimated that Japan regularly pays twice as much per unit for the American-designed weapons (e.g., F-15 fighters, P-3C anti-submarine aircraft, Patriot missiles and Aegis destroyers) it manufacturers under license than would be the case with direct imports (Chinworth, 2000: 382–384). Even South Korea, which has the advantage of a highly qualified and comparatively lower-cost workforce, pays cost-penalties of at least 20 percent for licensed-production (Bitzinger and Kim, 2005: 192–193). Thus, by opting to manufacture arms under license, nations give up one of the major benefits of acquiring U.S. weaponry, superior cost -effectiveness, that is a product of the United States' long production runs and high volumes of production.

While licensed-production forgoes certain of the economic advantages of procuring armaments from the United States, it also fails to provide licensees with a high degree of defense-industrial independence. While licensed manufacturing transfers the know-how to build products, it rarely transfers the tacit knowledge required to design or modify high technology weaponry. Moreover, since the 1970s the United States' policy of restricting technology transfers has further limited the benefits licensed-production confers on licensees. In general, the United States "black-boxes" sensitive components (i.e., supplies them only as end-items from U.S. contractors) when it agrees to the licensed production of weaponry.[11] For example, when it licenses combat aircraft designs, the United States systematically restricts the ability of licensees to master hot-section technologies for jet engines, electronic warfare suites, and certain software codes (Lorell, 1995: 77–79; Bitzinger and Kim, 2005: 192). Therefore, nations that produce U.S. weapons under license remain dependent on the delivery of components from the United States and face substantial shortfalls in their defense-industrial capabilities should they attempt to progress to the autonomous design *and* production of weapons systems.

The shortcomings of licensed-production have led many of the United States friends and allies to pursue a greater degree of defense-industrial autonomy via the development and production of weapons combining substantial U.S. technical inputs with a greater degree of indigenous design activity. The attraction of developing indigenous weapons with significant U.S. inputs is that it permits nations to maintain skilled design teams, develop systems integration skills and accrue experience in the management of complex programs. Since these defense-industrial skills are considered to be the most strategically important and also those that involve the greatest economic value-added, many nations are willing to go to significant lengths to develop and/or preserve them. In principle, a policy of developing indigenous weapons based on U.S. technologies should permit nations to achieve a greater degree of defense-industrial autonomy than permitted by licensed production, while incurring a fraction of the R&D costs necessary for the development of a wholly indigenous weapon.

To consider, for example, just the domain of combat aircraft, Japan's F-2 fighter, South Korea's T-50 training/strike aircraft, Taiwan's F-CK-1 fighter and Sweden's Gripen fighter are all present-day examples of this phenomenon. In the first three cases, Asian nations' "national champion" firms forged partnerships with U.S. prime contractors to develop aircraft for their domestic markets (Lorell, 1995; Bitzinger, 2011: 441–442; Steuer et al., 2011: 95–98). Japanese and South Korean aircraft for which information is available involved U.S. partners and sub-contractors contributing approximately 40 percent of the aircrafts' components (as measured by value) (Chinworth, 2000: 386; Bitzinger, 2011). In Sweden's case, the state's national champion, SAAB, acted as the Gripen's sole prime contractor and systems integrator but purchased many of the aircraft's sophisticated subsystems from the United States (as well as the United Kingdom and France) (Andersson, 1989: 54–55; Keijsper, 2003: 34–43). In certain respects, all of these projects *can* be qualified as successes because nations succeeded at domestically developing weapons systems that would normally have been technically and financially beyond their reach.

Nonetheless, the development of indigenous weapons systems based on U.S. technology has proven both significantly more costly than mere licensed-production and much less valuable from a defense-industrial perspective than building completely indigenous weapons systems. Because projects based on U.S. technology involve designing new weapons systems, nations must bear the economic burdens and technical risks inherent in defense-industrial R&D. However, because the United States government reserves the right to prohibit the export of products based on its technology, it is difficult for these nations to utilize arms exports to amortize their R&D

expenditures adequately over longer production runs. As a consequence, the indigenous production of armaments based on U.S. technology frequently results in weapons whose per unit program costs markedly exceed those of either direct imports or licensed-production. For example, Japan's F-2 fighter, which is based on the United States' F-16, reportedly provides only marginally better performance than its cheaper U.S. counterpart and costs Japanese taxpayers three times more per unit than if they had bought the latest models of F-16s directly from U.S. production lines (Chinworth, 2000; Steuer et al., 2011: 98).

Independent Arms Producers' Struggle to Survive

Because of the disadvantages inherent in designing weapons based on U.S. technological inputs, many nations consider it necessary to design and produce weapons systems on an entirely indigenous basis. The principle advantage of developing weapons indigenously lies in the superior degree of defense-industrial autonomy it confers. A state that is self-sufficient when it comes to the production of modern weaponry is: 1) not subject to the shifting policies of arms exporters, 2) comparatively immune to arms embargos, and 3) capable of exporting its wares to whomever it pleases (Krause, 1992). Because of these perceived benefits, the indigenous production of armaments remains a goal cherished by many nations. It receives additional support in certain nations because of their "techno-nationalist" identities held by political elites and populations alike (Samuels, 1994). As a consequence, in addition to China and Russia, which cannot reliably import U.S. weapons, nations as diverse as France, India, Israel, Italy and the United Kingdom strive to develop and produce at least some entirely indigenous weapons systems.

However, despite its advantages in terms of defense-industrial autonomy, the development and production of indigenous weapons systems is a challenging and costly process. Unlike other policies, developing weapons indigenously obliges nations to bear all of the considerable R&D costs and risks involved in high technology projects. In many instances, governments that embark on this course of action ultimately discover, to their chagrin that they have overreached in attempting to develop weapons whose complexity exceeds their available supply of highly-educated human capital and the capabilities of existing national R&D institutions. India is a case in point. Having launched a number of armaments projects in the 1980s and 1990s, India's ambitious effort to develop major weapons systems proved premature. While India's much vaunted Tejas fighter is already 12 years behind schedule and has already consumed an R&D budget two times larger than originally anticipated, the Arjun tank has not yet entered service more than three decades after the project began (Gupta, 1990; Bitzinger, 2011: 435–438).

Even when nations *can* indigenously develop weapons systems, the financial effort required to do so exceeds what is needed to produce weapons under license or even develop them based on U.S. technology. For example, France's indigenous development of the Rafale fighter and the exclusively French sub-systems comprising it required an R&D budget four times larger than Sweden needed to develop the comparable Gripen fighter, which incorporates many sub-systems purchased from the United States and other foreign suppliers (Hartley, 2001). The high R&D costs that governments must invest to indigenously develop and produce weapons renders it imperative that they produce sufficient quantities of the final product to amortize projects' high sunk costs. Invariably, because foreign arms producers lack internal defense markets comparable to the United States', they can attain adequate production runs only by exporting a disproportionately large share of the arms they produce.

However, the basic economics of arms production renders it difficult for foreign arms producers to achieve the volumes of arms exports needed to sustain their indigenous development and production. While the inadequate size of these nations' domestic markets necessitates their exporting a disproportionately large share of the arms they produce, the larger size of the United States' domestic market ensures that U.S. arms producers enjoy competitive advantages in terms of the quality and price of their equipment. As a consequence, under *ceteris paribus* conditions, U.S. arms producers can usually offer weapons that are more innovative and cost-effective than the competition.

Therefore, the question must be posed as to how certain foreign arms producers succeed at exporting a higher percentage of the arms they produce despite laboring under competitive disadvantages in terms of the quality and cost effectiveness of the products they can offer? The answers to this question lie in governmental policies of: 1) catering to markets where U.S. products are either unavailable or politically unwanted; 2) going to greater lengths to meet the military requirements and delivery schedules of export clients; 3) providing various indirect subsidies for arms exports, and 4) providing more advantageous options for clients to finance their purchases.

One of the principal reasons for the comparative success of certain arms exporters lies in the restrictiveness of the United States' own export procedures. At present, the United States' State Department explicitly prohibits lethal arms exports to 23 nations for humanitarian or political reasons, and the government regularly denies specific requests from many more. This group of nations, which includes China, Cuba, Iran, Libya, Myanmar, Syria and Venezuela, amongst others, comprises a sizeable arms export market for which U.S. arms producers cannot compete. (U.S. Department of State, 2011) Ultimately, the United States' unwillingness to export armaments to these nations provides other arms producers with a precious opportunity to sell their goods abroad. For example, a close examination of sales by the world's second largest arms exporter, Russia, reveals that the bulk of its business has been conducted with nations that cannot import U.S. (and, in many cases, European) weapons (Kalinina and Kozyulin, 2010: 33–34). Although its export volumes are more modest, China's burgeoning role in the arms trade is likewise based on the United States' unwillingness to export weapons to nations such as Iran, Myanmar, North Korea, Sri Lanka, Sudan and Zimbabwe (Medeiros and Gill, 2000; Michel and Beuret, 2008: 221–235; Boutin, 2009). Israel, which is perhaps the world's most export-dependent arms producer, has also partly built its success on sales to nations that the United States has blacklisted (e.g., Angola, China, Congo/Zaire, Ivory Coast and Sri Lanka) (Clarke, 1995; Berghezan and Richard, 2002; Hofnung, 2006: 88–89; Reyntjens, 2009: 211).

Even when the United States does not categorically ban nations from importing U.S. arms, its perceived unreliability as an arms exporter has provided other arms producers with valuable commercial opportunities. Such is particularly the case when arms importers adopt a policy of limiting their dependence on the United States by acquiring weapons systems from multiple foreign suppliers. In this fashion, the decisions of Greece, Saudi Arabia, Taiwan and Pakistan to supplement their purchases of U.S. weapons by buying equivalent equipment from other suppliers has provided otherwise unattainable commercial opportunities for the British, French and Chinese producers fortunate enough to become the "second-source" for the categories of arms in question (Phythian, 2000: 188–258; Carlier, 2002: 243–261).

While the United States' self-imposed restrictions on arms exports are one of the reasons other producers can export disproportionately large shares of the arms they manufacture, another lies in the greater lengths that other arms producers will go in satisfying the military requirements and delivery schedules of export clients. U.S. weapons are built to meet the exacting specifications of the United States' armed forces and enough weapons must be produced to meet U.S. military demands before foreign clients can expect the delivery of weapons produced in the United States. Because of the primacy accorded to the technical and delivery needs of the United States' armed forces, would-be customers frequently discover that U.S. products are either too specialized for their needs or cannot be delivered when clients desire. One of the reasons for this phenomenon is that most major arms importers cannot afford or maintain the entire panoply of specialized equipment deployed by the United States and other great powers, but need simpler multi-purpose weaponry (Kaldor, 1983). As a consequence, lightweight fighters, multi-purpose frigates and simple armored personnel carriers tend to sell better in international markets than stealthy strike aircraft, anti-air warfare (AAW) destroyers and infantry fighting vehicles.

Although the United States can afford *not* to take export markets into consideration when procuring equipment for its armed forces, other arms producers do not have the luxury of behaving in a similar manner. For example, since the British government's adoption of the recommendations contained in the so-called Stokes Report of 1965, the United Kingdom's official policy has been to develop arms with the demands of export markets in mind and to satisfy foreign clients' delivery timetables by diverting arms, if necessary, from Britain's armed forces (Phythian, 2000: 58–69). Although France has never officially announced a policy equivalent to the United Kingdom's, an examination of its activities reveals that its government relies on similar principles to boost arms sales. For example, not only have export possibilities repeatedly shaped the specifications of the weapons France procures, but the state also provides special subsidies (estimated at $22 to 26 million [current monetary values] per annum in 1998) for French defense industries to develop or adapt products destined *exclusively* for export (Kolodziej, 1987; Sandler and Hartley, 1995: 251–53; Hébert, 1998). When the rapid delivery of equipment has proven crucial to the signing of export contracts, France's government has not hesitated to supply clients with equipment originally destined for France's armed forces. The most blatant example of this phenomenon occurred in 1983 when it diverted 10 percent of the French Navy's inventory of Super Etendard attack aircraft to Iraq (Hébert, 1998: 73).

In addition to going to great lengths to meet the military requirements and delivery schedules of export clients, many arms producing nations also provide a range of indirect subsidies to aid their industries in the struggle for export markets. One form of export subsidy lies in the administration of government-to-government sales. Whereas the United States' Foreign Military Sales (FMS) program operates on a no profit, no loss basis and charges foreign clients 3.8 percent of a contract's value to administer the sale, both the United Kingdom and France operate equivalent programs (run respectively by Britain's Defence Exports Sales Organisation [DESO] and France's General Delegation for Armament [DGA]) that *do not* charge customers for their services. Consequently, British and French taxpayers ultimately pay for their governments to administer foreign arms sales (Hébert, 1998: 40–41; Hartley, 2000: 449).

In addition to subsidizing the administration of sales, governments in countries with state-owned defense industries can also boost arms exports by permitting public corporations to sell weapons at a financial loss. France, for example, has sold South Africa Puma helicopters and the United Arab Emirates Leclerc tanks at a net economic loss (Hébert, 1998: 72–73). Although the precise magnitude of indirect arms export subsidies has never been precisely calculated, estimates for nations such as France and the United Kingdom plausibly range in hundreds of millions of dollars per annum (Hébert, 1998: 35–42; Martin, 1999: 34–35).

One particularly important form of subsidy that arms producers employ to encourage exports lies in the mechanisms whereby arms sales are financed. Major weapons purchases represent large and long-term investments on the part of nations. Within this context, many nations, especially developing ones, cannot afford to pay for weapons upon delivery and require financing to spread payments over the life span of the equipment they intend to purchase. As a result, these nations attach equal weight to the availability and quality of the credit they are offered as to the price and performance of weapons in deciding which arms to import. When an arms producer cannot offer adequate financing options, its products are generally excluded from competitions, regardless of their technical merits (Johnson, 1994).

In many instances, commercial banks are unwilling to offer the unsecured loans for weapons purchases that clients demand. The reasons for banks' uneasiness with these types of financial transactions are easy to understand. Many large arms purchasers have historically been nations where a high degree of political instability coincides with the precarious management of state finances (e.g., Argentina, Indonesia, Iran, Iraq, Nigeria, Libya and Pakistan). As a result, the risks of arms purchasers defaulting on their loans is comparatively high. To make matters worse, banks and other commercial lending institutions fear the potential negative publicity that could result from financing arms purchases to repressive or expansionist governments (Johnson, 1994).

To compensate for commercial banks' hesitancy to provide credit for arms exports, many arms exporters offer financing through official (i.e., state managed and/or insured) credit institutions that either provide loans directly or guarantee commercial loans extended to would-be arms purchasers. Britain's Export Credits Guarantee Department (ECGD), France's Compagnie Française d'Assurance pour le Commerce Extérieur (COFACE) and the German HERMES Kreditversicherungs AG all serve such a role in providing financing for arms exports (Johnson, 1994; Hébert, 1998: 37–39; Phythian, 2000: 77–79). Although the United States' Export-Import Bank plays a similar role as do non-military exports, it has not provided financing for arms exports to developing countries since 1968, and has not offered credit for exports to developed countries since 1974 (Johnson, 1994: 114).

The absence of an official credit institution willing to provide financing for arms exports places U.S. defense industries at a competitive disadvantage against foreign rivals when it comes to winning contracts with many developing countries. Contrarily, other arms producing nations have exploited their ability to provide export financing to win orders in the face of more cost-effective U.S. products. For example, France won: a large helicopter contract with India because it provided an exceptionally low interest rate (2.5 percent) over an extremely long repayment period (28 years); a helicopter deal with Brazil based on its offer of export credit worth 185 percent the contract's value; and a naval deal with Saudi Arabia because of its willingness to accept a down-payment representing only 0.5 percent of the contracts' value (rather than the customary 15 percent) (Hébert, 1998: 52–53).

However, although effective at stimulating exports, governments' practice of extending loans or loan guarantees to arms importers has proven costly for taxpayers in nations that employ this practice. The arms sales financing activities of both France's COFACE and Britain's ECGD have consistently generated annual deficits measured in the tens or hundreds of millions of dollars (Hébert, 1998: 37–39; Phythian, 2000: 77–79). During the 1990s alone, British taxpayers were obliged to cover loan defaults on defense exports of: £253 million to Jordan; £98 million to Algeria; £46 million to Egypt; £16 million to Kenya; and £11 million to Indonesia (Phythian, 2000: 78–79).

In sum, while it is possible for certain nations to maintain a high degree of defense industrial autonomy through large-scale arms exports, such policies impose heavy direct and indirect costs on the nations that pursue them. In order to export a sizeable proportion of the armaments they produce, nations must: subordinate their own military requirements to the dictates of export markets; be willing to export weapons without political strings attached to virtually any state capable of paying for them; and provide a variety of (direct and indirect) subsidies for the export of weaponry. Even when nations are prepared to take all of the above steps, reliance on a high volume of arms exports exposes nations' defense-industrial bases to the risk that sufficient contracts simply might not be won in the face of stiff international (frequently U.S.) competition. As a consequence of the costs and risks of maintaining an export-based defense-industrial base, a number of once significant arms producers, such as Brazil and South Africa, have largely abandoned the indigenous development and production of major weapons systems (Conca, 1998; Bitzinger, 2003). Certain analysts have even raised questions about the sustainability of the export-based defense-industrial model for nations such as Russia and France (Fontanela and Hébert, 1997; Kalinina and Kozyulin, 2010).

THE UNITED STATES AND THE ARMS MARKET

Through its decisions about what weapons to develop and whom to export them to, the United States government plays a crucial role in structuring both the international arms market and the defense-industrial policies of other nations. The reason for the United States' overseas defense-industrial impact lies in its unrivaled domestic expenditures on defense R&D and procurement, which enables U.S. arms manufactures to offer new technologies for export earlier and more cost-effectively than other suppliers. However, although the United States is the world's largest arms exporter and accounts for 27 to 58 percent of the world market, its' market share understates its true impact on the international arms market and defense industries world-wide.

Unlike most other nations, the United States' arms export decision making process frequently denies sales to nations capable of paying the full cost of weapons for humanitarian or political reasons. Because the United States is perceived as a restrictive and unreliable exporter, other nations are frequently willing to sacrifice much in terms of the economic efficiency of their procurement activities in order to achieve a greater degree of autonomy from U.S. imports. Diversified purchasing, manufacturing U.S. weapons under license, and domestically developing weapons based on U.S. technology are all common, yet costly responses to U.S. arms export policies. Since the latter two policies involve substantial U.S. inputs, which are not generally included in arms export statistics, the United States' true share of the international production of armaments is far greater than export statistics suggest.

Ultimately, the fact that U.S. products possess decisive advantages in terms of cost and performance renders the path difficult for those nations that attempt to leverage arms exports as a means of sustaining autonomous defense-industrial bases. In general, only through comparatively unrestrained arms exports and a range of indirect export subsidies can most nations achieve the sales volumes needed to sustain the indigenous development and production of major weapons systems. However, while necessary, such policies cannot guarantee the success of an export-driven domestic defense-industrial base. Relying on exports to achieve adequate production runs is intrinsically risky because it involves achieving a volume of sales that is both large and predictable in a market that is highly competitive and where demand is extremely volatile.

U.S. DEFENSE BUDGETS AND THE INTERNATIONAL SYSTEM

Arms Races and Burden-Sharing

Although nations' defense budgets are established through domestic political and administrative processes, a variety of interactions can occur between different nations' budgeting decisions. In fact, there are powerful reasons why similar trends should be observable in all great powers' defense budgeting decisions. Because governments develop military forces in response to capabilities possessed by potentially hostile nations and allies collaborate in meeting threats to their mutual security, an increase (or decrease) in the defense expenditures of one great power might logically trigger a response from rivals and allies alike.

This section will examine the extent to which such dynamics can be observed in the present international environment and whether U.S. defense budgets can plausibly be characterized as either responses to or drivers for other nations' decisions about the proportion of their national resources that should be dedicated to defense. Within this context, two particular types of interactions between nations' defense budgets-arms races and burden-sharing amongst allies will be examined. After demonstrating that the evolution of U.S. defense budgets has been largely unconnected to those of its allies and rivals, this section will discuss plausible explanations, differing threat perceptions, allied free riding and domestic politics, for the absence of greater interactions between nations' defense budgeting decisions.

Historically, one of the most common forms of interaction between different nations' defense budgets has been so-called "arms races." Arms racing can best be conceptualized by an action-reaction dynamic wherein the decision of one state to invest more on its armed forces will trigger an equivalent response from other nations (Richardson, 1960; Buzan and Herring, 1998: 75–100). In theory, arms races can occur even when nations do not harbor hostile intentions towards one another. Such is the case because of the dynamic known as the "security dilemma," whereby one state's efforts to strengthen itself defensively relative to perceived threats makes other nations feel less secure. The tendency for even defensive military investments on one state's part to alarm others can be explained by the difficulty of distinguishing between offensive and defensive military preparations, and the impossibility of knowing another state's future behavior. As a consequence, a state will likely respond to another state's defensive military preparations by investing more in their own armed forces, which can result in the first state feeling more threatened than was initially the case (Schelling, 1966; Jervis, 1978).

Scholars have identified several historic arms races that closely correspond to this action-reaction model. Amongst the oft-cited examples are the pre-World War I Anglo-German naval competition, the Cold War arms race between the Soviet Union and the United States, and the Arab-Israeli arms race prior to the 1978 Camp David Accords (Weir, 1992; Friedman, 2000; Freedman, 2003). In each case, nations' defense budgeting decisions, weapons acquisitions and force structures were driven by rival nations' actual and anticipated actions. In principle, it is even possible for asymmetric or "offense-defense" arms races to occur in which one state attempts to establish a plausible offensive option against an opponent, which responds by striving to maintain a credible defensive posture (Wolfson, 1968). Both the strategic relationship between neutral Scandinavian nations (Finland and Sweden) and the Soviet Union during the Cold War and the hypothetical competition between one state's ballistic missile defenses and another's nuclear deterrent forces are examples of this phenomenon (Roberts, 1976; Yanarella, 2002). Because arms races are frequently asymmetric in nature, the best evidence that an arms race is in progress lies in the similarity of two nations' defense budgeting trends (i.e., whether their budgets rise and fall at approximately the same time), rather than similar absolute levels of expenditure (Wolfson, 1968).

Since the end of the Cold War, speculation and official statements (particularly the United States' 2002 National Security Strategy and the 2008 National Military Strategy) alike have focused on China, Russia and, to a lesser degree, India as potential participants in and/or instigators of arms races with the United States (Office of the President, 2002; Department of Defense, 2008). If such is the case, then these nations' decisions about what proportion of national resources to spend on defense should correlate closely with defense budgeting trends in the United States, which would suggest that there is a cause-effect relationship between each party's defense budgeting decisions.

While arms racing is one way the United States' defense budgeting decisions might be linked to those of other nations, alliance burden-sharing is another. Military alliances, whereby nations combine forces to further their mutual security, date back to the earliest chronicles of international relations (e.g., the Amarna letters, Homer's Iliad and Herodotus) (David, 2000). Because alliances aim to supply a collective good shared by all members—security—their proper functioning depends on agreements to share the burden of providing for a common defense. Both the Franco-Russian defense consultations resulting from the nations' 1893 bilateral alliance and North Atlantic Treaty Organization's (NATO) activities during the Cold War resulted in tangible examples of burden-sharing amongst allies. In the latter case, individual nations' commitments to provide specific numbers of army divisions and air wings and, after the 1978 North Atlantic Council (NAC) meeting, the agreement of all member nations to implement a three percent increase in defense spending constitute notable instances of NATO burden-sharing (Sandler, 1987).

At present, the United States stands at the center of an unprecedentedly broad and complex network of alliances. Within this context, the United States' alliances in Europe and Asia are particularly important because they tie the United States to many of the world's other great powers and are best situated for containing potentially revisionist nations. In Europe, NATO remains the lynchpin of an alliance structure that the United States has helped pioneer and lead since 1949. Remarkable for both its duration and degree of institutionalization, NATO today counts 28 member nations, including all of Europe's militarily significant nations (e.g., France, Germany, Italy, Poland, Spain, Turkey and the United Kingdom), with the sole exception of Russia. Compared with other alliances, NATO possesses sophisticated institutions for promoting burden-sharing, including regular NAC meetings, a collectively-financed NATO infrastructure program and, most importantly, an annual review process that subjects each member's defense program to the scrutiny and criticism of its allies and NATO's international staff (Sandler and Hartley, 1999: 24–41).

Although the United States does not possess an overarching alliance organization equivalent to NATO for Asia, it nonetheless has long maintained alliances with many of the region's key nations. Within this context, the United States' security relationships with Australia, Japan and the Philippines date back to 1951, South Korea's to 1953, Thailand's to 1954 and Taiwan's (in its present form) to 1979. If the United States' European and Asian allies share its threat perceptions and are actively sharing the burden of providing collective security, then these nations' defense budgeting decisions should mirror the United States.

The following pages will first examine defense expenditures in Europe and then Asia to ascertain whether U.S. defense budgeting decisions can be linked to arms races and/or burden-sharing in either region.

United States and European Defense Budgets

Throughout the second half of the 20th century, Europe became the focus of the United States' longest and most intense peacetime security commitment ever. In effect, through the creation of NATO, the United States assumed the military leadership of a coalition of Western European nations dedicated to containing the Soviet Union. Compared to alliances preceding it, NATO achieved an unprecedented degree of political and military integration, which has contributed to the alliance's remarkable longevity. As a consequence, this transatlantic alliance enabled the United States, Western Europe and Canada to achieve a reasonable level of security relative to the Warsaw Pact at a political and economic cost that was acceptable to modern democratic nations (Kaplan, 1999). Because of Europe's past centrality to U.S. security, as both a source of committed allies and significant threats, many observers expect Europe to continue to be the region where arms races with rivals and burden-sharing amongst allies will be most common.

In terms of arms races, certain journalists and policy analysts alike have recently highlighted Russia's authoritarian political system, willingness to use force and conflicting interests with the west as proof that a new cold war is in the offing (Brzezinski, 2007; Lucas, 2008; McLaughlin and Mock,

2009). To make matters worse, Russia's leaders have repeatedly threatened that a new arms race or cold war would ensue if NATO did not acquiesce to Russia's policies in the Caucasus or accommodate it in terms of ballistic missile defense (Harding, 2007; Blomfield and McElroy, 2008). Given a combination of this posturing and Russia's suspension since 2007 of its participation in the Conventional Forces in Europe (CFE) Treaty, it is natural to examine both whether Russia's current defense budgeting decisions are driven by a desire to compete with the United States and whether the United States' decisions are still motivated by the need to contain Russia.

While some observers believe Russia could spark an arms race in Europe, others view NATO's European members as partners in sharing the defense burden needed to render the world secure for the United States and Europe alike. Believers in the reality of transatlantic burden-sharing can point to NATO's remarkable resilience since the end of the Cold War. In effect, far from dissolving after the collapse of the Soviet menace it was designed to combat, NATO has expanded both its membership and missions. In terms of membership, the alliance has grown from 15 members nations in 1989 to 28 today—an accomplishment that largely consisted of incorporating into NATO nations that had formerly belonged to NATO's former rival, the Warsaw Pact (DeHart, 2008). In terms of mission, NATO has gradually transitioned from an organization dedicated to the defense of its members' territory to a broader agenda of peacekeeping and crisis management (North Atlantic Council, 1999). Over the course of successive interventions in Bosnia-Herzegovina (1995), Kosovo (1999), Afghanistan (2001) and Libya (2011), NATO demonstrated its capacity to undertake the new missions.

In certain respects, NATO appears more solid today than at any time in the past. Within this context, experts have observed a notable convergence in the published national strategies and defense policies of NATO member nations (Serfaty and Biscop, 2009). One of the most notable developments has been Germany's slow emergence from the pacifist shell it forged after the Second World War to play a more active role in NATO's foreign interventions. Progressing incrementally from providing medical aid in Bosnia to conducting counterinsurgency operations in Afghanistan while increasing the professional component of its armed forces, Germany has gradually become a full partner in NATO's new missions (Kümmel, 2006; Noetzel, 2010). Likewise, by choosing to rejoin NATO's integrated force structure in 2008, France's government put an end to the diffident stance that President Charles de Gaulle adopted towards NATO in 1966 and, thereby, re-incorporated one of Europe's premier armed forces into this American-led alliance (Cameron and Maulny, 2009; de Russé, 2010). As a sign of the growing strategic concord between the United States and its European allies, NATO's heads of government adopted a new joint strategic concept (i.e., "Active

Engagement: Modern Defense"), highlighting their agreement on key strategic issues (North Atlantic Council, 2010).

Part and parcel to the strategic beliefs shared by NATO governments on both sides of the Atlantic has been the regular participation by NATO's European members in American-led military interventions. Since the end of the Cold War, *all* of the United States significant military interventions (e.g., the 1991 Gulf War, Somalia, Haiti, Bosnia-Herzegovina, Kosovo, Afghanistan, the 2003 Iraq War, the Horn of Africa and Libya) have featured the participation of at least some European nations. With nearly 70,000 military personnel deployed abroad at any given time, most of whom are participating in joint operations with the United States, Europe's NATO members are the United States' most significant allies when it comes to projecting power overseas (EDA, 2009; IISS, 2011). Considering that the United States' allies in the Asia-Pacific region collectively contribute fewer than 5,000 troops to United States-sponsored operations, Europe's role as the United States' principal purveyor of deployable military forces looms even larger (IISS, 2011). Given an ostensibly shared strategic vision and numerous joint military operations, the defense budgets of the United States' European allies should be examined to ascertain whether they exhibit the similar trends to the United States', which would demonstrate both effective burden-sharing and a shared threat perception.

Figure 12.4, below, illustrates the defense budgeting trends of Europe's six largest defense-spenders in terms of aggregate annual expenditures,. To ascertain whether arms races and/or burden-sharing is occurring, these budget trends must be compared to the United States', which was illustrated in Figure 12.1 towards the beginning of this chapter. To recapitulate, the United States defense budget declined gradually between 1992 and 1998; increased gradually between 1999 and 2001, and has increased dramatically since 2002.

As may be seen in this figure, America's European allies *do not* exhibit similar trends in defense budgeting to the United States. Within this context, the key divergence between defense budgeting trends in the United States and its European allies can be traced back to 2002. Although NATO's European members invoked Article 5 of NATO's founding Washington Treaty, in solidarity with the United States, following Al Qaeda's September 11, 2001 attacks on the United States, the United States' European allies have proven unwilling to increase their own defense budgets in a manner consonant with the United States'. In fact, when defense spending is examined as a proportion of nations' gross national products (GNP), as is done in Figure 12.5 below, it becomes apparent that the defense efforts of most of America's European allies continued to steadily decline, even as the United States' experienced massive increases. As a consequence, it can be concluded that the United States' European allies are not contributing

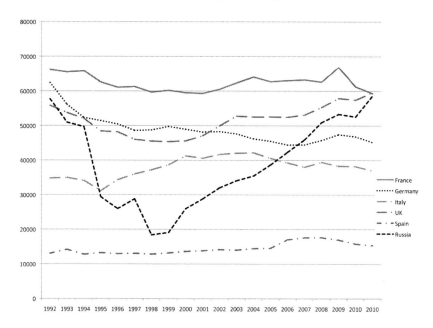

Figure 12.4 European Defense Expenditures Compared. (Figures in Billions of Constant 2010 Dollars)

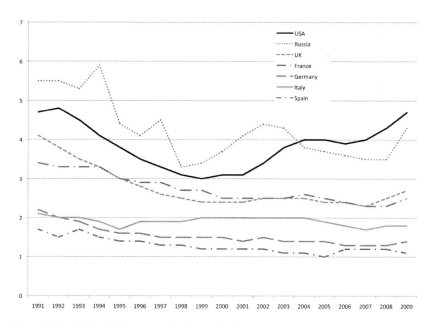

Figure 12.5 US and European Defense Budgets as a Percentage of GNP.

additional resources to share a security burden that U.S. policymakers contend has become more onerous since September 2001.

While Europe's NATO members are clearly not sharing the United States' increased defense burden, is Russia engaging in arms racing behavior against the United States? Is the Cold War arms race reemerging? Superficially, such a case might be made because both U.S. and Russian defense budgets experienced sharp upswings at approximately the same time. However, the beginning of large Russian budgetary increases can be traced to 2000, which preceded the United States' defense build-up by two years. Therefore, if there is a causal link between U.S. and Russian military build-ups, then the former must be a result of the latter. However, such a link would be improbable because U.S. spending was nearly 14 times larger than Russian spending even before the United States' build-up began with the 2002 budget.

To further examine the relationship between U.S. and European defense budgets, Figure 12.5 examines budgeting trends in terms of the percentage of nations' national wealth, as measured by GNP, which has been devoted to defense.

By analyzing defense expenditures as a proportion of GNP, rather than in absolute terms, this figure demonstrates the absence of arms racing behavior between Russia and the United States. Rather than reflecting a desire to match or counteract U.S. military capabilities, much of Russia's arms buildup since 2000 can be explained by the changing fortunes of Russian state finances, which are highly dependent on oil and gas prices.[12] Therefore, even though Russia's defense budget has increased significantly in monetary terms since 2000, the long-term trend has actually been one of Russia devoting a smaller percentage of its national resources to defense. In fact, the years since 2004 represent the first period in recorded history when the United States has dedicated a larger proportion of its national resources to defense than Russia (or the Soviet Union preceding it).

A detailed analysis of how Russia is spending its defense budgets further proves the absence of a present-day Russo-U.S. arms race. After a decade of chaotic defense budgeting following the Soviet Union's collapse, the bulk of Russia's growing defense budget is now dedicated to reestablishing Russia's status as the preeminent power within the regions its leaders consider Russia's historic sphere of influence. Within this context, maintaining credible military options for the Caucasus, Central Asia and the Soviet Union's former European possessions (e.g., Belarus, Moldova and Ukraine) constitute Russian defense planners' primary objective (Rukshin, 2005). Although it may be debated whether Russia has any right to hegemony in its self-described "near abroad," preserving this state of affairs is a modest ambition and one compatible with the maintenance, rather than modification, of the international status quo. Those resources that have not been dedicated to reasserting Russian preeminence in these regions have been allocated to the urgent task of re-capitalizing a defense-industrial base that was starved of resources for over a decade (Bjelakovic, 2008).

In parallel with these prosaic, albeit needed investments, Russia's attitude towards strategic weapons, which drove the Cold War arms race, has been remarkable for its restraint. In fact, Russia has proven far more proactive than the United States in pushing for further Russo-U.S. arms control agreements. Thus, it was the United States Senate rather than the Russian Duma that prevented the implementation of the second Strategic Arms Reduction Treaty (START II) by refusing to ratify an addendum to the agreement (Woolf, 2006). After the collapse of START II, it was Russia rather than the United States that pushed for additional talks and proposed a dramatic reduction of nuclear forces to a level of 1,550 warheads per state (Woolf, 2010). Compared with these initiatives, the United States' own actions, such as its 2002 release of an offensive "Nuclear Posture Review" and withdrawal from the Anti-Ballistic Missile (ABM) Treaty, have been far less conducive to the continued reduction of Russian and U.S. nuclear forces (Evstafiev, 2007; Hildreth and Woolf, 2010). Nevertheless, Russia has not responded to its arms control disappointments by embarking on an arms race, but has rather proceeded with a modest modernization of its nuclear forces through the slow introduction of Topol-M (SS-27) ICBMs and Project 955 ballistic missile submarines (SSBN) (Podvig et al., 2004).

In sum, no direct relationship is apparent between U.S. and European decisions about how many resources societies should expend on their armed forces. Within this context, America's allies have made no discernable effort to share the larger defense burden that the United States has imposed upon itself since September 2001 and the only regional great power not aligned with the United States—Russia—has not attempted to either match or counter U.S. investments in military power. Given the absence of either arms racing or burden-sharing in Europe, U.S. and European decisions about how much to spend on defense appear to be fundamentally unconnected to one another.

THE UNITED STATES AND ASIAN DEFENSE BUDGETS

If U.S. decisions about how many resources to dedicate to defense are not linked to those of European great powers through either arms racing or burden-sharing, does the same hold true for Asia? While U.S. policymakers long considered Europe the most important region for the United States' security, more U.S. military personnel have fought and been killed in Asia since 1945 than on any other continent. Given this legacy of U.S. military engagement—spanning the Korean, Vietnam and Afghan Wars—it is only natural to examine whether the United States' defense budgeting decisions may be linked more closely to those of significant Asian, rather than European, nations.

In many respects, a comparison of international relations in Europe and Asia provides additional reasons to suspect that the latter region may witness a greater degree of arms racing and burden-sharing than the former. Unlike Europe, which has benefited from a pacifying process of regional integration culminating in the formation of the European Union and a common currency, Asia is still subject to traditional great power rivalries and unbridled nationalism. Because Asia has also recently experienced rapid economic growth, its nations possess both greater resources for waging war and face greater needs for natural resources than was hitherto the case. For these reasons, numerous scholars have argued that Asia will, in the future, likely endure dynamics of inter-state conflict equivalent to those Europe experienced prior to 1945 (Friedburg, 2000; Mearshimer, 2001).

Given this state of affairs, incidents since the end of the Cold War highlight the potential for great power conflict in Asia. These include: China's occupation of Mischief Reef in the disputed Spratley Islands (1994); provocative Chinese missile tests into the waters surrounding Taiwan (the so-called Third Taiwan Straits Crisis of 1995–96); the Kargil War between India and Pakistan (1999); Japan's sinking of a North Korean spy ship in its territorial waters (2001); provocative North Korean missiles tests into the Sea of Japan (2005 and 2007); North Korean nuclear tests (2006 and 2009); and artillery duels between the two Koreas over Yeonpyeong Island (2010). If Asia is more conflict prone today than Europe, then one might expect both a significant degree of arms racing between the United States and its potential Asian rivals, and an elevated level of burden-sharing between the United States and allies eager to collectively achieve a high degree of security.

Scholars and policymakers alike focus on China as the state most likely to engage the United States in an arms race. Having experienced rapid economic growth over the course of three decades, China today possesses both the world's second largest economy and second largest defense budget. However, along with these resources, China is also a state which many scholars characterize as dissatisfied with its current position in international affairs. Resentful of the "unequal treaties" foisted upon it during the 19th century, frustrated with the province of Formosa's (Taiwan) escaping Beijing's control since the communist victory of 1949, possessing 22,000 kilometers of disputed borders, and ruled by an undemocratic elite dependent on nationalism to compensate for its lack of other forms of legitimacy, China allegedly possesses powerful motivations for challenging the status quo in Asia (Wan, 2005; Hongyi, 2009; Buza 2010). Because of the United States' alliances with many of the nations surrounding China (e.g., Japan, South Korea, the Philippines and Taiwan), many observers predict that China's rise will result in an intense Sino-American military competition, if not war (Mearshimer, 2010).

As if to emphasize this possibility, China's biannual defense White Papers single-out the United States for criticism and obliquely state that the Chinese armed forces' principle challenge is preparing to fight a high technology war with the United States (People's Republic of China 2004; 2007; 2009; 2011). For its part, the United States' 2002 National Security Strategy condemned China's pursuit of advanced military capabilities and its 2008 National Military Strategy characterized China as an "ascendant state with the potential for competing with the United States" (Office of the President, 2002; Department of Defense, 2008: 3). To meet this challenge, the latter document emphasized the "need to hedge against China's growing military modernization and the impact of its strategic choices upon international security." Given the fact that U.S. and Chinese armed forces acknowledge one another as potential adversaries, the question should be posed as to whether or not the two nations' defense budgeting decisions are linked by an arms racing dynamic.

While a Sino-American arms race is one way that the defense budgeting decisions of the United States and Asian nations might be linked, burden-sharing between the United States and its allies in the Asia-Pacific region constitutes another. As already mentioned, the United States is connected to Australia, Japan, the Philippines, South Korea, Taiwan and Thailand through bilateral security agreements. Faced with China's growing power and the danger posed by a nuclear-armed North Korea, many of these nations have sought to reaffirm their ties with the United States in recent years. The United States' most powerful Asian ally, Japan, has: collaborated with the United States on ballistic missile defenses since 1998; participated in an ongoing security dialogue with the United States since 2002; deployed troops to Iraq in 2003; and embarked in 2005 on a process designed to improve the ability of U.S. and Japanese armed forces to operate together as an integrated fighting force. As part of deepening its strategic partnership with the United States, Japan has also committed itself to supporting the United States in the event of fighting in either Korea or the Taiwan Straits (Samuels, 2007).

While Japan is exemplary in its pursuit of more robust security options in conjunction with the United States, America's other regional partners have also expressed their growing appreciation for the value of their long-standing alliances with the United States. Australia, for instance, prioritized improving interoperability with the U.S. armed forces in its 1997 Strategic Policy and declared its objective to remain "a highly valued ally of the United States" in its 1998 Defense Review (Department of Defence [Australia], 1997; 1998). Since then, it has contributed credible contingents to American-led coalitions in Iraq and Afghanistan, announced plans to expand its high technology naval and air forces, and reaffirmed in its 2009 Defence White Paper the United States' centrality to Australian security (Department of Defence [Australia], 2009; Ayson, 2010).

South Korea, too, has reaffirmed and expanded its security ties with the United States, gradually transforming the two nations' alliance from a pact designed exclusively to protect against North Korea into a strategic partnership with broader ramifications. This process became apparent in 2000 when U.S. and South Korean leaders declared that their alliance "will serve to maintain peace and stability in Northeast Asia and the Asia-Pacific region as a whole" (Suh, 2009: 129). Since then, South Korea committed itself in 2003 to transforming its own military forces to remain interoperable with the United States, contributed forces to the United States' wars in Afghanistan and Iraq, and has worked with the Pentagon to develop new mechanisms for coordinating how the two nations will militarily respond to crises (Chang-hee, 2007; Suh, 2009). Given the value that certain Asian nations ostensibly place in their alliances with the United States, it is worth examining whether they are also sharing the additional defense burden that the United States has assumed.

To ascertain whether arms races and/or burden-sharing links the United States to the Asia-Pacific region, Figure 12.6, below, illustrates the defense budgets of the United States' potential rivals in this region (China and India)

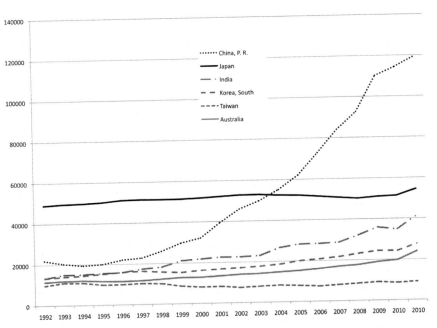

Figure 12.6 Asia-Pacific Defense Expenditures Compared. (Figures in Billions of Constant 2010 Dollars).

and its most significant allies (Australia, Japan, South Korea and Taiwan). To determine how the defense budgeting trends of these nations compare to the United States, readers should re-examine Figure 12.1 presented at the beginning of this chapter.

As may be judged from this figure, two of the United States' key regional allies—Japan and Taiwan—are clearly *not* sharing the burden of increasing U.S. defense expenditures. Although both nations have expressed their attachment to and attempted to build upon their alliances with the United States, neither has increased its defense budgets, which have remained essentially flat (in real terms) since the end of the Cold War. Although none of the other four nations in question exhibit budgeting trends comparable to the United States', what they share in common with the United States is that all have experienced significant growth to their defense budgets. However, whereas the United States' budget expanded dramatically from September 2001 onwards, budgetary growth was comparatively linear and occurred over the course of two decades in Australia, India and South Korea. Of all the Asia-Pacific nations examined, only China's defense budget exhibits a growth curve of equal or greater magnitude to the United States. However, the fact that China's period of spectacular budgetary growth preceded the United States' by four years renders it difficult to deduce a connection between the budgetary developments in the two nations.

To ascertain whether any of the growth in defense budgets in the Asia-Pacific region can be linked to U.S. budgeting trends through either arms racing (for China and India) or burden-sharing (for Australia and South Korea), Figure 12.7 below compares defense expenditures as a proportion of GNP, rather than in absolute terms. By capturing the share of nations' wealth that governments are willing to dedicate to their armed forces, this measure can better reveal the lengths to which nations are willing (or unwilling) to go to strengthen themselves militarily.

By analyzing defense expenditures as a proportion of GNP, rather than in absolute terms, this figure demonstrates the absence of either burden-sharing or arms racing between the United States and great powers in the Asia-Pacific region. As may be seen from this figure, increasing Australian and South Korean defense do not reflect a greater willingness on behalf of these nations to share the United States' defense burden, but rather economic growth, which has permitted these nations to produce greater levels of military force for a degree of national effort that has actually declined over the past two decades.

Likewise, when judged in terms of the proportion of national resources dedicated to defense, the trend in Taiwan's defense effort diverges even more sharply from the United States'. While the United States' defense effort stabilized in 1999 and increased dramatically beginning with the 2002 budget, Taiwanese spending declined steadily as a proportion of GNP until 2007, at

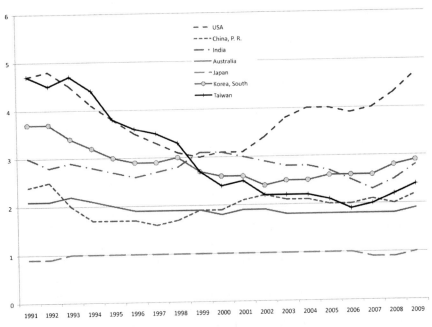

Figure 12.7 US and Asia-Pacific Defense Budgets as a Percentage of GNP

which time Taiwan was spending less than half as much as proportion of GNP on defense as the United States. This reduction in Taiwanese defense spending at a time of rising Chinese defense budgets is arguably rendering Taiwan even more dependent on U.S. assistance than what had hitherto been the case (Liu, 2011; Willner, 2011). In short, as in Europe, the United States' allies in the Asia-Pacific region are *not* sharing the larger defense burden that the U.S. government has imposed upon itself since September 2001.

Likewise, an analysis of defense budgets as a proportion of GNP proves the absence of an arms race between the United States and China (or India for that matter). Although China's defense budget has increased dramatically over two decades, this is a product of China's economic development, rather than a willingness to invest a larger share of national resources to overtake U.S. military developments. An in-depth analysis of Chinese defense policy reinforces this finding. Ever since Chinese Communist Party Chairman Deng Xiaoping articulated his "Four Modernizations" policy in 1978, which relegated defense to the fourth place amongst China's development priorities, the strengthening of China's armed forces has been considered tributary to and dependent on the development of the country's economy as a whole (Pollack, 1980; Deng, 1985).[13]

Having experienced the stifling economic impact of excessive military preparations during the regime of Chairman Mao Zedong, Deng and his successors have kept defense expenditures within such limits that Chinese economic growth would not suffer. Most recently, this concern for not overtaxing the economy was encapsulated in the inclusion of the slogan "prosperous nation with a strong military" into the Chinese Communist Party's platform in 2007 (Lai, 2010). Even though the Chinese government acknowledged in its 2004 defense White Paper that the United States' post-9/11 military build-up was widening the existing military imbalance, it refrained from dedicating a greater share of China's national resources to offset U.S. actions (People's Republic of China, 2004).

In keeping with this policy, China has sought to develop counters to U.S. capabilities that are both asymmetric and affordable, rather than attempting to compete more directly with America's advantages in conventional high technology warfare. It is within this context that China is developing: anti-satellite weapons; "cyber war" capabilities; a large submarine force; anti-ship ballistic missiles; and a substantial conventionally-armed medium-range ballistic missile force (Cole, 2007; Scobell, 2010; Cliff, 2011; Libicki, 2011). However, this concentration on so-called "anti-access / area denial" capabilities must be recognized as one of leveraging a fixed budget to improve China's ability to deter and, if necessary, fight a war with the United States in the Taiwan Straits or Korean Peninsula, rather than an attempt to achieve any form of across-the-board military advantage. That China's current and planned measures are probably insufficient even for this limited objective is openly acknowledged in the pages of Chinese military publications, as is the fact that it will take several decades of uninterrupted economic growth before China can confidently engage in an arms race with the United States (Lai, 2010: 8–11).

In sum, there is no direct relationship between U.S. and Asian decisions about how many resources societies should expend on their armed forces. While America's allies in the Asia-Pacific region have made no additional effort since 2002 to share the larger defense burden that the United States has imposed upon itself, the two regional great powers not aligned with the United States—China and India—have not drawn deeper on their national resources to match America's investments in military power.

EXPLAINING DISCONNECTEDNESS

Considering the pervasive impact of the United States' defense budgeting policies on so many facets of the international security environment, it may appear surprising the that United States' decisions about how many resources to dedicate to its armed forces appear fundamentally unconnected to those of other great powers in either Europe or Asia. However, the findings are quite stark. The United States' potential great power adversaries are not basing their decisions on how many resources to dedicate to their armed forces on a desire to offset (even at a much lower absolute level of expenditure) U.S. budgeting trends. Likewise, the United States' allies have not followed the United States' lead in either lowering or raising their level of defense expenditures. How does one explain the absence of either a measurable degree of arms racing or burden-sharing amongst the United States and the other principal nations in the international system?

For the great powers not aligned with the United States—a category including China, India and Russia—the unwillingness to engage in an arms race with the United States can largely be explained by the size of the U.S. economy, the overwhelming nature of its military expenditures and the value of the arsenal of weaponry it has already accumulated. In effect, it is assumed that the United States' economy is and will likely remain stable (compared to most nations) for some time and hence remain the world's largest. Combined with the United States' fundamental economic strength is the size of its ongoing defense expenditures, which are over five times China's, 11 times Russia's and 17 times India's. Attempting to achieve anything comparable with the United States' degree of military power would require an economically crippling increase in the proportion of national resources these nations dedicate to defense.

Such is especially the case, as from the perspective of a potential arms race, all three nations would be beginning from a position of substantial weakness. With Russia's armed forces having been significantly under-capitalized since the end of the Cold War, many of China's units still being equipped with weaponry of 1960s vintage, and India seemingly incapable of piloting any one of its ambitious defense-industrial projects to a successful conclusion, all three nations face a long road ahead before they will be capable of wholly equipping their existing armed forces with modern weaponry. Given the huge challenges they would face, none of these nations appears eager to repeat the Soviet Union's error of engaging in an open-ended arms race with an economically larger United States. In China, the preoccupation with affordability has prompted investments in asymmetric anti-access and area-denial capabilities for use against the United States in the unlikely eventuality of war. In Russia and India, similar worries about affordability have led to investments designed to enable both nations to assert themselves as regional powers, rather than efforts to offset the United States advantages. Thus, no state appears likely to challenge the United States to an arms race for either the present or the next couple decades.

While a lack of realistic prospects for success explains the unwillingness of the United States' potential rivals to engage it in an arms race, the question must be asked as to why the United States' allies do not exhibit a greater willingness to share its defense burden. Given the rhetorical attachment that all of these nations have expressed for their alliances with the

United States, the demonstrable absence of burden-sharing in budgeting decisions must be attributable to either a desire to *free-ride* on the United States' provision of security or *differing perceptions* about the acuteness of security threats and how to deal with them.

From an economic point of view, alliances should be viewed as providers of collective goods, wherein two or more nations collaborate to provide a good—security in this case—that is shared equally by all members. Because the common security provided by an alliance is the sum of the contributions provided by each ally, members' contributions are, at least to a degree, substitutable. As Mancur Olson and Richard Zeckhauser first argued in 1966, it is this substitutability of contributions that makes free-riding a problem for alliances. The reason for this is that, given a static threat environment, an increase in one member's defense effort creates incentives for its alliance partners to reduce their own contributions as doing so will not result in a net decline in security (Olson and Zeckhauser, 1966; Sandler and Hartley, 1995: 19–51). Empirical studies demonstrate that certain past increases in United States defense spending have produced precisely this phenomenon. For example, high levels of U.S. defense spending during the administrations of Dwight Eisenhower and Ronald Reagan permitted European allies to devote fewer resources to defense, thereby free-riding on the U.S. public's willingness to spend more on defense (Sandler and Hartley, 1999: 37–41). Thus, an ally can under certain circumstances rely on the defense provision of partners to underwrite its own national security.

When analyzing the current security environment, it becomes apparent that the scale of U.S. defense expenditures provides ample scope for allies to free-ride on the United States' provision of security (Kapstein, 1992: 169–72). At a time when the United States is investing nearly $700 billion per annum on defense, the marginal impact of lower spending by allies is comparatively small. It is within this context that high and growing United States defense budgets since the end of the Cold War may have encouraged nations such as Germany, South Korea and Taiwan to economize on the proportion of their national wealth dedicated to security. Over the longer term, Japan has also quite explicitly benefited from a "cheap ride" on security, at least since its government's 1967 decision to freeze defense spending at one percent of GNP (Samuels, 2007: 39–45). Given this state of affairs, if the United States government wants its allies to contribute more to the common defense, it must be prepared to contribute less (Kapstein, 1992: 171–72; Gholz, Press and Sapolsky, 1997).

The practical limit on allies' ability to free-ride on the United States' provision of security is their fear of abandonment. Because an alliance with the United States offers many nations a considerable gain in security compared to what they could achieve on their own, America's allies are generally loath to act in ways that would alienate the U.S. government to such a degree as to jeopardize an alliance's continued existence. For this reason, allies cannot openly shirk burden-sharing, but instead seek to determine the minimum amount they must contribute to preserve their ties with the United States. To this end, allies generally favor those policies that generate a maximum amount of political capital in Washington D.C. at minimum expense over alternatives that may be less appreciated by the United States government. At some level, this is one of the reasons why so many allies are willing to contribute forces to American-led interventions, sign-up for high profile projects such as ballistic missile defense, and place interoperability with the United States high on their agendas (Samuels, 2007: 86–108; Suh, 2009). Given the countervailing pressures of allies' desire to minimize defense expenditures and their fears of abandonment should they contribute too little, certain authors have argued that America's allies are not pursuing a "free-ride" on defense, but rather a "cheap ride" wherein they shift as much of the defense burden as possible to the United States (Samuels, 2007).

Besides a tendency to free-ride on the United States' provision of security, another reason America's allies have not shared its increasing defense burden lies in differing perceptions of threats and how best to meet them. Although the United States' allies ostensibly share its concerns about terrorism, the proliferation of weapons of mass destruction and the emergence of rival great powers, different nations clearly exhibit different degrees of alarm at these developments. For example, while the United States' East Asian allies (i.e., Japan, South Korea and Taiwan) appear comparatively unconcerned with terrorism originating in the Middle East, its European allies are much less worried about the emergence of China as a peer competitor to the United States (Howorth, 2007; Samuels, 2007; Suh, 2009). Even when the United States' allies share U.S. perceptions of a threat, different strategic preferences can generate different patterns of defense budgeting.

A good example of this phenomenon lies in U.S. and allied responses to the danger of so-called "rogue nations" acquiring weapons of mass destruction. Over the course of two decades, the use of preventative military attacks to halt the proliferation of weapons of mass destruction—known as "proactive counter-proliferation"—remains a viable policy option in the United States whose merits have been repeatedly discussed with reference to nations such as Iran and North Korea. In the 2002 National Security Strategy, proactive counter-proliferation was even elevated to the status of an official policy of the United States government (Office of the President, 2002). However, to be successful, a strategy of counter-proliferation must be backup-up by a panoply of specialized high technology weaponry, including: long-range precision weapons; "bunker busting" munitions able to destroy hardened facilities; stealth aircraft capable of evading detection by enemy air defenses; and theater ballistic missile defense capable of fending off retaliatory missile strikes against either U.S. bases or allies.

In contrast to continuing, albeit fluctuating support for counter-proliferation in the United States, many of America's allies have either explicitly or implicitly rejected counter-proliferation as a viable policy. Most explicitly, the 27 member nations of the European Union (23 of whom are also NATO members) ruled out counter-proliferation as an option in their 2003 European Security Strategy, whose publication followed closely on the heels of the United States' 2002 National Security Strategy (European Council, 2003; Howorth 2007: 199–214; Stritzel and Schmittchen, 2011). More subtly, Japan and South Korea have quietly opposed counter-proliferation as a means of halting North Korea's nuclear program (Samuels, 2007: 171–176; Suh, 2009: 121–22). By rejecting the United States' policy of counter-proliferation, its allies also dispensed themselves from needing to acquire the costly military means to enact such a policy.

Thus, despite the United States' broad international role and its alliance connections with a large number of nations, U.S. decisions about how many resources to expend on defense appear to exercise little impact on those of other great powers. Rather than share the United States' defense burdens, allies are happier free-riding on the United States' provision of security and, in some instances, do not agree with either the necessity or wisdom of certain U.S. categories of expenditure. For their part, those great powers not aligned with the United States accept as economically counter-productive any attempt to imitate U.S. trends in defense expenditure. Fundamentally, the reasons for the disconnectedness of the United States' defense expenditures with those of other nations—whether allied or not with the United States—can be traced back to the high level of U.S. expenditures relative to all other nations. In principle, if the United States spent comparatively less on defense, then its allies could be predicted to engage in a greater degree of burden-sharing and its potential adversaries would be more tempted to engage in arms races.

CONCLUSIONS

As this book has demonstrated, defense budgeting and financial management processes in the United States are a product of several distinct, often-times conflicting, imperatives. Because the United States' political system is based on the separation of power amongst different branches of government and civilian control of the armed forces, U.S. defense budgeting procedures should accord adequate voice to a plurality of political and bureaucratic actors. However, since the ultimate purpose of defense budgeting is providing the military capabilities, generated by a combination of weapons systems and the trained personnel to operate them, needed to keep America safe, the defense budgeting process must also produce outcomes that are both efficient and well-integrated with the country's foreign policy. Finally, because U.S. weapons acquisition and arms transfer policies exercise a powerful impact on the military opportunities and constraints facing other nations, the United States' government should approach procurement and export decisions with an understanding of all the direct and indirect effects their actions might have. Having evolved gradually since the 1960s, the United States' PPBES system and arms transfer procedures are sophisticated, albeit imperfect, responses to all of these conflicting demands.

Within this context, it is in their effects on international relations that the impact of the United States' defense budgeting is probably least understood. This final chapter has attempted to fill this analytical lacuna and provide future managers of the United States defense budget with useful insights as to how their actions affect global politics in the broadest sense. To this end, this chapter has successively examined issues such as: the United States' impact on the development and diffusion of new military technologies; the pros and cons of arms exports; the United States' impact on the international arms trade; the question of burden-sharing amongst allies; and the presence or likelihood of arms races. As has already been shown, the impact of U.S. defense budgeting and arms transfer decisions is significant, complex and, at times, counterintuitive. Now, this concluding section will explore in greater depth the policy implications of the findings presented earlier in this chapter.

Because of the scale and nature of its defense budgets, the United States plays a crucial role in the development and diffusion of new military technologies. However, the United States pays a heavy premium, in terms of higher R&D budgets and numerous failed projects, to generate this level of innovation. Other arms producers, by way of contrast, achieve substantial economies through "sub-optimization," which means relying only on proven technologies and pursuing more moderate performance goals in the development of new weapons systems. While the United States' across-the-board drive for innovation was essential during the Cold War, when the Soviet Union contested America's technological lead, is it still necessary today at a time when no other state is either pushing the technological frontier in weapon design or engaging in an arms race with the United States? Or might the United States itself sub-optimize on some future weapons systems? The inability to fund all promising procurement programs to completion (e.g., the Crusader artillery system, Comanche helicopter and DD-21 destroyer) under even the generous post-9/11 defense budgets suggests that such a policy might actually improve America's military strength.

The fact that the United States is the originator of much of the world's new military technology should draw fresh attention to the United States' technology transfer policies. In order to economically develop their own military capabilities, America's allies and rivals alike seek to access its military technology. By transferring such technology to allies, the United States can strengthen nations whose security is, to some degree, interdependent with the United States'. However, every transfer of U.S. technology to a friend or ally comports a risk that the technology will be retransferred to potential rivals. Given a historic record that has witnessed even such close allies as Israel and the United Kingdom (deliberately or inadvertently) transfer U.S. technology to the Soviet Union and China, it can never be known for certain that the intended recipient of a technology transfer will also be its final recipient.[14] Within this context, if technological superiority is so important to U.S. security that its citizens should pay a heavy premium for military innovation, then should not the United States government restrict to a maximum *any* state's access to its recently developed military technology?

While national security concerns militate against arms sales or defense technology transfers, the very scope of the United States' defense-industrial effort creates powerful incentives for an economically liberal approach towards the sale and licensed production of armaments. Not only does the size of U.S. production runs provide its arms producers with cost advantages when it comes to competing for export markets, but the United States' defense R&D effort represents a sizeable diversion of human and financial resources away from the civilian economy. At present, over half of the Federal government funding for technological R&D is devoted to defense and defense contractors employ a sizeable proportion of U.S. engineering talent (Lockheed Martin, for example, is the United States' largest recruiter of graduating engineers) (CBO, 2007; Denney, 2011). Given the concentration of science and engineering resources on defense, the United States should logically maximize its economic return on this defense-industrial investment through a liberal arms export policy. Moreover, partisans of arms exports make the oftentimes correct argument that U.S. export restrictions only serve to create commercial opportunities for other producers, which, in certain cases, are geopolitical rivals of the United States.

In keeping with the United States' pluralistic political system, the task of striking the proper balance between the security arguments against exports and the economic reasons for arms sales is not entrusted to any single body. Rather, both the executive and legislative branches of government, as well as a multiplicity of officials at the State and Defense Departments, determine the merits of each sales requirement on a case-by-case basis. Because the United States' arms export decision making process is comparatively restrictive and can produce unpredictable results, foreign nations are willing to sacrifice much in terms of the economic efficiency of their procurement activities in order to achieve a greater degree of defense-industrial autonomy from the United States. Within this context, diversified purchasing from multiple exporters, manufacturing American-designed weapons under license, and developing indigenous weapons based on U.S. technologies all represent different forms of foreign hedging against unpredictable interruptions in U.S. arms sales.

Motivated, in part, by uncertain access to U.S. defense technologies, certain nations elect to maintain the highest possible degree of defense-industrial autonomy, which consists of the indigenous development and production of weapons systems. Unlike the United States, the survival of these nation's domestic defense-industrial bases depends largely on their ability to export a large proportion of the arms they produce. However, the cost and performance advantages of American-produced armaments render this task fundamentally difficult in those markets where U.S. defense contractors compete for sales.

As a consequence, the world's other armaments producers have a powerful incentive to cater to markets that the United States has embargoed regardless of the political or humanitarian concerns that such sales might generate. When foreign arms producers are obliged to compete directly with the United States for export markets, they must rely on subsidies, a generous provision of financing and a greater flexibility in meeting clients' delivery requirements to compensate for the superior cost effectiveness of U.S. contractors. Thus, the United States government's decisions about when and where to export armaments shapes other arms producers' ability (or lack thereof) to achieve the sales volumes needed to sustain domestic defense-industrial bases.

While America's allies covet its defense technologies and frequently import a large proportion of their armaments from the United States, they have systematically resisted calls to follow the United States example in consecrating a larger proportion of their national wealth to defense. Because the security provided by a military alliance is a collective good whose benefits are shared by all members, a unilateral decision by one state to increase its investment in defense creates opportunities for its partners to reduce their contributions, provided that the aggregate amount of security generated by the alliance is still considered sufficient. It is within this context that the United States' post-9/11 military build-up has encouraged its allies to free-ride (or "cheap-ride") on its provision of security, rather than increase their own contributions as a form of burden-sharing. Thus, many of the United States' allies have cut their defense budgets either in real terms or as a proportion of GNP even as the United States has spent more. This has resulted in such seemingly paradoxical situations as Taiwan reducing its defense effort even as the United States implemented budget increases designed, in part, to contain Chinese ambitions in Asia. Likewise, even as the United States invested significant sums in fighting terrorism across the globe, two allies whom immigration has rendered more potentially vulnerable to Salafi terrorism—Germany and Spain—cut their defense budgets.

The current prevalence of free-riding on the part of many United States' allies has counterintuitive implications for U.S. defense budgeting. Because any cut in U.S. defense budgets reduces its allies' ability to free-ride, lower levels of defense spending in the United States will not necessarily generate a commensurate decline in national security. In principle, allies will compensate, at least in part, for cuts in U.S. defense spending by increases in their own. Therefore, the real security impact of U.S. defense budget cuts will logically be somewhat less than the magnitude of the cuts would suggest. Of course, before using the collective goods theory to justify defense budget cuts, it is necessary to closely examine to what extent U.S. defense spending actually provides a collective good (i.e., security) *shared* by allies. If allies hold fundamentally divergent threat perceptions or are normatively attached to different strategies than the United States, then they will be less likely to substitute their own contributions for a reduction in U.S. expenditures. Thus, the degree of burden-sharing between the United States and its allies will always likely be a function of both the scale of U.S. defense spending and the level of strategic concord prevailing between its allies and itself.

While high levels of U.S. defense spending currently encourage allies to free-ride on its provision of security, it also appears to have dissuaded potential rivals from engaging in an arms race. Although China and Russia view the United States' global presence as a challenge to their own regional ambitions, the United States' existing military advantages and the magnitude of its defense spending is such as to deter either state from increasing the proportion of its national resources dedicated to defense in a vain effort to compete with the United States. Within this context, today's great powers appear to have learnt the lesson of the Cold War's arms race that it is counterproductive to engage in an open-ended defense-industrial competition with a nation far wealthier than one's own. As a result, neither contemporary Russia nor post-Maoist China are nations likely to replicate the Soviet Union's mistakes. Rather, both are keeping their defense expenditures within reasonable limits and are striving to acquire the best mixtures of military capabilities commensurate with their budgets and foreign policy ambitions. In theory, arms races should only reemerge as a characteristic of the United States' relations with other great powers once economic growth and/or a decline in U.S. defense spending permits other nations to compete with the U.S..

Considering the many ramifications of U.S. budgetary choices on the international system, defense budgeting should no longer be viewed as a purely domestic process for funding the equipment, training and high-level of operational readiness needed for the United States' armed forces to effectively enact the government's foreign policy. Rather, U.S. budgeting and arms transfer policies have direct and indirect effects on other nations' decisions about; how many resources to dedicate to their armed forces; what types of capabilities they should develop; and whether nations should import weapons, build them under license or attempt to develop them domestically. At a more fundamental level, U.S. defense budgeting and transfer decisions shape how the international market for armaments functions as well as how new military technologies are developed and diffused throughout the international state system. Only by adapting the United States' defense budgeting and financial management policies and processes to take into account these frequently unanticipated or, at least, underappreciated national security threats and associated risks, can other nations achieve the foreign policy outcomes desired by their governments and citizens.

NOTES

1. The six figures presented in this chapter were produced using the defense spending and export data sets compiled by the Stockholm International Peace Research Institute. For countries such as China and Russia, where defense budgets are not published in a transparent manner, SIPRI employs estimates generated with the use of purchasing power parity (PPP) as a means of correcting for these states' lower labor costs. Overall, SIPRI's budgetary estimates for these states are comparatively close to those of the International Institute of Strategic Studies (IISS), the other major public source for information on world defense expenditures (IISS 2011). See SIPRI Military Expenditure Database 2011, http://milexdata.sipri.org (accessed August 2011).
2. Japan's actual R&D spending may be higher than the budgeted figure because of the tendency of private firms to conduct R&D and then seek reimbursement through production contracts.
3. Data in these sources is in 2010 constant U.S. dollars.
4. The figures for China and Russia include defense R&D expenditures. Those for all other states (including the United States) do not.
5. Paradoxically, the U.S. Expeditionary Force (AEF) of 1918 was largely equipped with weapons bought or borrowed from its European allies. However, the reason for this was not a lack of defense-industrial capacity, but rather the inadequacy of U.S. procurement policies prior to its entry in the war. Since 1915, the United States had exported vast quantities of war material-including European-designed weapons manufactured under license (British Enfield rifles and French Model 1897 artillery pieces), ammunition, explosives, motor vehicles (nearly 3,000 in total), high-quality steel and gas masks-to the Entente Powers. (Porte, 2005: 178–85).
6. In practice, the DOD might lose money on smaller contracts and profit from larger ones because administrative costs are not necessarily proportional to contract values (GAO, 1999).

7. One factor motivating the Air Force was the apprehension of a Japanese officer who had been transmitting Aegis blueprints (a system which the United States had sold to Japan) to China (Sorenson 2009, 134).

8. Specifically, the law prohibits "the use of appropriated funds to approve or license the sale of the F-22 to any foreign government (Bolkcom and Chanlett-Avery 2009, 1)."

9. In certain categories of weapons system, U.S. products face the additional problem of being too costly and sophisticated for the needs of many export clients. Mary Kaldor (1983) has referred to the development of weapons where the marginal cost of additional sophistication is high and its value low as "baroque." A good example of U.S. weapons being "baroque" compared to the needs of many customers lies in warships. In recent years, one of the smallest surface combatants procured by the United States Navy has been the Arleigh Burke-class destroyer, which displaces between 8,000 and 10,000 tons (depending on the model). However, most international demand for surface combatants is concentrated on frigates displacing between 3,000 to 4,000 tons. Therefore, although the Arleigh Burke is a cost-effective weapons system *for its size*, the smaller frigates produced by Germany and France appear to better suit the needs of many foreign navies (Todd and Lindberg, 1996; Sadler, 2007).

10. In terms of restrictiveness, Germany and Japan constitute the two notable exceptions as they both possess arms export legislation that is more restrictive than the United States'. In Germany's case, firms and government officials have developed several subterfuges to circumvent export regulations that normally prohibit lethal arms sales to states that are not members of either the North Atlantic Treaty Organization (NATO) or the European Union (EU). One such subterfuge has been to license foreign companies (frequently in Turkey or the United Kingdom) to manufacture German-designed arms for export (Abel 2000). Another such subterfuge has been to build unarmed platforms in Germany that are then equipped with weapons systems in another state (Sadler, 2007: 63–94). A third and final means of avoiding stringent export regulations has been European collaboration, whereby Germany's partners are allowed to negotiate export contracts that Germany firms cannot (Vetter and Vetter, 2009: 169–79). In contrast to Germany, Japan has scrupulously abided to its post-1967 arms export ban. This has proven exceptionally costly as poor scale- and learning-economies have rendered Japanese-designed weapons amongst the least cost-effective in the world (Chinworth, 2000; Samuels, 2007: 143–48).

11. The evolution of the United States' policy towards licensed-production can be observed in the changing degree of U.S. content in weapons manufactured under license in Japan. The 1960s license for production of the F-4 Phantom II permitted the Japanese to produce 90 percent of the aircraft's content. Later deals, concluded in the 1970s and 1980s, reduced Japan's share of weapons produced under license to 70 percent for the F-15 and 60 percent for Patriot missiles (Chinworth, 1992).

12. The reason for the high degree of volatility in Russia's defense expenditures as a proportion of GNP lies in both the nature of Russian state revenues and

defense budgeting. On the one hand, Russian state revenues are subject to substantial annual variations because they depend on value of raw material (oil and gas) exports. On the other hand, Russia's system of defense budgeting, which was inherited from the Soviet Union and is termed the Arms Program, operates on a five year planning cycle that allows for neither annual updates to the program nor a moveable time horizon. (Zatsepin, 2005)

13. To understand the relationship between economic development and military power that has prevailed in post-Mao China, it is worth citing Deng Xiaoping at length. According to Deng, "The four modernizations include the modernization of defense. But the four modernizations should be achieved in order of priority [i.e., with defense last]. Only when we have a good economic foundation will it be possible to modernize the army's equipment.... If the economy develops, we can accomplish anything. What we have to do now is to put all our efforts into developing the economy. That is the most important thing, and everything else must be subordinated to it" (Deng, 1985).

14. During the 1950s, a Soviet spy ring operating at the Admiralty Underwater Weapons Establishment at Portland, England, passed U.S. anti-submarine warfare technology to the Soviet Union. In the 1960s, the United Kingdom sold Vickers Viscount aircraft to China. Although the Viscount was a British-designed aircraft, the exported aircraft were equipped with U.S. avionics, whose export to China was proscribed by both the United States government and COCOM (the international organization created to control high technology exports to the Communist states) (Bulloch and Miller, 1961; Engel, 2007: 216–51).

BIBLIOGRAPHY

Abel, Pete. 2000. "Manufacturing Trends: Globalising the Source," in Lore Lumpe, ed., Running Guns: The Global Black Market in Small Arms. London: Zed Books: 81–104.

Adelman, K., and N. Augustine. 1990. The Defense Revolution: Strategy for the Brave New World. San Francisco, CA: Institute for Contemporary Studies Press.

Andersson, Hans. 1989. Saab Aircraft Since 1937. London: Putnam.

Annals of Congress. No date.1: 929.

Annals of Congress. No date. 3: 1, 259.

Anonymous. 1911. "Will Congress Put our Navy on the Sea? The Story of Secretary Meyer's Fight Against Waste and Bureaucracy," McClure's Magazine XXXVI: 523–535.

Anonymous. 2010. Head Count: Tracking Obama's Appointments. June 23. The Washington Post http://projects.washingtonpost.com/2009/federal-appointments/

Ashdown, Keith. 2004. "Defense Spending Pork Fest," The Waste Basket, Taxpayers for Common Sense IX/45, 12 November.

Ashdown, Keith. 2005. "Emergency Spending Bill Pays for Political Pork," News Release, Washington DC: Taxpayers for Common Sense.

Ashdown, Keith. 2005. "Emergency Spending Bill Pays for Political Pork," News Release, Taxpayers for Common Sense, Washington, DC.

Asher, Harold. 1956. RAND Memorandum R-291. Santa Monica: RAND.

American Society for Military Comptrollers (ASMC). 2010. Financial Management on the Battlefield: Annual Survey of Defense and Military Department Financial Management Executives. Alexandria, VA: American Society for Military Comptrollers And Grant Thorton LLP.

Anthony, Robert N., 2002. "Federal Accounting Standards Have Failed," International Public Management Journal, 5/3: 297–312.

Financing National Defense: Policy and Process, pages 455–479
Copyright © 2012 by Information Age Publishing
All rights of reproduction in any form reserved.

Anthony Robert N., and David Young. 1988. Management Control in Nonprofit Organizations. 4th edition. Homewood, IL: Irwin.

Appleby, Paul. 1957. "The Role of the Budget Division." Public Administration Review 17 Summer: 156–158.

Augustine, Norman R. 1982. Augustine's Laws. New York: American Institute of Aeronautics and Astronautics.

Ayson, Robert. "Australia's Defense Policy: Medium Power, Even Bigger Ambitions?" Korean Journal of Defense Analysis 22/2 2010: 183–96.

Bartle, J. R. 2001. "Seven Theories of Public Budgeting," in J. R. Bartle, Evolving Theories of Public Budgeting, New York: JAI: 1–11.

Belasco, A. 2010. The Cost of Iraq, Afghanistan, and Other Global War on Terror Operations Since 9/11. CRS Report for Congress, RL33110. Washington, DC: CRS.

Bennett, John T., and Vago Muradian. 2007. Interview Kenneth Krieg, Pentagon Acquisition Chief. Defense News March 26: 30.

Berghezan, Georges, and Pierre Richard. 2002. "Israël, terre promise des mafias ex-soviétiques?" in Georges Berghezan, ed., Trafics d'armes vers l'Afrique: Pleins feux sur les réseaux français et le "savoir-faire" belge. Brussels: Groupe de recherche de d'information sur la paix et la sécurité GRIP: 109-16.

Berman, Larry. 1979. The Office of Management and the Budget and the Presidency, 1921–1979, Princeton, NJ: Princeton University Press.

Bitzinger, Richard, and Mikyoung Kim. 2005. "Why Do Small States Produce Arms? The Case of South Korea," Korean Journal of Defense Analysis 17/2: 183–205.

Bitzinger, Richard 2011. "China's Defense Industrial Base in a Regional Context: Arms Manufacturing in Asia," Journal of Strategic Studies 34/3 June: 425–50.

Bitzinger, Richard. 2003. Towards a Brave New Arms Industry? Oxford: Oxford University Press.

Bjelakovic, Nebojsa. 2008. "Russian Military Procurement: Putin's Impact on Decision-Making and Budgeting,» The Journal of Slavic Military Studies 21/3: 527-42

Black, G. 1971. "Externalities and Structure in PPB." Public Administration Review, 31/6: 637–643.

Blakeman, S. T., A. R. Gibbs, J. Jeyasingam and L. R. Jones. 2010. Arming America at War. Charlotte, NC: Information Age.

Blechman, M. 1990. The Politics of National Security: Congress and U.S. Defense Policy, New York: Oxford University Press.

Blomfield, Adrian, and Damien McElroy. 2008. "Russia 'ready for new Cold War' over Georgia: Russia has said it is prepared for a new Cold War, after President Dmitry Medvedev defied the West by formally declaring the independence of two Georgian rebel regions," The Telegraph August 27: 24.

Bolkcom, Christopher, and Emma Chanlett-Avery. 2009. CRS Report for Congress– Potential F-22 Raptor Export to Japan. Washington, DC: CRS.

Bolles, Albert S. 1896 and 1969. The Financial History of the United States from 1774 to 1789. New York: D. Appleton.

Boutin, J.D. Kenneth. 2009. "Arms and Autonomy: The Limits of China's Defense-Industrial Transformation," in Richard Bitzingerd, ed., The Modern Defense

Industry: Political, Economic, and Technological Issues. Santa Barbara: Praeger Security International: 212–26.

Braddon, Derek. 1995. "Regional Impact of Defense Expenditure," in Todd Sandler and Keith Hartley, eds. Handbook of Defense Economics: Volume 1. Amsterdam: Elsevier: 491–522.

Brook, D. A., and P. J. Candreva. 2007. "Business Management Reform in the Department of Defense in Anticipation of Declining Budgets." Public Budgeting and Finance 27/3: 50–70.

Brook, D. A. 2001. Audited Financial Statements: Getting and Sustaining "Clean" Opinions. Washington: The PricewaterhouseCoopers Endowment for the Business of Government.

Brook, D. A. 2010. Audited Financial Statements in the Federal Government: Intentions, Outcomes and On-Going Challenges for Management and Policy Making. Journal of Public Budgeting, Accounting and Financial Management 22/1: 52–83.

Browne, Vincent J. 1949. The Control of the Public Budget. Washington, D C: Public Affairs Press.

Brzezinski, Zbigniew. 2007. "How to Avoid a New Cold War," Time June 7: 15.

Budget of the United States for Fiscal Year 2006. Historical Tables www.whitehouse.gov/omb/budget/fy2006/pdf/hist.pdf.

Bulloch, John, and Henry Miller. 1961. Spy Ring : The Full Story of the Naval Secrets Case. London: Secker and Warburg.

Burk, James. 1993. "Morris Janowitz and the Origins of Sociological Research on Armed Forces and Society," Armed Forces and Society 19/2 Winter: 167–185

Burkhead, Jesse.1959. Government Budgeting. New York: John Wiley and Sons: 340–356. .

Burlingham, D.M. 2001. "Resource Allocation: A Practical Example," Marine Corps Gazette, 85 January: 60–64.

Bush, George W. 2007. Veto Message from President of the United States. House Document 110–31, May 2.

Business Transformation Agency (BTA). 2009. December. Enterprise Transition Plan. http://www.bta.mil/products/BEA_6.2/ETP/home.html

Business Transformation Agency (BTA). 2010,. Business Enterprise Architecture. March 12 http://www.bta.mil/products/bea.html

Buzan, Barry, and Eric Herring. 1998. The Arms Dynamic in World Politics. Boulder: Lynn Reiner.

Buzan, Barry. 2010. "China in International Society: Is Peaceful Rise Possible," The Chinese Journal of International Politics 3: 5–36

Caiden, Naomi, and Aaron Wildavsky. 1974. Planning and Budgeting in Poor Countries. New York: Wiley.

Caldwell, L. K. 1944. "Alexander Hamilton: Advocate of Executive Leadership," Public Administration Review 4, Spring: 41–64.

Cameron, Alastair, and Jean-Pierre Maulny. 2009. France's NATO Reintegration: Fresh Views with the Sarkozy Presidency. London: RUSI.

Candreva, P. J. 2008. "Financial Management" in Rene Rendon and Keith Snider, eds., Management of Defense Acquisition Projects. Reston, VA: AIAA: 106–121.

Candreva, P. J., and D. A. Brook. 2008. Transitions in Defense Management Reform: A Review of the Government Accountability Office's Chief Management Officer Recommendation and Comments for the New Administration. Public Administration Review, 68/2: 1043–1049.

Candreva, P. J., and L. R. Jones. 2005. "Congressional Delegation of Spending Power to the Defense Department in the Post 9-11 Period," Public Budgeting and Finance 25/4:1–19.

Carlier, Claude. 2002. Serge Dassault: 50 ans de défis. Paris: Perrin.

Carter, Linwood B., and Thomas Coipuram, Jr. 2005. Defense Authorization and Appropriations Bills: A Chronology, FY1970-FY2006. CRS Report for Congress, 98-756C, May 23, Washington, DC.

Caverley, Jonathan. 2007. «United States Hegemony and the New Economics of Defense,» Security Studies 16/4 October-December: 598–614.

Chan, James L. 2003. Government Accounting: An Assessment of Theory Purposes and Standards. Public Money and Management. January: 13–20.

Chang-Hee, Nam. 2007. "Realigning the U.S. Forces and South Korea's Defense Reform 2020," Korean Journal of Defense Analysis 19/1: 165–89.

Cheung, Tai Ming. 2009. Fortifying China: The Struggle to Build a Modern Defense Economy. Ithaca: Cornell.

Chiarelli, Peter W. 1993. "Beyond Goldwater-Nichols," Joint Forces Quarterly, Autumn: 71 –81 http://www.dtic.mil/doctrine/jel/jfq_pubs/index.htm

Chinworth, M.W. 2000."Country Survey XIII: Japan's Security Posture and Defense Industry Prospects," Defence and Peace Economics 11/2: 369–401.

Chinworth, M.W. 1992. Inside Japan's Defense. Washington, DC: Brassey's.

Churchman, C. W., and A. H. Schainblatt.1969. "PPB: How Can It be Implemented?" Public Administration Review 29/2: 178–189.

Chwastiak, M. 2001. "Taming the Untamable: Planning, Programming and Budgeting and the Normalization of War." Accounting, Organizations, and Society 26: 501–519.

Citizens Against Government Waste Pig Book (CAGW). 2005. "2005 Pig Book Exposes Record $27.3 Billion in Pork" Citizens Against Government Waste, Washington, DC, Press Release, 6 April.

Clarke, Duncan. 1995. "Israel's Unauthorized Arms Transfers," Foreign Policy 99 Summer: 89–109.

Clemens, Austin. 2005. "Defense Pork Reaches Record High," Taxpayers for Common Sense Defense Appropriations Database. 2005. Washington, DC: DOD.

Defense Appropriations Act. 2006: SEC 8105. (b), Washington, DC: DOD.

Cliff, Roger. 2011. Anti-Access Measures in Chinese Defense Strategy. Santa Monica: RAND.

Cohen, M. D., J. G. March, J. G., and J. P. Olsen. 1972. "A Garbage Can Model of Organizational Choice." Administrative Science Quarterly 17/3: 1–25.

Cohen, Sharon. 2007. "Hurricane Katrina Cuts A Wide Swath of Fraud-600 Charged." Monterey County Herald, April 2: A3.

Cole, Bernard. 2007. "China's Maritime Strategy," In China's Future Nuclear Submarine Force. Edited Andrew Erickson. Annapolis: Naval Institute Press: 22–42.

Collet, André. 1993. Armements et conflits contemporains de 1945 à nos jours. Paris: Armand Colin.

Comptroller General of the U.S. 1927. 6: 619, 621.

Conca, Ken. 1998. «Global Markets and Domestic Politics: Brazil's Military Industrial Collapse,» Review of International Studies 24/4 October: 499–513.

Congressional Budget Office. 1999. "Emergency Funding Under the Budget Enforcement Act: An Update," Washington, DC: CBO.

Congressional Budget Office. 2001. "Supplemental Appropriations in the 1990s," Washington, DC: CBO, March, www.cbo.gov

Congressional Budget Office. 2007. Federal Support for Research and Development. Washington, DC: USGPO.

Congressional Budget Office. 2010a. The Budget and Economic Outlook: Fiscal Years 2010 to 2020. CBO, January 26 http://www.cbo.gov/ftpdocs/108xx/doc10871/BudgetOutlook2010_Jan.cfm

Congressional Budget Office, 2010b. CBO Data on Supplemental Budget Authority for the 2000's. Washington, DC: CBO.

Congressional Budget Office. 2011. Linking the Readiness of the Armed Forces to DOD's Operation and Maintenance Spending. Washington, DC: CBO.

Congressional Quarterly Almanac (CQA). 1975–2001. Washington, DC: CQ.

Congressional Quarterly Almanac (CQA). 2001: 7.4

Congressional Quarterly Almanac (CQA). 2002: 2-40.

Congressional Quarterly Almanac (CQA). 2003: 2-83, 2-27.

Congressional Quarterly Almanac (CQA). 2004. 16, Washington, DC: CQ.

Congressional Quarterly Almanac (CQA). 2005. 17, Washington, DC: CQ: 2–14, 18.

Congressional Quarterly Weekly. 2001. Feb. 10, Washington, DC: CQ: 7.3, 337.

Congressional Quarterly Weekly. 2006. War and Disaster Supplementals, 18, Washington, DC: CQ: 3331, 3336.

Congressional Research Service. 2011. Duration of Continuing Resolutions in Recent Years, CRS Report RL32614, Washington, DC: Congressional Research Service.

Cottey, A., T. Edmunds, and A. Forster. 2002. "The Second Generation Problematic: Rethinking Democracy and Civil-Military Relations," Armed Forces and Society, 29, 1, Fall: 31–56.

Craig, Chris. 1995. Call for Fire: Sea Combat in the Falklands and the Gulf War. London: John Murray.

Daggett, Stephen. 2006. Military Operations: Precedents for Funding Contingency Operations in Regular or in Supplemental Appropriations Bills. CRS Report for Congress, RS22455. Washington, DC: CRS.

Daniels, Mitchell, 2001. "A-11 Transmittal Letter from the Director of OMB to Agencies," Washington, DC: OMB, February 14.

David, Steven. 2000. "Realism, Constructivism, and the Amarna Letters," in Amarna Diplomacy: The Beginnings of International Relations. Raymond Cohen and Raymond Westbrook, eds. Baltimore: Johns Hopkins University Press: 54–67.

Davis James W., and Randall B. Ripley. 1967. "The Bureau of the Budget and Executive Branch Agencies: Notes on their Interaction," Journal of Politics 29: November: 749–769.

Defense News. 2006. February 26, 106:1

Defense Aid Supplemental Appropriation Act. 1941. Mar. 27, 1941, ch. 30, 55 Stat. 53.

Defense Aid Supplemental Appropriation Act. 1942. Oct. 28, 1941, ch. 460, title I, 55 Stat. 745.

Defense Aid Supplemental Appropriation Act. 1943. June 14, 1943, ch. 122, 57 Stat. 151.

de Russé, Anne-Henry. 2010. France's Return into NATO: French Military Culture and Strategic Identity in Question. Paris: IFRI.

DeHart, James. 2008. The Burden of Strategy: Transatlantic Relations and the Future of NATO Enlargement. Washington, DC: Institute for the Study of Diplomacy, Georgetown University Press.

Deng Xiaoping. 1985. Selected Works of Deng Xiaoping, Vol. III. Beijing: The People's Daily Online www.people.com.cn/English/dengxp/

Denney, Jason. 2011. Innovation Pump: America Needs More Scientists, Engineers, And Basic Research. Washington, DC: Defense Acquisition University.

Department of Defence. [Australia]. 1997. Australia's Strategic Policy. Canberra: Australian Government.

Department of Defence. [Australia]. 1998. Building Combat Capability. Canberra: Australian Government.

Department of Defence. 2009. [Australia]. Defence Department Budget Request, Canberra: Australian Government.

Department of Defense. 2003. "Greenbook: National Defense Budget Estimates for FY 2004," Office of the Principal Deputy Under Secretary of Defense Comptroller, Washington, DC: DOD, March, www.dod.mil/comptroller/defbudget/fy2004

Department of Defense. 2004. Future Years Defense Program FYDP Structure: Codes and Definitions for All DOD Components. DOD 7045.7-H. Washington, DC: DOD.

Department of Defense. 2007. FY 2007 Supplemental Request for the Global War on Terror. Washington, DC, February, 2007.

Department of Defense. 2008. National Defense Strategy: June. Washington, DC: USGPO.

Department of Defense. 2009. Agency Financial Report for Fiscal Year 2009: Addendum A: Strategic Management Plan. Washington, DC: DOD.

Department of Defense. 2010. Defense Appropriation Act of 2010. Washington, DC: DOD.

Department of Defense. 2010. Financial Improvement and Audit Readiness FIAR Plan Status Report. Washington, DC: Undersecretary of Defense Comptroller.

Department of Defense. 2011a. DOD Financial Management Regulation, 7000.14-R. Washington, DC: DOD, 17 May, http://comptroller.defense.gov/fmr/

Department of Defense. 2011b. Fiscal Year 2012 Defense Department Budget Request Roll Out Briefing Slides. February. Washington, DC: DOD http://comptroller.defense.gov/defbudget/fy2012/fy2012_BudgetBriefing.pdf.

Department of the Air Force. 2002a. "Budget Process," Deputy Assistant Secretary Budget, November, http://www.saffm.hq.af.mil

Department of the Air Force. 2002b. "The Planning, Programming, and Budgeting System and the Air Force Corporate Structure Primer," November, http://www.saffm.hq.af.mil

Department of the Army. 2002. "Budget Process," Deputy Assistant Secretary of the Army, Budget Office, http://www.asafm.army.mil/budget/budget.asp

Department of the Navy. 2011. DON Budget Guidance Manual. Washington, DC: DON. 17 May 2011 http://www.finance.hq.navy.mil/fmb/guidance/bgm/bgm.htm.

Department of the Treasury. 2010. Citizens Guide to the 2009 Financial Report of the U.S. Government. Washington, DC: Department of Treasury.

DeVore, Marc. 2009. "A Convenient Framework: The Western European Union in the Persian Gulf, 1987–1988 and 1990–1991," European Security 18/2: 227–43.

DeVore, Marc. 2010. "The Military Revolution: A Dangerous Utopia from the Cold War to the War on Terror," in Michael Cox and George Lawson, eds., The Global 1989: Continuity and Change in World Politics, 1989–2009. Cambridge: Cambridge University Press: 219–242.

Dewey, Davis R.1968. Financial History of the United States, New York: Longmans, Green.

Defense Finance and Accounting Service (DFAS). 2010. DFAS: Our History. June 17, Washington, DC: Defense Finance and Accounting Service http://www.dfas.mil/about/OurHistory.html

Defense Logistics Agency (DLA). 2010. DLA at a Glance. June 23: http://www.dla.mil/ataglance.aspx

DOD. 2001. Financial Management Regulations, vol. 12: 23–26. Washington, DC: DOD.

Doyle, Richard. 2011. "Real Reform or Change for Chumps: Earmark Policy Developments, 2006–2010." Public Administration Review 71/1: 34–44.

Dörfer, Ingemar. 1973. System 37 Viggen: Arms, Technology and the Domestication of Glory. Oslo: Universitetsforlaget.

Dreher, Rod. 2007. "War Wrecking National Guard," Monterey County Herald, April 2: A-8.

Dror, Y. 1967. "Policy Analysts: A New Professional Role in Government Service." Public Administration Review 27/3: 197–203.

Drury. and Tayles, No date. "Misapplication of Capital Investment Appraisal Techniques," unpublished paper.

Duma, David. 2001. "A Cost Estimation Model for CNAP TACAIR Aviation Depot Level Repair Costs," Master Degree Thesis, Monterey, CA: Naval Postgraduate School.

Eisenhower, Dwight D. 1961. President's Farewell Speech, January 17, Washington, DC: Office of the President.

European Defence Aagency (EDA). Defence. 2009. Data of EDA Participating Member States in 2009. Brussels: European Union.

Engel, Jeffrey. 2007. Cold War at 30,000 Feet: The Anglo-American Fight for Aviation Supremacy. Cambridge: Harvard University Press.

England, Gordon. Deputy Secretary of Defense, 2005. Memorandum for Service Secretaries, CJCS, VCJS, and Service Chiefs, Subject: FY07 Budget, October 19.

European Council. 2003. A Secure Europe in a Better World. Brussels: European Union.

Evans, Amanda. 2006. Long-term Military Contingency Operations: Identifying the Factors Affecting Budgeting in Annual or Supplemental Appropriations. Masters Thesis, Naval Postgraduate School, March: 11.

Evstafiev, Gennady. 2007. "Disarmament Returns," Security Index 13/2: 19–30.

Feaver, Peter. 1996. "The Civil-Military Problematique: Huntington, Janowitz and the Question of Civilian Control," Armed Forces and Society 23/2 Winter: 149.

Feltes, L.A. 1976. "Planning, Programming, and Budgeting: A Search for a Management Philosopher's Stone." January-February. Air University Review www.airpower.maxwell.af.mil/airchronicles/aureview/1976/jan-feb/feltes.html

Fenno, Richard. 1966. The Power of the Purse. Boston: Little, Brown.

Ferrari, Giovanni. 1995. "NATO's New Standardization Organization Tackles an Erstwhile Elusive Goal," NATO Review 43/3: 33–35.

Fesler, James, W. 1982. American Public Administration: Patterns of the Past. Washington, D C: American Society for Public Administration: 71-89.

Fisher, Louis. 1975. Presidential Spending Power. Princeton, NJ: Princeton University Press.

Fisher, Richard. 2007. "The Impact of Foreign Technology on China's Submarine Force and Operations," in Andrew Erickson, et al., eds., China's Future Nuclear Submarine Force. Annapolis: Naval Institute: 135–161.

Fontanela, Jacques, and Jean-Paul Hébert. 1997. «The End of the 'French Grandeur Policy," Defence and Peace Economics 8/1: 37–55.

Fox, Ronald. 1988. The Defense Management Challenge. New York: Harper and Row.

Fox, Ronald with J. L. Field. 1988. The Defense Management Challenge: Weapons Acquisition. Boston: Harvard Business School Press.

Frame, J. D. 2002. The New Project Management: Tools for an Age of Rapid Change, Complexity, and Other Business Realities. San Francisco: Jossey-Bass.

Francis, D. B., and R. J. Walther. 2006. A Comparative History of Department of Defense Management Reform from 1947 to 2005. Monterey, CA: Center for Defense Management Reform.

Frank, J. E. 1973. "A Framework for Analysis of PPB Success and Causality." Administrative Science Quarterly 18/4, 527–543.

Freedman, Lawrence. 2003. The Evolution of Nuclear Strategy. Third Edition. Basingstoke: Palgrave.

Frenzel, Bill. 2005. "Budgeting in Congress: How the Budget Process Functions," Testimony before the U.S. House of Representatives, Budget Committee, Washington, DC, 22 June.

Friedburg, Aaron. 2000. "Will Europe's Past Be Asia's Future," Survival 42/3 Autumn: 147–59.

Friedman, Norman. 2000. The Fifty Year War: Conflict and Strategy in the Cold War. Annapolis: Naval Institute Press.

Friedman, Stephen. 2002. Testimony to the Subcommittee on National Security, Veterans Affairs, and International Relations of the House Government Reform Committee, U.S. House of Representatives, June 4.

Friedman, Thomas L. 2007. U.S. Soldiers Shortchanged. Monterey County Herald, March 8,: A-8.

Gansler, Jacques. 1989. Affording Defense. Cambridge, MA: MIT Press.

GAO. Government Accountability Office. 2004. Principles of Federal Appropriations Law Red Book http://www.gao.gov/legal/redbook.html

GAO. Government Accountability Office. 2005. CFO Act of 1990: Driving the Transformation of Federal Financial Management. Washington, DC: Government Accountability Office.

GAO. Government Accountability Office. 2006. Contract Management: DOD Vulnerabilities to Fraud, Waste, and Abuse. Washington, DC: Government Accountability Office.

GAO. Government Accountability Office. 2007a. Defense Business Transformation: Achieving Success Requires a Chief Management Officer to Provide Focus and Sustained Leadership, GAO-07-1072, September.

GAO. Government Accountability Office. 2007b. Fiscal Year 2006 U.S. Government Financial Statement: Sustained Improvement in Federal Financial Management is Crucial to Addressing Our Nation's Accountability and Fiscal Stewardship Challenges. Washington, DC: GAO.

GAO. Government Accountability Office. 2008a. Contingency Contracting. Washington, DC: Government Accountability Office.

GAO. Government Accountability Office. 2008b. DOD Business Systems Modernization: Military Departments Need to Strengthen Management of Enterprise Architecture Programs. Washington, DC: Government Accountability Office.

GAO. Government Accountability Office. 2009a. Defense Acquisitions: Actions Needed to Ensure Value for Service Contracts. Washington, DC: Government Accountability Office.

GAO. Government Accountability Office. 2009b. DOD Business Systems Modernization: Recent Slowdown in Institutionalizing Key Management Controls Needs to be Addressed. Washington, DC: Government Accountability Office.

GAO. Government Accountability Office. 2009c. High-Risk Series: An Update. Washington, DC: Government Accountability Office.

GAO. Government Accountability Office. 2009d. Understanding the Primary Components of the Financial Statements of the United States Government. Washington: GAO.

GAO. Government Accountability Office. 2010. Defense Inventory: Defense Logistics Agency Needs to Expand on Efforts to More Effectively Manage Spare Parts. Washington, DC: Government Accountability Office.

Gareev, Makhut. 1998. If War Comes Tomorrow?: The Contours of Future Armed Conflict. London: Frank Cass.

General Accounting Office. 1991 ."Principles of Federal Appropriations Law," 2nd edition, OBC-91-5, I, Chapter 2, July, Washington, DC: GAO.

General Accounting Office. 1999. "Managing for Results: Opportunities for Continued Improvements in Agencies' Performance Plans," GGD/AIMD-99-215, Washington, DC: GAO, July 20.

General Accounting Office. 2000. FY2000 Contingency Operations Costs and Funding Audit. GAO/NSIAD-00-168, June: Washington, DC: GAO: 2.

General Accounting Office. 2002. "Canceled DOD Appropriations: Improvements Made but More Corrective Actions are Needed," Washington, DC: GAO, July 31.

Gholz, Eugene, Daryl G. Press, and Harvey M. Sapolsky. 1997. "Come Home America: The Strategy of Restraint in the Face of Temptation," International Security 21/4 Spring: 5–48.

Godek, Paul. 2000. "Emergency Supplemental Appropriations: A Department of Defense Perspective." Master's Degree Thesis, Monterey, CA: Naval Postgraduate School, December.

Goldsmith, Ronald E., Leisa R. Flynn and Elizabeth B. Goldsmith. 2003. "Innovative Consumers and Market Mavens," Journal of Marketing Theory and Practice 11/4: 54–65.

Goldwater-Nichols Act (reforms). 1986. U.S. Congress, Washington, DC: U.S. Congress.

Government Accounting Standards Board (GASB). 2006. Why Governmental Accounting and Financial Reporting Is—and Should Be—Different. Government Accounting Standards Board, March http://gasb.org/jsp/GASB/Page/GASBSectionPageandcid=1176156741271

Grimmett, Richard. 2009. CRS Report for Congress–U.S. Arms Sales to Pakistan. Washington, DC: CRS.

Grimmett, Richard. 2010. CRS Report for Congress–Arms Sales: Congressional Review Process. Washington, DC: CRS.

Gupta, Amit. 1990. "The Indian Arms Industry: A Lumbering Giant?" Asian Survey 30/9 September: 846–61.

Hackett, Clifford. 1988. "Congress and Greek American Relations: The Embargo Example," Journal of the Hellenic Diaspora 15/1–2: 5–31.

Hale, Robert F. 1987. "Statement of the Assistant Director National Security Division Congressional Budget Office," Subcommittee on Conventional Forces and Alliance Defense and the Subcommittee on Defense Industry and Technology Committee on Armed Services United States Senate. Washington, DC: USGPO, March 17.

Hale, Robert F. 2011. P. Candreva, personal interview, Monterey, CA, May 12.

Halperin, Morton H., and Kristen Lomasney. 1999. "Playing the Add-On Game in Congress: The Increasing Importance of Constituent Interests and Budget Constraints in Determining Defense Policy," in The Changing Dynamics of U.S. Defense Spending, Leon V. Sigal, ed., Westport: Praeger Press: 85–106.

Hamilton, Alexander. 1790. 1791. Reports to Congress on the Public Credit, January 14, 1790; December 13, 1790; The Report on a National Bank, December 14, 1790; The Report on Manufactures, December 5, 1791. Washington, DC: U.S. Congress.

Harding, Keith. 2007. "Russia Threatening New Cold War Over Missile Defence: Kremlin Accuses US of Deception on East European Interceptor Bases," The Guardian April 11: 24.

Hartley, Keith, and Stephen Martin. 1993. "The Political Economy of International Collaboration," in Richard Coopey, et al., eds., Defence Science and Technology: Adjusting to Change. Chur, Switzerland: Harwood: 172–179.

Hartley, Keith. 1997. Towards a European Weapons Procurement Process. Paris: Institute of Security Studies of WEU.

Hartley, Keith. 2000. «The Benefits and Costs of the UK Arms Trade,» Defence and Peace Economics 11/3: 445–459.

Hartley, Keith. 2001. The Common European Security and Defence Policy: An Economic Perspective. University of York, Centre for Defence Economics Paper www.york.ac.uk/depts/econ/rc/cde.htm#current

Hartley, Thomas and Bruce Russett. 1992. "Public Opinion and the Common Defense: Who Governs Military Spending in the United States?", *The American Political Science Review* 86/4:905–915.

Hartung, W. D. 1999. "The Shrinking Military Pork Barrel: The Changing Distribution of Pentagon Spending, 1986–1996" in Leon V. Sigal, ed., The Changing Dynamics of U.S. Defense Spending. Westport: Praeger Press: 29–84.

Hébert, Jean-Paul. 1995. Production d'Armement: Mutation du Système Français. Paris: Documentation Français.

Hébert, Jean-Paul. 1998. Les Exportations d'armement: A Quel Prix? Paris: Documentation Français.

Heniff Jr., Bill, and Robert Keith. 2004. "Federal Budget Process Reform: A Brief Overview," CRS Report for Congress," CRS RS21752, Washington, DC, 8 July www.opencrs.com

Heniff, Bill. 2005. "Congressional Budget Resolutions: Selected Statistics and Information Guide," CRS Report for Congress, 25 January, Washington, DC

Henrotin, Joseph. 2010. 'Entre héritage sunzien et révolutionnisme: Quel avenir pour la PLA?' Défense and Sécurité Internationale Hors-Série 15: 54–61.

Hewson, Robert. 2008."Chinese J-10 'Benefited From The Lavi Project'," Jane's Defence Weekly May 16: 12.

Higgs, R., and A. Kilduff. 1993. "Public Opinion: A Powerful Predictor of U.S. Defense Spending," Defense Economics 4: 227–238.

Hildreth, Steven, and Amy Woolf. 2010. Ballistic Missile Defense and Offensive Arms Reductions: A Review of the Historical Record. Washington, DC: CRS.

Hill, T. F. 2008. An Analysis of the Organizational Structures Supporting PPBE within the Military Departments. Masters Thesis. Naval Postgraduate School. Monterey, CA.

Hinricks, H., and G. Taylor, eds. 1969. Program Budgeting and Benefit-Cost Analysis. Pacific Palisades, CA: Goodyear.

Hitch, C., and R. McKean. 1960. The Economics of Defense in the Nuclear Age. Cambridge: Harvard University Press.

Hofnung, Thomas. 2006. La Crise en Côte-d'Ivoire: Dix Clés pour Comprendre. Paris: La Découverte.

Hoge, J., and E. Martin. 2006. Linking Accounting and Budgeting Data: A Discourse. Public Budgeting and Finance 26/2: 121–142.

Hongyi, Nie. 2009. Explaining Chinese Solutions to Territorial Disputes with Neighboring States," Chinese Journal of International Politics 2: 487–523.

Hormats, Robert. 2007. The Price of Liberty. New York: Henry Holt and Company.

Hood, Christopher. 1991."A Public Management for All Seasons," Public Administration 69/1: 3–20.

House of Representatives. 1912. Doc. 854, 62-2: 138

Howorth, Jolyon. 2007. Security and Defence Policy in the European Union. Basingstoke: Palgrave.

Huertas, Salvador Mafé. 1996. Dassault Mirage: The Combat Log. Atglen, Pennsylvania: Schiffer.

Hughes, Robin. 2003. «Israeli Defence Industry: In the Lion's Den,» Jane's Defence Weekly February 26: 13.

Human Rights Watch. 2008. Flooding South Lebanon: Israel's Use of Cluster Munitions in Lebanon in July and August 2006. New York: Human Rights Watch, February.

Huntington, Samuel P. 1957. The Soldier and the State: The Theory and Politics of Civil-Military Relations, Cambridge: Harvard University Press.

Interview with Hilmar Linnenkamp. 2010. Former Deputy Chief Executive of EDA, March 17.

IISS–International Institute of Strategic Studies. 2011. The Military Balance. Abington: Routledge.

Janowits, M. 1960. The Professional Solider: A Social and Political Portrait. Chicago, IL: Glencoe.

Jervis, Robert. 1978. "Cooperation Under the Security Dilemma," World Politics 30/2:167–214.

Johnson, Joel.1994. "Financing the Arms Trade," Annals of the American Academy of Political and Social Science 53/5 September: 110–21.

Joint DOD/GAO Working Group on PPBS. 1983. "The Department of Defense Planning, Programming and Budgeting System," Washington, DC: DOD/GAO, May 26.

Jonas, Tina. 2002. Statement before the House Government Reform Committee, Subcommittee on Government Efficiency, Financial Management and Intergovernmental Relations, 20 March. Washington, DC: GPO.

Jones, B. D. 2001. Politics and the Architecture of Choice. Chicago: University of Chicago Press.

Jones, David C. 1982. "Why the Joint Chiefs Must Change," Armed Forces Journal International March, 64: 24.

Jones, L. R. 2010. "Restructuring Public Organizations in Response to Global Economic and Financial Stress," International Public Management Review 11/1: 1–14.

Jones, L. R., and Glenn C. Bixler. 1992. Mission Budgeting to Realign National Defense. Greenwich, CT: JAI Publishers.

Jones, L. R., and Kenneth Euske. 1991. "Strategic Misrepresentation in Budgeting," Journal of Public Administration Research and Theory 3/3: 37–52.

Jones, L. R., and J. L. McCaffery. 1989. Government Response to Financial Constraints. New York: Greenwood Press.

Jones, L. R., and J. L. McCaffery. 1997. Implementing the Chief Financial Officers Act and the Government Performance and Results Act in the Federal Government. Public Budgeting and Finance 17/1: 35–55.

Jones, L. R., and J. L. McCaffery. 2005. "Reform of PPBS and Implications for Budget Theory," Public Budgeting and Finance 25/3: 1–19.

Jones, L. R., and J. L. McCaffery. 2008. Budgeting, Financial Management and Acquisition Reform in the U.S. Department of Defense. Charlotte, NC: Information Age Publishing, 2008.

Jones, L. R., and F. Thompson. 1999. Public Management: Institutional Renewal for the 21st Century. New York: Elsevier Science.

Jones, L. R., and F. Thompson. 2007. From Bureaucracy to Hyperarchy in Netcentric and Quick Learning Organizations. Charlotte, NC: Information Age Publishing

Jones, L. R., and Aaron Wildavsky. 1995. "Budgetary Control in a Decentralized System: Meeting the Criteria for Fiscal Stability in the European Union," Public Budgeting and Finance 14/4: 7–21.

Joyce, Philip G. 2002. "Federal Budgeting After September 11th: A Whole New Ballgame or Deja Vu All Over Again?" Paper presented at the conference of the Association for Budgeting and Financial Management, Kansas City, MO, October 10.

Joyce, Philip G. 2003. Linking Performance and Budgeting: Opportunities in the Federal Budget Process. Washington, DC: IBM Center for the Business of Government.

Joyce, Philip G. 2011. "The Obama Administration and PBB: Building on the Legacy of Federal Performance-Informed Budgeting." Public Administration Review 71/3: 356–367.

Kadish, R., G. Abbott, F. Cappuccio, R. Hawley, P. Kern, D. Kozlowski, et al. 2006. Defense Acquisition Performance Assessment Report for the Deputy Secretary of Defense http://www.acq.osd.mil/dapaproject/

Kaldor, Mary. 1983. The Baroque Arsenal. London: Sphere, 1983.

Kalinina, Natalia and Kozyulin Vadim. 2010. "Russian Defense Industry: Feet of Clay," Security Index 1/90, Winter: 31–46.

Kan, Shirley. 2002. CRS Report for Congress–Taiwan: Major U.S. Arms Sales Since 1990. Washington, DC: CRS.

Kanter, A. 1983. Defense Politics: A Budgetary Perspective. Chicago, IL: University of Chicago Press.

Kaplan, Lawrence. The Long Entanglement: NATO's First Fifty Years. Wesport: Praeger, 1999.

Kapstein, Ethan B. 1994. "America's Arms-Trade Monopoly: Lagging Sales Will Starve Lesser Suppliers," Foreign Affairs 73/3 May–June: 13–19.

Kapstein, Ethan B. 2004. "Capturing Fortress Europe: International Collaboration and the Joint Strike Fighter," Survival 46/3 Autumn: 137–60.

Kapstein, Ethan B.1992. The Political Economy of National Security: A Global Perspective. New York: McGraw-Hill.

Karp, J. 2007. January "Pentagon Redefines 'Emergency'" The Wall Street Journal January 3: 12.

Keijsper, Gerard. 2003. Saab Gripen: Sweden's 21st Century Multi-Role Aircraft. Hinckley, England: Aerofax.

Kemp, Geoffrey. 1994. "The Continuing Debate Over U.S. Arms Sales: Strategic Needs and the Quest for Arms Limitations," Annals of the American Academy of Political and Social Science 535 September: 146–157.

Key, V. O. 1940. The Lack of a Budgetary Theory. The American Political Science Review 3/4, 1137–1144.

Klamper, Amy. 2005. Subcommittee Voices Dismay Over Defense Supplemental Spending, Congress Daily 4 March: 2.

Kollias, Christos, George Manolas and Suzanna-Maria Paleologou. 2004."Defence Expenditure and Economic Growth in the European Union: A Causality Analysis." Journal of Policy Modeling 26/5 (July): 553–569.

Kolodziej, Edward. 1987. Making and Marketing Arms: The French Experience and its Implications for the International System. Princeton: Princeton University Press.

Korb, L. 1977. "Department of Defense Budget Process: 1947–1977." Public Administration Review 37/4: 247–264.

Korb, L. 1979. The Rise and Fall of the Pentagon. Westport, CT: Greenwood Press.

Kozar, M. J. 1993. "An Analysis of Obligation Patterns for the Department of Defense Operations and Maintenance Appropriations." Master's Degree Thesis, Monterey, CA: Naval Postgraduate School.

Krause, Keith. 1992. Arms and the State: Patterns of Military Production and Trade. Cambridge: Cambridge University Press.

Krepinevich, Andrew. 1992. The Military-Technical Revolution: A Preliminary Assessment. Washington, DC: Office of Net Assessment.

Kutz, Gregory, 2002b. "Defense Department Financial Management," Director, GAO Financial Management and Assurance Team, Hearing of the National Security, Veterans Affairs, and International Relations Subcommittee of the House Government Reform Committee, U.S. House of Representatives, June 4.

Kutz, Gregory, 2002b. "Defense Department Financial Management," Director, GAO Financial Management and Assurance Team, Hearing of the National Security, Veterans Affairs, and International Relations Subcommittee of the House Government Reform Committee, U.S. House of Representatives, June 25.

Kümmel, Gerhard. 2006. "An All-Volunteer Force in Disguise: On the Transformation of the Armed Forces in Germany," in Curtis Gilroy and Cindy Williams, eds., Service to Country: Personnel Policy and the Transformation of Western Militaries. Cambridge: MIT Press: 203–232.

Kyle, Keith.1991. 2011. Suez: Britain's End of Empire in the Middle East. London: I.B. Taurus.

Labaree, Leonard W. 1958. Royal Government in America: A Study of the British Colonial System before 1783, New Haven, CT: Yale University Press.

Lai, David. 2010. "Introduction," in Roy Kamphausen, et al., The PLA at Home and Abroad: Assessing the Operational Capabilities of China's Military. Carlisle, PA: U.S. Army Strategic Studies Institute: 1–44.

Le Roy, François. 2002. "Mirages over the Andes: Peru, France, the United States, and Military Jet Procurement in the 1960s," The Pacific Historical Review 71/2 May: 269–300.

Lee, R. D., and R. W. Johnson, 1983. Public Budgeting Systems. Baltimore: University Park Press.

Lee, Jr., R. D., R. W. Johnson, and P.G. Joyce. 2004. Public Budgeting Systems. 8th Edition. Boston: Jones and Bartlett Publishers.

LeLoup Lance, and William Moreland. 1978. "Agency Strategies and Executive Review: The Hidden Politics of Budgeting." Public Administration Review 38/1: 232–239

Library of Congress. 1997. Congressional Record S11817-8, Washington, DC: Congressional Research Service, November 6.

Library of Congress. 2000. Exhibit: Defense Appropriation Bill in 2000–Passed on Time www.thomas.gov.

Lieberman, Robert. 2002. Testimony by the Deputy Inspector General, Department of Defense to the Subcommittee on National Security, Veterans Affairs, and International Relations of the House Government Reform Committee, U.S. House of Representatives, June 4.

Liebman, J. B., and N. Mahoney. 2010. Do Expiring Budgets Lead to Wasteful Year-End Spending? Evidence from Federal Procurement. Unpublished working paper http://www.hks.harvard.edu/jeffreyliebman/LiebmanMahoneyExpiringBudgets.pdf

Lindblom, Charles E. 1959. "The Science of Muddling Through," Public Administration Review 19/2: 79–88.

Lindsay, James, M. 1987. "Congress and Defense Policy: 1961–1986," Armed Forces and Society 13/3, Spring: 371–401.

Lindsay, James, M. 1990. "Congressional Oversight of the Department of Defense: Reconsidering the Conventional Wisdom," Armed Forces and Society 17, Fall: 7–33.

Lipicki, Martin. 2011. Chinese Use of Cyberwar as an Anti-Access Strategy: Two Scenarios. Santa Monica: RAND.

List, Friedrich. 2005. French Deltas: The Dassault Mirage 2000 over Europe–Part 1. Erlangen, Germany: AirDoc.

Liu, Kuo Liu. 2011. "Taiwan's Long-Term Challenges and Strategic Preparations," In New Opportunities and Challenges for Taiwan's Security. Roger Cliff, et al., eds. Santa Monica: RAND: 89-98.

Locher, James R.1996. "Taking Stock of Goldwater-Nichols," Joint Forces Quarterly Autumn: 10–17, http://www.dtic.mil/doctrine/jel/jfq_pubs/index.htm

Locher, James R. 2002.Victory on the Potomac: The Goldwater-Nichols Act Unifies the Pentagon, College Station, Texas: Texas A&M University Press.

Lorell, Mark. 1995. Troubled Partnership: A History of U.S.-Japan Collaboration on the FS-X Fighter. Santa Monica: RAND.

Lucas, Edward. 2008. The New Cold War: Putin's Russia and the Threat to the West. Basingstoke: Palgrave.

Lungu, Sorin. 2004. "Military Modernization and Political Choice: Germany and the US-Promoted Military Technological Revolution During the 1990s," Defense and Security Analysis 20/3 September: 261–272.

Mann, G. J. 1988. "Reducing Budget Slack," Journal of Accountancy 166/2: 118–23.

March, J. G. 1994. A Primer on Decision Making. New York: The Free Press.

Marks, Robert A. 1989. "Program Budgeting within the Department of Navy," Master's Degree Thesis, Monterey, CA: Naval Postgraduate School.

Martin, Steven. 1999. "The Subsidy Savings from Reducing UK Arms Exports," Journal of Economic Studies 26/1 1999: 15–37.

Mathews, William. 2007. "U.S. House Passes $124.3B Supplemental," Defense News March 26,: 4

Mayer, Kenneth R. 1993. "Policy Disputes as a Source of Administrative Controls: Congressional Micromanagement of the Department of Defense," Public Administration Review 53/4: 293–302.

McCaffrey, J. L., and P. Godek. 2003. "Defense Supplementals and the Budget Process," Public Budgeting and Finance 23/2: 53–72.

McCaffery, J. L., and L. R. Jones. 2001. Budgeting and Financial Management in the Federal Government. Greenwich, CT: Information Age Publishers.

McCaffery, J. L., and L. R. Jones. 2004. Budgeting and Financial Management for National Defense. Greenwich, CT: Information Age Publishers.

McCaffery, J. L., and J. E. Mutty. 2003. "Issues in Budget Execution" in A. Khan and W. B. Hildreth, eds., Case Studies in Public Budgeting and Financial Management, 2nd Edition New York: Marcel Dekker.

McCain, J. L. 1997. List of 95 Press Releases and Talking Points about Pork, Office of Senator John L. McCain website www.mccain.senate.gov/index.cfm?fuseaction=Issues.ViewIssue&Issue_id=27

McLaughlin, Daniel, and Vanessa Mock. 2009. "New Cold War in Europe as Russia Turns Off Gas Supplies," The Independent January 7: 2.

Mearshimer, John. 2001. The Tragedy of Great Power Politics. New York: W.W. Norton.

Medeiros, Evan, and Gill Bates. 2000. Chinese Arms Exports: Policies, Players, and Process. Fort Leavenworth, Kansas: U.S. Army Strategic Studies Institute.

Merchant, K. A. 1981. "The Design of the Corporate Budgeting System, Influences on Managerial Behavior and Performance," The Accounting Review 56/4: 813–829.

Merewitz, L., and S. H. Sosnick. 1972. The Budget's New Clothes. Chicago: Markham.

Mĕrtl, Miroslav. 1998. Army of the Czech Republic in Achieving Interoperability with NATO. Monterey: Naval Postgraduate School, Master's Degree Thesis, December.

Meyers, Roy. 1994. Strategic Budgeting. Ann Arbor: University of Michigan Press, 1994.

Meyers, Roy. 1997. "Late Appropriations and Government Shutdowns," Public Budgeting and Finance 17/3: 23–42.

Meyers, Roy. 2002. "Comments on The Federal Budget 2002," Presentation at the conference of the Association for Budgeting and Financial Management, Kansas City, MO, October 10.

Michel, Serge, and Michel Beuret. 2008. La Chinafrique: Pékin à la Conquête du Continent Noir. Paris: Bernard Grasset.

Micheletti, Eric. 1999. COS, le Commandement des Opérations Spéciales. Paris: Histoire and Collections.

Migdalovitz, Carol. 2008. CRS Report for Congress–Israel: Background and Relations with the United States. Washington, DC: CRS.

Miller, David. 2006. The Fiscal Blank Check Policy and its Impact on Operation Iraqi Freedom. Masters Thesis, Monterey, CA: Naval Postgraduate School, December.

Miller, Gene. 1990. Security Assistance to Saudi Arabia: Should the United States Provide Combat Aircraft Replacements? Carlisle Barracks, Pennsylvania: U.S. Army War College.

Miller, John C. 1959. Alexander Hamilton: Portrait in Paradox, New York: Harper.

Miller, G. J., W. B. Hildreth, and J. Rabin. 2001. Performance-Based Budgeting. Boulder, CO: Westview Press.

Moravcsik, Andrew. 1992. "Arms and Autarky in Modern European History," in Raymond Vernon and Ethan Kapstein, eds., Defense and Dependence in a Global Economy. Washington, DC: Congressional Quarterly: 23–45.

Morrison, Blake, Tom Vanden Brook and Peter Eisler. 2007. "When the Pentagon Failed to Buy Enough Body Armor, Electronic Jammers and Hardened Vehicles to Protect U.S. Troops from Roadside Bombs In Iraq, Congress Stepped In." USA Today Tuesday, September 4: 23.

Musgrave, R. A. 1959. The Theory of Public Finance: A Study in Public Economy. New York: McGraw-Hill.

National Audit Office (NAO). 2009. Support to High Intensity Operations. London: HMSO.

National Audit Office (NAO). 2001. Maximizing the Benefits of Defence Equipment Co-Operation. London: HMSO.

National Academy of Public Administration (NAPA). 2006. Moving from Scorekeeper to Strategic Partner: Improving Financial Management. Washington: National Academy of Public Administration.

Neuman, Stephanie. 2009. "Power, Influence, and Hierarchy: Defense Industries in a Unipolar World," The Modern Defense Industry: Political, Economic, and Technological Issues. Richard Bitzinger, ed. Santa Barbara: Praeger Security: 60–94.

New York Times–MSNBC. 2011. Public Opinion Survey, reported September 19, MSNBC.com.

Noetzel, Timo. 2010. "Germany's Small War in Afghanistan: Military Learning amid Politico-strategic Inertia," Contemporary Security Policy 31/3: 486–508.

North Atlantic Council. 2010. "Active Engagement: Modern Defence," Strategic Concept for the Defence and Security of the Members of the North Atlantic Treaty Organisation." Brussels: NATO.

Organization for Economic Co-operation and Development (OECD). 2010. Economic Policy Reforms: Going for Growth 2010, Summary of Chapter 1 "Responding to the Crisis While Protecting Long-term Growth." OECD, March 10 http://www.oecd.org/document/51/0,3343,en_2649_34325_44566259_1 _1_1_1,00.html

Odom, William. 1998. The Collapse of the Soviet Military. New Haven: Yale University Press.

Office of Management and Budget. 1993. Objectives of Federal Financial Reporting: Statement of Federal Financial Accounting Concepts, Number 1. Washington, DC: Office of Management and Budget.

Office of Management and Budget. 2005. Budget of the United States for Fiscal Year 2006, Historical Tables, Table 4.1: Outlays by Agency: 1962–2010: 76 www.whitehouse.gov/omb/budget/fy2006/pdf/hist.pdf.

Office of Management and Budget. 2008. Table 1.1, U.S. Budget FY 08 summary of receipts and expenditures. Washington, DC: OMB.

Office of Management and Budget. 2011. Budget of the United States Government, FY2011. Analytical Perspectives, Chart 12.1: Relationship of Budget Authority to Outlays. Washington, DC: OMB.

Office of the President. 2002. The National Security Strategy of the United States of America. Washington DC: USGPO, 2002.

Office of the President. 2010. National Security Strategy. Washington DC: USGPO.

Office of the Secretary of Defense. 2003. Justification for FY2004 Component Contingency Operations and the OCOTF, February.

Office of the Secretary of Defense. 2010. "Fiscal Year FY 2011 Overseas Contingency Operations OCO Request." Office of the Secretary of Defense. February http://comptroller.defense.gov/defbudget/fy2011/budget_justification/pdfs/16_Overseas_Contingency/oco_request_fy2011.pdf.

Olson, Mancur, and Richard Zeckhauser. "An Economic Theory of Alliances," Review of Economics and Statistics 48/3 1966: 266–79.

Osborne, David, and Ted Gaebler.1993. Reinventing Government. New York: Penguin Books.

Ostrom, E., R., Gardner, J. Walker. 1994. Rules, Games, and Common-Pool Resources. Ann Arbor, MI: University of Michigan Press.

Overy, Richard. 1995. Why the Allies Won. New York: W.W. Norton.

Owens, Mackubin T. 1990. "Micromanaging the Defense Budget," The Public Interest 100. Summer: 131–146.

Peckman, Joseph A.1983. Setting National Priorities: The 1984 Budget, Washington, D C: The Brookings Institution.

People's Republic of China. 2004. China's National Defense in 2004. Beijing: State Council Information Office.

People's Republic of China. 2007. China's National Defense in 2006. Beijing: State Council Information Office, 2007.

People's Republic of China. 2009. China's National Defense in 2008. Beijing: State Council Information Office.

People's Republic of China. 2011. China's National Defense in 2010. Beijing: State Council Information Office.

Peters, K. M. 2010. "Defense Budget Tackles Major Management Issues." February 2 http://www.govexec.com/story_page.cfm?filepath=/dailyfed/0210/020210kp1.htmandoref=search

Philips, William E. 2001. "Flying Hour Program (FHP) Cash Management at Commander Naval Air Forces Pacific (CNAP)", Appendix B in J. L. McCaffery and L. R. Jones. 2001. Budgeting and Financial Management in the Federal Government. Greenwich, CT: Information Age Publishers: 423–440.

Phythian, Mark. 2000. The Politics of British Arms Sales Since 1964. Manchester: Manchester University Press.

P.L. 97-25.

P.L. 101-576.

P.L. 103-62.

P.L. 103-356.

P.L. 104-208.

Podvig, Pavel, et al. 2004. Russian Strategic Nuclear Forces. Cambridge: MIT University Press.

Pollack, Jonathan. 1980. China's Military Modernization, Policy, and Strategy. Santa Monica: RAND.

Pollpeter, Kevin. 2010. "Towards an Integrative C4ISR System: Informatization and Joint Operations in the People's Liberation Army," in Roy Kamphausen, et al., eds., The PLA at Home and Abroad: Assessing the Operational Capabilities of

China's Military. Edited by Carlisle, Pennsylvania: U.S. Army Strategic Studies Institute: 193–235.

Porte, Rémy. 2005. La Mobilization Industrielle: "Premier Front" de la Grande Guerre? Paris: Soteca, 14–18 Éditions.

Porten, R. E., D. L. Cuda, A. C. Yengling, C. V. Fletcher, and D. A. Drake. 2003. Exploring a New Defense Resource Management System. Alexandria, VA: Institute for Defense Analysis.

Posner, Paul L., and Denise L. Fantone. 2008. "Performance Budgeting: Prospects for Sustainability," in F. S. Redburn, R. J. Shea and T. F. Buss, eds., Performance Measurement and Budgeting: How Government Can Learn from Experience. Armonk, NY: M. E. Sharpe.

Posner, Paul. 2002. "Performance-Based Budgeting: Current Developments and New Prospects." Paper presented at the conference of the Association for Budgeting and Financial Management, Kansas City, MO, October 10.

Powell, Fred W. 1939. "Control of Federal Expenditures." Washington, D C: The Brookings Institution.

Puritano, V. 1981. "Streamlining PPBS." Defense August: 20–28.

Reinhard, Scott. 2006. Joint Strike Fighter Across the Atlantic: To Unite or Divide. Monterey: Naval Postgraduate School MA Thesis.

Reynolds, Gary K. 2000. "Defense Authorization and Appropriation Bills: A Chronology," Washington, DC: CRS, November 21.

Reyntjens, Filip. 2009. The Great African War: Congo and Regional Geopolitics, 1996–2006. Cambridge: Cambridge University Press.

Rich, Michael, et al. 1984. Cost and Schedule Implication for Multinational Coproduction. Santa Monica: RAND.

Richardson, Lewis. 1960. Arms and Insecurity; A Mathematical Study of the Causes and Origins of War. Pittsburgh: Boxwood Press.

Roberts, Adam. 1976. Nations in Arms: The Theory and Practice of Territorial Defense. New York: Praeger.

Robinson, Dan. 2006. Democrats Want More Accountability for Iraq Spending, Capitol Hill, 15 February.

Robinson, M. 2007. Performance Budgeting: Linking Funding and Results. New York: Palgrave Macmillan.

Rogers, David. 2007. "Democrats Bill to Fund Surge, Within Limits," Wall Street Journal, Thursday, March 3: A2.

Roosevelt, Franklin D. 1933. Inaugural Speech to the Nation, March 4, Washington, DC: Office of the President, as quoted in Samuel Rosenman, ed., 1938. The Public Papers of Franklin D. Roosevelt, Volume Two: The Year of Crisis, 1933. New York: Random House: 11–16.

Rose, J. Holland. 1911a. William Pitt and National Revival. London: G. Bell and Sons.

Rose, J. Holland. 1911b. William Pitt and the Great War. London: G. Bell and Sons.

Rose, J. Holland, 1912. Pitt and Napoleon. London: G. Bell and Sons.

Roum, C. 2007. The Nature of DOD Reprogramming. Master's Degree Thesis. Monterey, CA: Naval Postgraduate School.

Rubin, I. S. 1993. The Politics of Public Budgeting: Getting and Spending, Borrowing and Balancing. Second Edition. Chatham, NJ: Chatham House Publishers.

Rukshin, Alexander. 2005. "The Russian Armed Forces in an Era of New Threats and Challenges," Digest of the Russian Journal Yaderny Control 10/1–2 Winter/Spring: 30–37.

Sadler, Karl-Otto. 2007. Ein Leben für den Marine-Schiffbau. Hamburg: Mittler.

Salmanov, G.I. 1995. "Soviet Military Doctrine and Some Views on the Nature of War in Defense of Socialism," in David Glantz, ed., The Evolution of Soviet Operational Art, 1927–1991: Volume II, Operational Art, 1965–1991. London: Frank Cass: 310–23.

Sampson, Anthony. 1977. The Arms Bazaar. London: Hodder and Stoughton.

Samuels, Richard. 1994. Rich Nation, Strong Army: National Security and the Technological Transformation of Japan. Ithaca: Cornell University Press.

Samuels, Richard. 2007. Securing Japan: Tokyo's Grand Strategy and the Future of East Asia. Ithaca: Cornell University Press.

Sander, Todd, and Keith Hartley. 1995. The Economics of Defense. Cambridge: Cambridge University Press.

Sandler, Todd, and Keith Hartley. 1999. The Political Economy of NATO: Past, Present, and into the 21st Century. Cambridge: Cambridge University Press.

Sandler, Todd. 1987. "NATO Burden Sharing: Rules or Reality," in Christian Schmidt and Frank Blackaby, eds., Peace, Defence, and Economic Analysis. London: Macmillan: 363–83.

Savage, James D. 1999. Funding Science in America: Congress, Universities, and the Politics of the Academic Pork Barrel. Cambridge: Cambridge: University Press.

Scardaville, Michael. 2002. "Congress Must Reform its Committee Structure to Meet Homeland Security Needs," Executive Memorandum #823, Heritage Foundation, Washington, DC; 12 July www.heritage.org/Research/Homeland-Defense/EM823.cfm

Schatz, Joseph J. 2005. "Urgency Driving Army Line Item." CQ Weekly February 28: 510.

Schelling, Thomas. 1966. Arms and Influence. New Haven: Yale University Press.

Schick, Allen. 1966. "The Road to PPB: The Stages of Budget Reform." Public Administration Review, 26/4: 243–258.

Schick, Allen. 1970. "The Budget Bureau that was: Thoughts on the Rise, Decline and Future of a Presidential Agency," Law and Contemporary Problems 35 Summer: 519–539.

Schick, Allen. 1973. "A Death in the Bureaucracy: The Demise of Federal PPB." Public Administration Review 33/2: 146–156.

Schick, Allen. 1980. Congress and Money: Budgeting, Spending, and Taxing, Washington, D C: Urban Institute.

Schick, Allen. 1985. "University Budgeting: Administrative Perspective, Budget Structure, and Budget Process." The Academy of Management Review 10/4: 794–802.

Schick, Allen. 1990. The Capacity to Budget. Washington, DC: The Urban Institute Press.

Schick, Allen, 2005. "Statement," Testimony before the U.S. House of Representatives, Budget Committee, Washington, DC, 22 June.

Schick, Allen. 2009. "Budgeting for Fiscal Space." OECD Journal on Budgeting 9/2: 7–24.

Schlecht, Eric V. 2002. "Dodging Pork Missiles," National Review Online, November 6: 1.

Schroeder, Matt. 2005. "Transparency and Accountability in Arms Export Systems: the United States as a Case Study," Disarmament Forum 3: 29–37.

Scobell, Andrew. 2010. "Discourse in 3-D: The PLA's Evolving Doctrine, Circa 2009," in Roy Kamphausen, et al., eds., The PLA at Home and Abroad: Assessing the Operational Capabilities of China's Military. Carlisle, Pennsylvania: U.S. Army Strategic Studies Institute: 99–134.

Secretary of Defense. 1990. Interviews by L. R. Jones with DOD Comptroller officials, Washington, DC: DOD, January 17–26.

Secretary of Defense. 2003. "Management Initiative Decision 913," Washington, DC: DOD, May 22.

Secretary of the Navy. 2002. Navy Budget Manual Instruction 7102.2a. Washington, DC: DON.

Secretary of the Navy. 2003. Interviews by Authors with Navy FMB Officials, Washington, DC: DON, January.

Seiko, Daniel T. 1940. The Federal Financial System, Washington, D C: The Brookings Institution.

Senate Armed Services Committee. 2005. Hearing on the Defense Authorization Request for Fiscal Year 2006 and the Future Years Defense Program, February 17.

Serfaty, Simon, and Sven Biscop. 2009. A Shared Security Strategy for a Euro-Atlantic Partnership of Equals: A Report on the Global Dialogue between the European Union and the United States. Washington, DC: CSIS.

Shalal-Esa, Andrea. 2005. Pentagon Plays Games with War Funding Requests, Capitol Hill Blue February 9.

Sharkansky, Ira. 1969. The Politics of Taxing and Spending, New York: Bobbs Merril.

Shays, Christopher. 2002. Statement to the Hearing of the Subcommittee on National Security, Veterans Affairs and International Relations of the House Government Reform Committee, U.S. House of Representatives, June 25.

Stanton, Shelby. 1987. The 1st Cav in Vietnam: Anatomy of a Division. Novato: Presidio.

Sherman, J. 2010. "Lynn Overturns Bush Administration DOD Budget Reforms, Directs Major Changes." InsideDefense.com: April 14 http://insidedefense.com/Inside-the-Pentagon/Inside-the-Pentagon-04/15/2010/lynn-overturns-bush-dod-budget-reforms-directs-major-changes/menu-id-148.html

Simon, H. A. 1993. Administrative Behavior. 3rd ed., New York: The Free Press

Simon, H. A. 1997. Administrative Behavior. 4th ed., New York: The Free Press.

SIPRI. 2011. Military Expenditure Database 2011 http://milexdata.sipri.org

SIPRI Yearbook. 1979. World Armaments and Disarmament. Oxford: Oxford University Press.

SIPRI Yearbook. 1987. World Armaments and Disarmament. Oxford: Oxford University Press.

SIPRI Yearbook. 2010. World Armaments and Disarmament. Oxford: Oxford University Press.

SIPRI Yearbook. 2011. World Armaments and Disarmament. Oxford: Oxford University Press.

Smith, Hedrick.1988. The Power Game: How Washington Works New York Press. Random House.

Smith, K. A., and Chen, R. H. 2006. Assessing Reforms of Government Accounting and Budgeting. Public Performance and Management Review 30/1:14–34.

Smith, Adam. 1776. An Inquiry into the Nature and Causes of the Wealth of Nations (The Wealth of Nations). Original unpublished manuscript.

Smithies, Arthur. 1955. The Budgetary Process in the United States.

Sorenson, David. 2009. The Process and Politics of Defense Acquisition: A Reference Handbook. Westport: Praeger.

Soutou, Georges-Henri. 1989. L'or et le Sang: Les Buts de Guerre Economiques de la Première Guerre Mondiale. Paris: Fayard.

Spinney, Franklin C. 1985. Defense Facts of Life: The Plans/Reality Mismatch. Boulder, CO: Westview.

Spinney, Franklin C. 2002. "Defense Department Financial Management," Testimony by Tactical Air Analyst, Department of Defense to the Subcommittee on National Security, Veterans Affairs, and International Relations Subcommittee of the House Government Reform Committee, June 4.

Steuer, Guillaume, et al. 2011. Air et Cosmos: Le Guide des Avions de Combat. Paris: Air and Cosmos.

Stevenson, James. 2000. The $5 Billion Misunderstanding: The Collapse of the Navy's A-12 Stealth Bomber Program. Annapolis: Naval Institute.

Strayer, Robert. 1998. Why Did the Soviet Union Collapse? Armonk, New York: M.E. Sharpe.

Stockman, David. 1986. The Triumph of Politics. New York: Avon Books.

Stritzel, Holger and Dirk Schmittchen. 2011."Securitization, Culture and Power: Rogue States in US and German Discourse," in Thierry Balzacq, ed., Securitization Theory: How Security Problems Emerge and Dissolve. London: Routledge: 170–85.

Suh, Jae-Jung. 2009. "Allied To Race? The U.S.-Korea Alliance and Arms Race," Asian Perspective 33/4: 101–127.

Sullivan, Sean C. 2010. The Four Year Integrated Defense Planning Cycle: The Formal Processes in U.S. Defense Planning, A Desk Reference. Newport, RI: Naval War College.

Taft, William H. 1912. President of the United States. Message on Economy and Efficiency in the Government Service, H. Doc 458 62–2, January 27.

Taylor, Brian. 2002. "An Analysis of the Departments of the Air Force, Army, and Navy Budget Offices and Budget Processes," Master's Degree Thesis, Monterey, CA: Naval Postgraduate School.

Taylor, Leonard B. 1974. Financial Management of the Vietnam Conflict. Department of the Army.

Thompson, F., and L. R. Jones. 1994. Reinventing the Pentagon. San Francisco: Jossey-Bass.

Thomson, Mark A. 1938. A Constitutional History of England, Vol. IV, London: Methuen and Co.

Thurmaier, K. M., and K. G. Willoughby. 2001. Policy and Politics in State Budgeting. Armonk, NY: M. E. Sharpe.

Tirpak, John A. 2003. Legacy of the Air Blockades, Air Force Magazine February: 46–52.

Todd, Daniel, and Michael Lindberg. 1996. Navies and Shipbuilding Industries: The Strained Symbiosis. Westport: Praeger.

Toll, I. W. 2006. Six Frigates: The Epic History of the Founding of the U.S. Navy. New York: W.W. Norton and Company.

Tolson, John. 1973. Vietnam Studies: Air Mobility, 1961–1971. Washington, DC: Department of the Army.

Tyszkiewicz, Mary, and Stephen Daggett. 1998. "A Defense Budget Primer." Washington, DC: CRS, December.

U.S. Code. 1921. 42 Stat. 20.

U.S. Code. 31: §1502.

U.S. Code. 31: 1501

U.S. Code. 31: 1341 and 1517.

U.S. Code. 31: 3511 (a).

U.S. Code. 253a 1 A.

U.S. Congress. 1789. I Statutes at Large, Ch. XXIII, Sept. 29: 95

U.S. Constitution. 1789. www.usconstitution.net.

U.S. Constitution. 1789. Article 1, Sec. 8.

U.S. Department of State. 2011."Country Policies and Embargos," Directorate of Defense Trade Controls www.pmddtc.state.gov/embargoed_countries/index.html

U.S. Marine Corps. 2002. "Program Objective Memorandum 2004 Guide," November http://www.usmcmccs.org/Director/POM/home.asp

U.S. Senate. Budget Committee, 2011.

Vetter, Bernd, and Frank Vetter. 2009. Der Alpha Jet. Stuttgart: Motorbuch Verlag.

Walker, David, 2001. Testimony by the Comptroller General to the Subcommittee of the National Security, Veterans Affairs, and International Relations Committee of the House Government Reform Committee, U.S. House of Representatives, March 7.

Wall Street Journal, 2007. "A Triumph for Pelosi," Review and Outlook, Saturday, March 24: A-10.

Walmart. 2010. About Us. Walmart Corporate http://walmartstores.com/AboutUs/

Wang, Dong. 2005. China's Unequal Treaties: Narrating National History. Lanham: Lexington Books.

Webb, N. J., and P. J. Candreva. 2010. "Diagnosing Performance Management and Performance Budgeting Systems: A Case Study of the U.S. Navy," Public Finance and Management 10/3, 524–555.

Weir, Gary. 1992. Building the Kaiser's Navy: The Imperial Naval Office and German Industry in the von Tirpitz Era: 1890–1919. Annapolis: Naval Institute Press.

Weisman, Jonathan. 2005. President Requests More War Funding, The Washington Post 15 February: B2.

Wescott, C., et al., eds. 2009. The Many Faces of Public Management Reform in the Asia-Pacific Region. Oxford, UK: Emerald Publishing.

Wheeler, Winslow, 2002. "Mr. Smith is Dead: No One Stands in the Way as Congress Laces Post-September 11 Defense Bills with Pork," Government Executive December 9 http://www.d-n-i.net/fcs/spartacus_mr_smith.htm

White, Leonard D. 1951. The Jeffersonians. New York: Macmillan.

White, Leonard D. 1954. The Jacksonians. New York: Macmillan

White, Leonard D. 1958. The Republican Era. New York: Macmillan

Wildavsky, Aaron. 1964. The Politics of the Budgetary Process. Boston: Little, Brown

Wildavsky, Aaron. 1969. "Rescuing Policy Analysis from PPBS." Public Administration Review, 292, 189–202.

Wildavsky, Aaron. 1975. Budgeting: A Comparative Theory of Budgetary Processes, Boston: Little, Brown.

Wildavsky, Aaron. 1979. The Politics of the Budgetary Process, Boston: Little, Brown.

Wildavsky, Aaron, 1988. The New Politics of the Budgetary Process. Glenview, IL: Scott, Foresman.

Wildavsky, Aaron, and Naomi Caiden. 1997. The New Politics of the Budgetary Process. New York: Addison Wesley Longman.

Wildavsky, Aaron, and Naomi Caiden. 2001. The New Politics of the Budgetary Process, 4th edition. New York: Longman.

Williams, M. J. 2000. "Resource Allocation: A Primer," Marine Corps Gazette, 84, February: 14–15.

Willner, Albert. 2011. "Implications of Recent and Planned Changes," in Roger Cliff, et al., eds., Taiwan's Defense Posture: New Opportunities and Challenges for Taiwan's Security. Santa Monica: RAND: 81–88.

Wilmerding, Jr., Lucius. 1943. The Spending Power: A History of the Efforts of Congress to Control Expenditures. New Haven, CT: Yale University Press.

Wlezien, Christopher. 1996. "The President, Congress, and Appropriations, American Politics Quarterly" 24/1 January: 62.

Wolf, Jr., Charles, and Derek Leebaert. 1978. "Trade Liberalization as a Path to Weapons Standardization in NATO," International Security 2/3 Winter: 136–159.

Wolfowitz, Paul. 2005. DOD Business Transformation. February 7 http://www.dtic.mil/whs/directives/corres/pdf/dsd050207transform.pdf

Wolfson, Murray. 1968. "A Mathematical Model of the Cold War," Peace Research Society: Conference Vol. 9: 107–23.

Wood, Gordon S. 2006. Revolutionary Characters: What Made the Founding Fathers Different. Penguin Press: New York, NY.

Woolf, Amy. 2006. Strategic Arms Control after START: Issues and Options. Washington, DC: CRS.

Woolf, Amy. 2010. Nuclear Arms Control: The U.S.-Russian Agenda. Washington DC: CRS.

World Bank. 1998. Public Expenditure Management Handbook. Washington, DC: The World Bank.

Yanarella, Ernest. 2002. The Missile Defense Controversy: Technology in Search of a Mission. Lexington: Kentucky University Press.

Yuen, D. C. Y. 2004. "Goal Characteristics, Communication and Reward Systems, and Managerial Propensity to Create Budgetary Slack," Managerial Accounting Journal 19/4: 517–532.

Zatsepin, Vasily. 2005. Performance-Oriented Defence Budgeting: A Russian Perspective. Moscow: Gaidar Institute for Economic Policy www.iet.ru/files/persona/zatsepin/Brno2005.pdf.

ABOUT THE AUTHORS

L. R. Jones

Lawrence R. Jones is Distinguished Professor and George F. A. Wagner Chair of Public Management in the Graduate School of Business and Public Policy, Naval Postgraduate School, Monterey, California. Professor Jones teaches and conducts research on a variety of topics including national defense budgeting and financial management, federal government budgetary and finance policy, national and international public sector management reform, regulatory policy and reform and international fiscal stress policy and management. He received his BA degree from Stanford University in political science and MA and PhD degrees with distinction from the University of California, Berkeley with emphasis on government planning, budgeting and financial management. He has authored more than one hundred twenty-five journal articles, book chapters and other publications on topics including international, national and state budgeting and financial management policy and practice, management control, government regulatory policy and practice, corporate environmental policy and management, government reform, and fiscal crisis and stress management. His work has appeared in a number of academic and practitioner journals including the *Journal of Policy Analysis and Management, Journal of Public Administration Research and Theory, International Public Management Journal, Policy Sciences, Public Administration Review, Academy of Management Review, Public Budgeting & Finance,* the *Journal of Comparative Policy Analysis,* the *International Journal of Public Administration, Public Administration and Development, Sloan Management Review, California Management Review, Canadian Business Review, Public Finance and Manage-*

Financing National Defense: Policy and Process, pages 481–484
Copyright © 2012 by Information Age Publishing
481

ment, Public Budgeting, Accounting and Financial Management, Government Executive, Canadian Public Administration, International Public Management Review, Azienda Publica, Revista do Servico Publico, the American Journal of Economics and Sociology, the Journal of Higher Education, Economics of Education Review, The Review of Higher Education, State and Local Government Review, the *Municipal Finance Journal, Armed Forces and Society, Armed Forces Comptroller, Defense Analysis, and the Journal of Health and Human Resource Administration.* Dr. Jones has published nineteen books including *Regulatory Policy and Practices* (1982), *University Budgeting for Critical Mass and Competition* (1985), *Government Response to Financial Constraints* (1989), *Mission Financing to Realign National Defense* (1992), *Reinventing the Pentagon* (1994), *Corporate Environmental Policy and Government Regulation* (1994), *International Perspectives on the New Public Management* (1997), *Public Management: Institutional Renewal for the 21st Century* (1999), *Budgeting and Financial Management in the Federal Government* (2001), *Learning From International Public Management Reform: Australia and New Zealand* (2001), *Learning From International Public Management Reform: International Experience* (2001), *Strategy for Public Management Reform* (2004), *International Public Financial Management Reform* (2005), *From Bureaucracy to Hyperarchy in Netcentric and Quick Learning Organizations* (2007), *The Many Faces of Public Management Reform in the Asia-Pacific Region* (2009), and *Arming America at War: A Model for Rapid Defense Acquisition in Time of War* (2010). Professor Jones has been a Fulbright Scholar and was honored in 2005 by the Association for Budgeting and Financial Management, American Society for Public Administration as recipient of the Aaron B. Wildavsky Award for Lifetime Achievement in the Field of Public Budgeting and Financial Management. Dr. Jones has consulted for a wide variety of national and international organizations including the Office of the Secretary of the Department Defense, the Ford Foundation, the Asian Development Bank, Harris Bank, the government of Canada, the government of Italy, the government of Brazil, the government of the Republic of China. He served as President (1997–2007) and co-founder of the International Public Management Network (IPMN at www. ipmn.net). He also serves as a public management book series editor for Information Age Publishing (US) and Emerald Press (UK), and is coeditor of the *International Public Management Journal.* For seven years he served as a budget officer for the State of California. He began his professional career as a production manager in the electronics industry in California. In September, 2011 Dr. Jones was honored by the NPS Provost and the President as Distinguished Professor for lifetime achievements and contributions to two sub-disciplines: public budgeting and financial management, and international public management reform.

Philip J. Candreva

Philip J. Candreva is Senior Lecturer of Financial Management in the Graduate School of Business and Public Policy (GSBPP) at Naval Postgraduate School in Monterey, California. He has over 25 years of practical experience and scholarship. He retired from the U.S. Navy at the rank of Commander. During his active duty he earned subspecialties in financial management, acquisition, and information technology management. In 2008, he served as special assistant to the Assistant Secretary of the Navy (Financial Management & Comptroller) in the Pentagon. A lifelong student, he is a 1984 graduate of Pennsylvania State University and in 1996 graduated with distinction from the Naval Postgraduate School (NPS) where he was honored as Conrad Scholar in financial management. He is currently enrolled in law school working on his J.D. degree. Philip has been part of the GSBPP public financial management faculty since 2002. In addition, he is a senior associate in the Center for Defense Management Research where he conducts research and consults with defense leaders on the department's management practices. Specifically, his work is in the areas of budgeting and accounting reform, performance management, acquisition, and governance. He has published more than two dozen articles, technical reports, case studies and book chapters. His work has appeared in *Armed Forces & Society*, the *International Public Management Review*, the *Journal of Government Financial Management*, *Public Administration Review*, and *Public Budgeting & Finance*.

Marc R. DeVore

Marc DeVore earned his Ph.D. in political science from the Massachusetts Institute of Technology (MIT) and is currently Jean Monnet Fellow at the European University Institute (EUI) in Florence. He has served as a lecturer and senior research fellow at the University of St. Gallen in the Centre for Security, Economics and Technology (CSET) and as a visiting research fellow at King's College London in the Department of War Studies, and at Sciences Po in Paris. He is also the youngest American to graduate from France's prestigious Institut des Hautes Etudes de Defense Nationale (Institute for Higher National Defense Studies–IHEDN). Dr. DeVore is a recipient of Fulbright and Truman scholarships and grants provided by the French government and both Harvard University and Columbia University centers for European Studies. He has conducted extensive research on national and international defense policy and management topics and related in areas and has published more than a dozen journal articles and book chapters in a variety of publications appearing in the United States,

Belgium, France, Germany, Spain, Switzerland and the United Kingdom. His work has been featured in BBC documentaries and has appeared in journals including *Security Studies, Cold War History, Guerres & Histoire, Small Wars & Insurgencies, European Security, Comparative Strategy, Swiss Political Science Review* and *Defense and Security Analysis.* Dr. DeVore's academic work has been complemented by experience as a policy advisor to political and military officials in the Balkans, Western Europe and Africa. In the Balkans he observed and participated in regional conflict resolution as a major part of an oral history research project in Bosnia-Herzegovina and as a participant in the Council of Europe's effort to enable members of the religious communities in the Balkans to gain better understandings of each other. In France and Switzerland, Dr. DeVore has been a consultant for government officials and representatives of industry in areas including military equipment acquisition processes and the evolving importance of the European Defense Agency. In 2002–2003 Dr. DeVore served as National Security Advisor to the President of the Central African Republic and has more recently provided advice to the government of Guinea-Bissau.

INDEX

Financing National Defense: Policy and Process, pages 485–489
Copyright © 2012 by Information Age Publishing
485

CPSIA information can be obtained at www.ICGtesting.com
Printed in the USA
LVOW071252211211

260495LV00002B/34/P

9 781617 356773